# The Role of Natural and Constructed Wetlands in Nutrient Cycling and Retention on the Landscape

Jan Vymazal
Editor

# The Role of Natural and Constructed Wetlands in Nutrient Cycling and Retention on the Landscape

*Editor*
Jan Vymazal
Czech University of Life Sciences Prague
Faculty of Environmental Sciences
Department of Applied Ecology
Praha
Czech Republic
and
ENKI, o.p.s.
Třeboň
Czech Republic

ISBN 978-3-319-08176-2     ISBN 978-3-319-08177-9 (eBook)
DOI 10.1007/978-3-319-08177-9
Springer Cham Heidelberg New York Dordrecht London

Library of Congress Control Number: 2014951701

© Springer International Publishing Switzerland 2015
This work is subject to copyright. All rights are reserved by the Publisher, whether the whole or part of the material is concerned, specifically the rights of translation, reprinting, reuse of illustrations, recitation, broadcasting, reproduction on microfilms or in any other physical way, and transmission or information storage and retrieval, electronic adaptation, computer software, or by similar or dissimilar methodology now known or hereafter developed. Exempted from this legal reservation are brief excerpts in connection with reviews or scholarly analysis or material supplied specifically for the purpose of being entered and executed on a computer system, for exclusive use by the purchaser of the work. Duplication of this publication or parts thereof is permitted only under the provisions of the Copyright Law of the Publisher's location, in its current version, and permission for use must always be obtained from Springer. Permissions for use may be obtained through RightsLink at the Copyright Clearance Center. Violations are liable to prosecution under the respective Copyright Law.
The use of general descriptive names, registered names, trademarks, service marks, etc. in this publication does not imply, even in the absence of a specific statement, that such names are exempt from the relevant protective laws and regulations and therefore free for general use.
While the advice and information in this book are believed to be true and accurate at the date of publication, neither the authors nor the editors nor the publisher can accept any legal responsibility for any errors or omissions that may be made. The publisher makes no warranty, express or implied, with respect to the material contained herein.

Printed on acid-free paper

Springer is part of Springer Science+Business Media (www.springer.com)

# Preface

The first workshop bearing the title "Nutrient Cycling in Natural and Constructed Wetlands" was organized at Třeboň, South Bohemia, in September 1995. The major organizing principle was to bring together scientists and researchers dealing with both natural and constructed wetlands; the majority of previous wetland conferences and seminars on these topics had been held separately. The first edition of this seminar was attended mostly by people dealing with constructed wetlands, but over the years, natural wetlands had become an increasingly substantial part of this workshop. The joint meeting has proven to be beneficial for both groups as interaction of natural and constructed wetland researchers is still limited. The International Water Association conferences on constructed wetlands and WETPOL conferences are usually not attended by many natural wetland researchers, while the Society of Wetland Scientists and INTECOL WETLANDS conferences mostly deal with natural wetlands. We believe that the interactions fostered by communication among all wetland researchers will continue to be a benefit for the full community.

The eighth meeting of this series took place again at Třeboň on May 17–22, 2013, this time entitled "The Role of Natural and Constructed Wetlands in Nutrient Cycling and Retention on the Landscape." The workshop was attended by participants from 17 countries in Europe, North America, and Australia. This volume contains a selection from the papers presented during the workshop —the first part of the book is mostly devoted to natural wetlands, while the second part deals with the use of constructed wetlands for wastewater treatment.

The organization of the workshop was partially supported by grant No. 5.1 PPO4/015 from the Ministry of Industry and Trade of the Czech Republic.

Praha, Czech Republic  Jan Vymazal
April 2014

Photo: Ketil Haarstad

# Contents

1 **Biomass Production in Permanent Wet Grasslands Dominated with *Phalaris arundinacea*: Case Study of the Třeboň Basin Biosphere Reserve, Czech Republic**........................................... 1
Hana Čížková, Jana Rychterová, Libuše Hamadejová, Karel Suchý, Monika Filipová, Jan Květ, and Neil O. Anderson

2 **Greenhouse Gas Fluxes from Restored Agricultural Wetlands and Natural Wetlands, Northwestern Indiana**..................... 17
Brianna Richards and Christopher B. Craft

3 **Assessment of Immobilisation and Biological Availability of Iron Phosphate Nanoparticle-Treated Metals in Wetland Sediments**... 33
Herbert John Bavor and Batdelger Shinen

4 **Spatial Variability in Sedimentation, Carbon Sequestration, and Nutrient Accumulation in an Alluvial Floodplain Forest**......... 41
Jacob M. Bannister, Ellen R. Herbert, and Christopher B. Craft

5 **Natural and Restored Wetland Buffers in Reducing Sediment and Nutrient Export from Forested Catchments: Finnish Experiences**........................................................ 57
Mika Nieminen, Annu Kaila, Markku Koskinen, Sakari Sarkkola, Hannu Fritze, Eeva-Stiina Tuittila, Hannu Nousiainen, Harri Koivusalo, Ari Laurén, Hannu Ilvesniemi, Harri Vasander, and Tapani Sallantaus

6 **Do Reflectance Spectra of Different Plant Stands in Wetland Indicate Species Properties?**...................................... 73
Katja Klančnik, Igor Zelnik, Primož Gnezda, and Alenka Gaberščik

| | | |
|---|---|---|
| 7 | **Global Boundary Lines of $N_2O$ and $CH_4$ Emission in Peatlands**...<br>Jaan Pärn, Anto Aasa, Sergey Egorov, Ilya Filippov, Geofrey Gabiri,<br>Iuliana Gheorghe, Järvi Järveoja, Kuno Kasak, Fatima Laggoun-Défarge,<br>Charles Kizza Luswata, Martin Maddison, William J. Mitsch,<br>Hlynur Óskarsson, Stéphanie Pellerin, Jüri-Ott Salm, Kristina Sohar,<br>Kaido Soosaar, Alar Teemusk, Moses M. Tenywa, Jorge A. Villa,<br>Christina Vohla, and Ülo Mander | 87 |
| 8 | **Distribution of Solar Energy in Agriculture Landscape: Comparison Between Wet Meadow and Crops**..................<br>Hanna Huryna, Petra Hesslerová, Jan Pokorný, Vladimír Jirka, and Richard Lhotský | 103 |
| 9 | **Surface Temperature, Wetness, and Vegetation Dynamic in Agriculture Landscape: Comparison of Cadastres with Different Types of Wetlands**........................<br>Petra Hesslerová and Jan Pokorný | 123 |
| 10 | **Agricultural Runoff in Norway: The Problem, the Regulations, and the Role of Wetlands**................<br>Anne-Grete Buseth Blankenberg, Ketil Haarstad, and Adam M. Paruch | 137 |
| 11 | **Subsurface Flow Constructed Wetland Models: Review and Prospects**.......................<br>Roger Samsó, Daniel Meyer, and Joan García | 149 |
| 12 | **Behaviour of a Two-Stage Vertical Flow Constructed Wetland with Hydraulic Peak Loads**......................<br>Guenter Langergraber, Alexander Pressl, and Raimund Haberl | 175 |
| 13 | **A New Concept of Multistage Treatment Wetland for Winery Wastewater Treatment: Long-Term Evaluation of Performances**..........................<br>Fabio Masi, Riccardo Bresciani, and Miria Bracali | 189 |
| 14 | **Polishing of Real Electroplating Wastewater in Microcosm Fill-and-Drain Constructed Wetlands**....................<br>Adam Sochacki, Olivier Faure, Bernard Guy, and Joanna Surmacz-Górska | 203 |
| 15 | **Relationship Between Filtering Material and Nitrification in Constructed Wetlands Treating Raw Wastewater**............<br>Georges Reeb and Etienne Dantan | 229 |
| 16 | **Single-Family Treatment Wetlands Progress in Poland**.........<br>Hanna Obarska-Pempkowiak, Magdalena Gajewska, Ewa Wojciechowska, and Arkadiusz Ostojski | 237 |

| | | |
|---|---|---|
| 17 | **Treatment Wetland for Overflow Stormwater Treatment: The Impact of Pollutant Particles Size**............................ Magdalena Gajewska, Marzena Stosik, Ewa Wojciechowska, and Hanna Obarska-Pempkowiak | 249 |
| 18 | **Treatment Wetlands in Rural Areas of Poland for Baltic Sea Protection**............................................................ Katarzyna Kołecka, Magdalena Gajewska, and Hanna Obarska-Pempkowiak | 259 |
| 19 | **Long-Term Performance of Constructed Wetlands with Chemical Dosing for Phosphorus Removal**............................. Gabriela Dotro, Raul Prieto Fort, Jan Barak, Mark Jones, Peter Vale, and Bruce Jefferson | 273 |
| 20 | **Use of the Macrophyte *Cyperus papyrus* in Wastewater Treatment**..................................................................... Njenga Mburu, Diederik P.L. Rousseau, Johan J.A. van Bruggen, and Piet N.L. Lens | 293 |
| 21 | **Does the Presence of Weedy Species Affect the Treatment Efficiency in Constructed Wetlands with Horizontal Subsurface Flow?**...... Jan Vymazal | 315 |
| **Index** | ............................................................ | 323 |

# Contributors

**Anto Aasa**  Institute of Ecology and Earth Sciences, University of Tartu, Tartu, Estonia

**Neil O. Anderson**  Department of Horticultural Science, University of Minnesota, Saint Paul, MN, USA

**Jacob M. Bannister**  School of Public and Environmental Affairs, Indiana University, Bloomington, IN, USA

**Jan Barak**  School of Applied Sciences, Cranfield University, Cranfield, Bedfordshire, UK

**Herbert John Bavor**  Centre for Water and Environmental Technology – Water Research Laboratory, University of Western Sydney – Hawkesbury, Penrith, NSW, Australia

**Anne-Grete Buseth Blankenberg**  Bioforsk, Norwegian Institute for Agricultural and Environmental Research, Aas, Norway

**Miria Bracali**  Casa Vinicola Luigi Cecchi e Figli, Castellina in Chianti, SI, Italy

**Riccardo Bresciani**  IRIDRA Srl, Firenze, Italy

**Hana Čížková**  Faculty of Agriculture, University of South Bohemia, České Budějovice, Czech Republic

**Christopher B. Craft**  School of Public and Environmental Affairs, Indiana University, Bloomington, IN, USA

**Etienne Dantan**  Atelier Reeb, Strasbourg, France

**Gabriela Dotro**  School of Applied Sciences, Cranfield University, Cranfield, Bedfordshire, UK

Waste Water Research and Development, Severn Trent Water, Coventry, West Midlands, UK

**Sergey Egorov** Institute of Ecology and Earth Sciences, University of Tartu, Tartu, Estonia

**Olivier Faure** GéoSciences and Environnement Département, Ecole Nationale Supérieure des Mines, Saint-Etienne Cedex 2, France

**Monika Filipová** Faculty of Agronomy, Mendel University, Brno, Czech Republic

**Ilya Filippov** UNESCO Chair of Environmental Dynamics and Climate Change, Yugra State University, Khanty-Mansiysk, Russia

**Raul Prieto Fort** School of Applied Sciences, Cranfield University, Cranfield, Bedfordshire, UK

**Hannu Fritze** Finnish Forest Research Institute, Vantaa, Finland

**Alenka Gaberščik** Department of Biology, Biotechnical Faculty, University of Ljubljana, Ljubljana, Slovenia

**Geofrey Gabiri** Department of Geography, Kenyatta University, Nairobi, Kenya

Department of Agricultural Production, College of Agricultural and Environmental Sciences, Makerere University, Kampala, Uganda

**Magdalena Gajewska** Faculty of Civil and Environmental Engineering, Gdansk University of Technology, Gdansk, Poznaň, Poland

**Joan García** GEMMA – Group of Environmental Engineering and Microbiology, Department of Hydraulic, Maritime and Environmental Engineering, Universitat Politècnica de Catalunya-BarcelonaTech, Barcelona, Spain

**Iuliana Gheorghe** Faculty of Ecology and Environmental Protection, Ecological University of Bucharest, Bucharest, Romania

**Primož Gnezda** Department of Biology, Biotechnical Faculty, University of Ljubljana, Ljubljana, Slovenia

**Bernard Guy** GéoSciences and Environnement Département, Ecole Nationale Supérieure des Mines, Saint-Etienne Cedex 2, France

**Ketil Haarstad** Bioforsk, Norwegian Institute for Agricultural and Environmental Research, Aas, Norway

**Raimund Haberl** Institute for Sanitary Engineering and Water Pollution Control, University of Natural Resources and Life Sciences, Vienna (BOKU), Vienna, Austria

**Libuše Hamadejová** Faculty of Agriculture, University of South Bohemia, České Budějovice, Czech Republic

**Petra Hesslerová** ENKI, o.p.s, Třeboň, Czech Republic

Contributors     xiii

**Hanna Huryna** Faculty of Science, University of South Bohemia, České Budějovice, Czech Republic

ENKI, o.p.s, Třeboň, Czech Republic

**Hannu Ilvesniemi** Finnish Forest Research Institute, Vantaa, Finland

**Järvi Järveoja** Institute of Ecology and Earth Sciences, University of Tartu, Tartu, Estonia

**Bruce Jefferson** School of Applied Sciences, Cranfield University, Cranfield, Bedfordshire, UK

**Vladimír Jirka** ENKI, o.p.s, Třeboň, Czech Republic

**Mark Jones** Waste Water Research and Development, Severn Trent Water, Coventry, West Midlands, UK

**Annu Kaila** Finnish Forest Research Institute, Vantaa, Finland

**Kuno Kasak** Institute of Ecology and Earth Sciences, University of Tartu, Tartu, Estonia

**Katja Klančnik** Department of Biology, Biotechnical Faculty, University of Ljubljana, Ljubljana, Slovenia

**Harri Koivusalo** Department of Civil and Environmental Engineering, Aalto University School of Science and Technology, Aalto, Finland

**Katarzyna Kołecka** Faculty of Civil and Environmental Engineering, Gdansk University of Technology, Gdańsk, Poznań, Poland

**Markku Koskinen** Department of Forest Sciences, University of Helsinki, Helsinki, Finland

**Jan Květ** Faculty of Science, University of South Bohemia, České Budějovice, Czech Republic

**Fatima Laggoun-Défarge** Institut des Sciences de la Terre d'Orléans, CNRS-Université d'Orléans, Orléans, France

**Günter Langergraber** Institute of Sanitary Engineering and Water Pollution Control, University of Natural Resources and Applied Life Sciences, Vienna (BOKU), Vienna, Austria

**Ari Laurén** Finnish Forest Research Institute, Joensuu, Finland

**Piet L. N. Lens** UNESCO-IHE Institute for Water Education, Delft, The Netherlands

**Richard Lhotský** ENKI, o.p.s, Třeboň, Czech Republic

**Charles Kizza Luswata** Department of Agricultural Production, College of Agricultural and Environmental Sciences, Makerere University, Kampala, Uganda

**Martin Maddison** Institute of Ecology and Earth Sciences, University of Tartu, Tartu, Estonia

**Ülo Mander** Institute of Ecology and Earth Sciences, University of Tartu, Tartu, Estonia

Hydrosystems and Bioprocesses Research Unit, National Research Institute of Science and Technology for Environment and Agriculture (Irstea), Antony Cedex, France

**Fabio Masi** IRIDRA Srl, Firenze, Italy

**Njenga Mburu** UNESCO-IHE Institute for Water Education, Delft, The Netherlands

Department of Civil and Structural Engineering, Masinde Muliro University of Science and Technology, Kakamega, Kenya

**Daniel Meyer** Freshwater Systems, Ecology and Pollution Research Unit, IRSTEA, Villeurbanne Cedex, France

**William J. Mitsch** Everglades Wetland Research Park, Kapnick Center, Florida Gulf Coast University, Fort Myers, FL, USA

**Mika Nieminen** Finnish Forest Research Institute, Vantaa, Finland

**Hannu Nousiainen** Finnish Forest Research Institute, Vantaa, Finland

**Hanna Obarska-Pempkowiak** Faculty of Civil and Environmental Engineering, Gdansk University of Technology, Gdańsk, Poznaň, Poland

**Hlynur Óskarsson** Faculty of Environmental Sciences, Agricultural University of Iceland, Borgarnes, Iceland

**Arkadiusz Ostojski** Faculty of Civil and Environmental Engineering, Gdansk University of Technology, Gdansk, Poznaň, Poland

**Jaan Pärn** Institute of Ecology and Earth Sciences, University of Tartu, Tartu, Estonia

**Adam M. Paruch** Bioforsk, Norwegian Institute for Agricultural and Environmental Research, Aas, Norway

**Stéphanie Pellerin** Institut de recherche en biologie végétale, Université de Montréal, Jardin botanique de Montréal, Montréal, QC, Canada

**Jan Pokorný** ENKI, o.p.s, Třeboň, Czech Republic

**Alexander Pressl** Institute for Sanitary Engineering and Water Pollution Control, University of Natural Resources and Life Sciences, Vienna (BOKU), Vienna, Austria

**Georges Reeb** Atelier Reeb, Strasbourg, France

Contributors

**Brianna Richards** School of Public and Environmental Affairs, Indiana University, Bloomington, IN, USA

**Diederik P. L. Rousseau** Department of Industrial Biological Sciences, Ghent University, Kortrijk, Belgium

**Jana Rychterová** Faculty of Agriculture, University of South Bohemia, České Budějovice, Czech Republic

**Tapani Sallantaus** Natural Environment Center, Biodiversity Unit, Finnish Environment Institute, Helsinki, Finland

**Jüri-Ott Salm** Institute of Ecology and Earth Sciences, University of Tartu, Tartu, Estonia

Estonian Fund for Nature, Tartu, Estonia

**Roger Samsó** GEMMA – Group of Environmental Engineering and Microbiology, Department of Hydraulic, Maritime and Environmental Engineering, Universitat Politècnica de Catalunya-BarcelonaTech, Barcelona, Spain

**Sakari Sarkkola** Finnish Forest Research Institute, Vantaa, Finland

**Batdelger Shinen** Centre for Water and Environmental Technology – Water Research Laboratory, University of Western Sydney – Hawkesbury, Penrith, NSW, Australia

**Adam Sochacki** Environmental Biotechnology Department, Faculty of Power and Environmental Engineering, Silesian University of Technology, Gliwice, Poland

GéoSciences and Environnement Département, Ecole Nationale Supérieure des Mines, Saint-Etienne Cedex 2, France

**Kristina Sohar** Institute of Ecology and Earth Sciences, University of Tartu, Tartu, Estonia

**Kaido Soosaar** Institute of Ecology and Earth Sciences, University of Tartu, Tartu, Estonia

**Marzena Stosik** Faculty of Civil and Environmental Engineering, Gdansk University of Technology, Gdańsk, Poznań, Poland

**Karel Suchý** Faculty of Agriculture, University of South Bohemia, České Budějovice, Czech Republic

**Joanna Surmacz-Górska** Environmental Biotechnology Department, Faculty of Power and Environmental Engineering, Silesian University of Technology, Gliwice, Poland

**Alar Teemusk** Institute of Ecology and Earth Sciences, University of Tartu, Tartu, Estonia

**Moses M. Tenywa** Department of Agricultural Production, College of Agricultural and Environmental Sciences, Makerere University, Kampala, Uganda

**Eeva-Stiina Tuittila** School of Forest Sciences, University of Eastern Finland, Joensuu, Finland

**Peter Vale** Waste Water Research and Development, Severn Trent Water, Coventry, West Midlands, UK

**Johan J. A. van Bruggen** UNESCO-IHE Institute for Water Education, Delft, The Netherlands

**Harri Vasander** Department of Forest Sciences, University of Helsinki, Helsinki, Finland

**Jorge A. Villa** Institut des Sciences de la Terre d'Orléans, CNRS-Université d'Orléans, Orléans, France

**Christina Vohla** Institute of Ecology and Earth Sciences, University of Tartu, Tartu, Estonia

**Jan Vymazal** Department of Applied Ecology, Faculty of Environmental Sciences, Czech University of Life Sciences, Praha 6, Czech Republic

ENKI, o.p.s, Třeboň, Czech Republic

**Ewa Wojciechowska** Faculty of Civil and Environmental Engineering, Gdansk University of Technology, Gdansk, Poznaň, Poland

**Igor Zelnik** Department of Biology, Biotechnical Faculty, University of Ljubljana, Ljubljana, Slovenia

# Chapter 1
# Biomass Production in Permanent Wet Grasslands Dominated with *Phalaris arundinacea*: Case Study of the Třeboň Basin Biosphere Reserve, Czech Republic

Hana Čížková, Jana Rychterová, Libuše Hamadejová, Karel Suchý, Monika Filipová, Jan Květ, and Neil O. Anderson

**Abstract** *Phalaris arundinacea* is a highly productive perennial grass which inhabits both natural and human-affected wetlands. Along with natural genotypes, there are a number of cultivars bred for fodder production, especially in cool climatic areas. At present *P. arundinacea* is being investigated as a potential energy crop. Use of seminatural and natural stands of *P. arundinacea* as an energy resource requires a knowledge of the variation of aboveground biomass production, which forms the agricultural yield. This work gives an overview of long-term investigation of the production of *P. arundinacea* on various types of natural biotopes. It also presents results of a detailed field experiment assessing the effects of various management (cutting frequency, mulching, fertilizing) on the production of aboveground biomass in a seminatural wetland dominated by *P. arundinacea*. The results confirm that monodominant stands of *P. arundinacea* attain a high production in Central Europe. The seasonal maximum of aboveground biomass of natural stands ranged from 4 to 14 metric tonnes dry weight per hectare (t.ha$^{-1}$) with an average of 9.5 t ha$^{-1}$. Among the management types, the lowest annual agricultural yield of 4.1 t ha$^{-1}$ (dry weight) was found in the treatment one cut per year and no fertilization. The maximum yield of 11 t ha$^{-1}$ was achieved under three cuts per year and fertilization with a double dose of N and single doses of P and K. Two cuts

---

H. Čížková (✉) • J. Rychterová • L. Hamadejová • K. Suchý
Faculty of Agriculture, University of South Bohemia, Branišovská 31, 370 05 České Budějovice, Czech Republic
e-mail: hcizkova@zf.jcu.cz

M. Filipová
Faculty of Agronomy, Mendel University, Zemědělská 1, 613 00 Brno, Czech Republic

J. Květ
Faculty of Science, University of South Bohemia, Branišovská 31, 370 05 České Budějovice, Czech Republic

N.O. Anderson
Department of Horticultural Science, University of Minnesota, 305 Alderman Hall, 1970 Folwell Avenue, 55108 Saint Paul, MN, USA

per year and fertilization by P and K seem to combine the production and non-production functions in an optimum way.

**Keywords** Aboveground biomass • Energy crop • Wetland • Yield

## 1.1 Introduction

*Phalaris arundinacea* L. (reed canary grass) is a highly productive, C3 cool season perennial grass naturally occurring in floodplains and other wetland biotopes of temperate climates (Merigliano and Lesica 1998). Due to its high biomass production and tolerance of fairly harsh climates, it has been used as a forage crop in areas of northern Europe and North America (Galatowitsch et al. 1999; Merigliano and Lesica 1998). The species is also used as a perennial cover in pastures (Casler et al. 1998; Hoveland 1992; Kading and Kreil 1990; Riesterer et al. 2000) and as a mixture or pure stand forage crop (Buxton et al. 1998; Ostrem 1988; Sheaffer and Marten 1992). Forage cultivars low in alkaloid content have been bred and introduced as commercial cultivars (Coulman 1995; Coulman et al. 1977; Ostrem 1987; Narasimhalu et al. 1995; Wittenberg et al. 1992).

Shoreline restoration and revegetation programmes have planted *Phalaris* (Figiel et al. 1995). Soil and water restoration projects have used *Phalaris* for phytoremediation: impoundment of acid slurries (Olsen and Chong 1991) and soil contaminant removal (Lasat et al. 1997; Samecka-Cymerman and Kempers 2001; Chekol et al. 2002). Wastewater treatment facilities have also employed *Phalaris* for removal of N forms (Groffman et al. 1991; Sikora et al. 1995; Vymazal 1995, 2001; Zhu and Sikora 1995).

Recently *P. arundinacea* has attracted attention also as a potential bioenergy crop (Burvall 1997; Hadders and Olsson 1997; Hallam et al. 2001; Lewandowski et al. 2003; Nilsson and Hansson 2001) and is currently in production for this purpose in Finland and Sweden (Lewandowski and Schmidt 2006). Its fresh and ensilaged biomass can be used for biogas production (Prochnow et al. 2009). Dry biomass can be used for burning both separately and in combination with coal, as feedstock for thermochemical processes such as pyrolysis and gasification to produce methanol, synthesis gas and pyrolysis oils and for biochemical processes (fermentation and anaerobic digestion) to produce ethanol or methane (Hallam et al. 2001). It can also be used as a short fibre raw material for the pulp and paper industry (Finell 2003; Hellqvist et al. 2003; Papatheofanous et al. 1995; Saijonkari-Pahkala 2001).

*Phalaris arundinacea* is native to Europe, North and Eastern Asia and, to a limited extent, in North America (Merigliano and Lesica 1998). Only a few, non-aggressive populations of *Phalaris* are native in North America – predating European settlements (Merigliano and Lesica 1998). *Phalaris arundinacea* was introduced to North America in the 1850s (Lavergne and Molofsky 2004), although Merigliano and Lesica (1998) noted herbarium specimens from 1825 that resembled the diploid *P. arundinacea* subsp. *rotgesii*. Both invasive and non-invasive populations have been discovered in Ontario, Canada (Dore and McNeill 1980), and elsewhere throughout the United States (Casler et al. 2009). It is a common dominant species of permanent grasslands in floodplains of the temperate zone. In its natural habitats it occurs primarily in floodplains on mineral substrates in which the running waters bring sufficient amounts of oxygen and nutrients (Gubanov et al. 1995; Lyons 1998; Casler et al. 2009). In such conditions it dominates a natural community Rorippo-Phalaridetum arundinacea Kopecký 1961 (called RPK in further text; Hrivnák and Ujházy 2003). *P. arundinacea* also dominates herbaceous wetlands with more stable water level on organic soils which are subjected to temporary flooding. Under such conditions it forms a community type described as Phalaridetum arundinacea Libert 1931 (PAL in further text; Hrivnák and Ujházy 2003). Both communities fall within the habitat reed canary-grass beds, listed under a code C3.26 in the European Nature Information System (http://eunis.eea.europa.eu/habitats/1398).

*Phalaris arundinacea* is regarded as an opportunistic species that is able to make use of suitable growing conditions. It is known to tolerate mechanical damage such as cutting, temporary flooding and readily spreads on wet sites well supplied with nutrients (Lamb et al. 2005; Merigliano and Lesica 1998). Reed canary grass has a high N requirement and growth/yield is significantly increased with N fertilizer applications, such as manure slurries (Lamb et al. 2005; Schmidt et al. 1999). Accordingly, stands dominated by *P. arundinacea* are usually not considered of great value by nature conservationists. Yet, regular management of such wet grasslands is recommended also by studies primarily focused on nature conservation as it prevents ruderalization and supports biodiversity. In North America, *P. arundinacea* is regarded as invasive species as it spreads in many herbaceous wetlands (Galatowitsch et al. 1999). This expansion is attributed to alien European genotypes.

Successful management, based on the use of biomass yield, requires understanding factors determining biomass production (Lamb et al. 2005; Lewandowski and Schmidt 2006). Biomass production depends on a variety of factors, among which climatic conditions, water and nutrient availability, cutting regime and genotype are the most important (Lewandowski and Schmidt 2006). While numerous studies are devoted to various aspects of production of *P. arundinacea* for North Europe and North America (Cherney et al. 2003; Jones et al. 1988; Lamb et al. 2005), few comprehensive studies have assessed the biomass production and its determinants in conditions of Central Europe (Lewandowski and Schmidt 2006). Nitrogen use efficiency (NUE) and energy use efficiency (EUE) in Southwestern Germany trials, at 100 kg ha$^{-1}$ year$^{-1}$ of N, are relatively low in reed canary grass (0.11 t dry biomass kg$^{-1}$ N and 13 GJ bioenergy/GJ energy input, respectively), when compared with *Miscanthus* × *giganteus* and *Triticosecale* Wittmack (Lewandowski and Schmidt 2006).

The aim of this paper is to assess the prerequisites of sustainable production of natural and seminatural wet grasslands dominated with *P. arundinacea* in climatic conditions of Central Europe. This study addresses the time and frequency of cutting, need for nutrient replenishment, vegetation changes associated with various management options and evidence for changes associated with long-term use. Four experiments involving *P. arundinacea* were conducted:

1. Seasonal course of aboveground biomass of a *P. arundinacea* stand
2. Variation in biomass production of natural communities
3. Effect of different fertilization and cutting regimes
4. Biomass production of a heavily fertilized wet grassland across a moisture gradient

## 1.2 Study Area

The UNESCO Třeboň Basin Biosphere Reserve (48°49′ to 49°20′N lat.; 14°39′ to 15°00′E long.) is a flat basin of 70,000 ha in the Třeboň basin of South Bohemia, Czech Republic, which has been subjected to human interactions for more than eight centuries (http://www.unesco.org/mabdb/br/brdir/directory/biores.asp?mode=all&code=CZE+02). Due to its flat topography and rich water availability, the Reserve is rich in wetlands of various types; their ecology and management have been the subject of numerous studies. Syntheses of this research are given by Prach et al. (1996) for biotopes of the Lužnice river floodplain and Květ et al. (2002) for biotopes of wet grasslands.

Both types of natural communities dominated with *P. arundinacea* (RPK, PAL) occur in the Třeboň Basin Biosphere Reserve. RPK covers large areas in the floodplains of main watercourses, the Lužnice and the Nežárka rivers, while the PAL occurs on occasionally flooded flat sites with organic soils.

## 1.3 Methods

### 1.3.1 Seasonal Course of Aboveground Biomass of a P. arundinacea *Stand*

The seasonal course of aboveground biomass production (all biomass above the soil line) was studied in a pure *P. arundinacea* stand occurring within the unmanaged part of the Mokré Louky wet meadows and estimated by the method of repeated small samples. Three replicate samples were harvested from $0.5 \times 0.5$ m$^2$ plots in two- to four-week intervals during the growing season of 2006. Aboveground plant material was taken to the laboratory, separated into live and dead parts, dried to a constant weight at 85 °C and weighed. The seasonal course of each biomass fraction was approximated using polynomial regression.

## 1.3.2 Variation in Biomass Production of Natural Communities

The annual production of aboveground biomass was approximated by estimating the aboveground biomass at the time of its seasonal maximum for the two typical naturally occurring community types (RPK, PAL) on various sites situated within the same catchment of the Upper Lužnice River, South Bohemia, Czech Republic. These localities included six RPK and five PAL sites. Aboveground biomass was estimated using the harvest method (Milner and Hughes 1968; Květ and Westlake 1998). At each site, all aboveground plant material was removed from five replicate plots of $0.5 \times 0.5$ m$^2$, taken to the laboratory and processed as described above for the study of seasonal development. The study lasted for two successive years, 1988–1989.

## 1.3.3 Effect of Different Fertilization and Cutting Regimes

The effect of various management regimes was tested in a field manipulative experiment (1995–1997) established in the Mokré Louky wet meadows. A total of 15 different treatments included combinations of mulching, up to four cuts over the season and up to four fertilization regimes. Each treatment was repeated four times in a completely randomized block design. Blocks of $5 \times 4$ m were surrounded by a border or protection strip.

Fertilizer treatments consisted of different additions of nitrogen, phosphorus and potassium in four different combinations: control (no NPK), PK, NPK and 2NPK (Table 1.1). Nutrients were added in the form of granulated superphosphate ($P_2O_5$), potassium salt ($K_2O$) and ammonium nitrate ($NH_4NO_3$) with limestone. Fertilizer was applied in spring at the beginning of the vegetation season (at the beginning of April, 1995–1997). In treatment 2NPK with a double dose of N, the second half of nitrogen fertilizer was applied after the first cut.

The experiment included six cutting regimes: uncut, one cut in June, one cut in August and two, three and four cuts per year. Uncut plots and plots cut once a year were not fertilized. All four fertilization treatments were applied to regimes with two to four cuts per year. The experiment lasted for 3 years (1995–1997).

In early summer each year, phytosociological relevés were taken at all plots, using the combined abundance-dominance scale (Dierschke 1994). The Simpson index $\lambda$ was calculated on the basis of the percentage cover $p_i$ of all species present, where $R$ is the total number of species:

$$\lambda = \sum_{i=1}^{R} p_i^2.$$

**Table 1.1** Amounts of pure nutrients: nitrogen (N), phosphorus (P) and potassium (K) added under four different fertilizer treatments in the field manipulative experiment on a wet grassland dominated with *Phalaris arundinacea*

| Treatment | Nutrient addition (kg ha$^{-1}$ year$^{-1}$) | | |
|---|---|---|---|
| | N | P | K |
| Control | 0 | 0 | 0 |
| PK | 0 | 6 | 25 |
| NPK | 17 | 6 | 25 |
| 2NPK | 33 | 6 | 25 |

Biomass yield (excluding a 6-cm high stubble; Lewandowski and Schmidt 2006) was determined by harvesting all matter from the whole plots, obtaining fresh weights and subsequent determination of the dry weight/fresh weight ratio for subsamples of the harvested material. The results are expressed as biomass dry weight.

Treatment effects on biomass yield were analysed using analysis of variance (ANOVA). Since the treatments did not include all possible combinations of factors, two groups of treatments were assessed separately: plots cut once a year and the non-fertilized control were subjected to a separate analysis, while all plots that were cut at least twice a year in combination with all fertilization regimes were subjected to another statistical analysis.

## 1.3.4 Biomass Production of a Heavily Fertilized Wet Grassland Across a Moisture Gradient

Biomass production in relation to plant species composition and soil moisture gradient was studied within the intensively managed part of the Mokré Louky wet meadow. The area is regularly mown two to three times a year. Nutrients have been applied to the site in the form of pretreated pig sewage since the 1980s, a common fertilizer for reed canary grass (Lamb et al. 2005). Three biotope types were selected along the moisture gradient on the basis of prevailing vegetation: (1) dry biotope with dominant *Elytrigia repens*, (2) moderately wet biotope dominated with *Alopecurus pratensis* and (3) wet biotope with dominant *P. arundinacea*. Aboveground biomass was estimated by the harvest method twice during the vegetation season, at times of regular cutting performed by the local agricultural firm (K + K Břilice). Within each biotope, the biomass was harvested from $0.5 \times 0.5$ m$^2$ plots replicated eight times. Biomass yield was determined by cutting all plant material at a height of 6 cm above the soil surface

(Lewandowski and Schmidt 2006). The remaining stubble was sampled separately. The plant material was separated into dead and live material of each species, dried to constant weight at 85 °C and weighed.

## 1.4 Results

### 1.4.1 Seasonal Course of Aboveground Biomass of a P. arundinacea Stand

Live, aboveground biomass of *P. arundinacea* approached two seasonal maxima in June (~3 t ha$^{-1}$) when flowering and September (~5 t ha$^{-1}$) (Fig. 1.1), although the biomass was a relatively flat line otherwise. The amount of live reed canary-grass biomass, however, did not change much as new shoots were formed from lower nodes of the dying shoots. Considerable amounts of dead biomass started accumulating from July (leaf decay) and exceeded 10 t ha$^{-1}$ by early October (die-off of shoots that finished flowering). Living and dead biomass of other species increased in a linear fashion, forming only a small percentage of the total live aboveground biomass (Fig. 1.1).

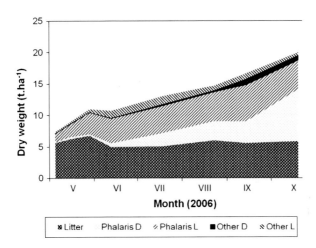

**Fig. 1.1** Seasonal maximum of aboveground biomass in natural stands of *Phalaris arundinacea* in the floodplain of the upper Lužnice River, Czech Republic. RPK, Rorippo-Phalaridetum arundinacea Kopecký 1961, found on well-flushed sites along river banks; PAL, Phalaridetum arundinacea Libert 1931, found on less well-flushed sites on organic soils. *Boxes* denote medians, quartiles and minimum/maximum values of five and separate sites with the RPK and PAL communities, respectively

### 1.4.2 Variation in Biomass Production of Natural Communities

In natural stands, *P. arundinacea* reached its maximum seasonal biomass in the second half of July (data not shown). Stands of *P. arundinacea* more closely associated with water flow (represented by the community type RPK) reached seasonal maxima of aboveground biomass in a range of 7–14 t ha$^{-1}$, with a pooled mean of 11 t ha$^{-1}$ (Fig. 1.2). Such values are higher than biomass production in stands on organic soils with less intensive water exchange (represented by the community type PAL), where most values of the seasonal maximum aboveground biomass were in a range of 5 to 9 t ha$^{-1}$, with a pooled mean of 7 t ha$^{-1}$ (Fig. 1.2). The only exceptions to this trend were (a) Nová Hlína (PAL) with 13.6 t ha$^{-1}$ in 1989 that rivalled the two sites in Lužnice village (RPK) and (b) Stará řeka bridge with 6.9 t ha$^{-1}$ in 1988 that was within the range of values for almost all PAL sites (Fig. 1.2). There were no consistent differences in seasonal maximum aboveground biomass among stands associated with the water flow, while there was a trend towards lower values in 1989, as compared with 1988, in the stands on less well-flushed sites.

### 1.4.3 Effect of Different Fertilization and Cutting Regimes on Biomass Yield

Within the plots mown once a year, annual mean yields of biomass ranged from 3.0 (June mowing, 1995) to 10.0 t ha$^{-1}$ (mulched, 1996; Fig. 1.3). June harvests had the lowest yields overall, which is understandable as the mowing time preceded the time of seasonal maximum for aboveground biomass. ANOVA indicated that both

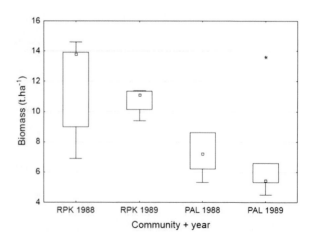

**Fig. 1.2** Seasonal course of cumulative living and dead aboveground biomass production (t ha$^{-1}$) in a stand dominated with *Phalaris arundinacea*

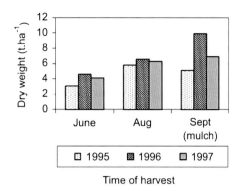

**Fig. 1.3** Mean aboveground biomass yield gained in treatments subjected to one cut per year in the field manipulative experiment on a wet grassland dominated with *Phalaris arundinacea*. Plots harvested in September were mulched. Values are based on four replicates within treatments

main effects (years, June vs. August cutting times) significantly affected the biomass yield ($p \leq 0.01$). The year x cutting time interaction was only barely significant at $p \leq 0.05$ (Table 1.2).

When multiple cuts were performed per year, the highest yield was gained in the treatment with three cuts and highest doses of fertilizer (Fig. 1.4). Further increase in cutting frequency to four cuts a year resulted in a decrease of yield regardless of fertilization regime. In most cases, the control (no NPK) resulted in the lowest comparative yield across all cutting cycles years when compared with fertilization;

**Table 1.2** ANOVA results for effects of years, cutting frequencies (1 vs. 2–4 cuts/year) and fertilizer treatments on aboveground biomass yield of stands dominated with *Phalaris arundinacea* in the field manipulative experiment

| Source of variation | | SS | DF | F | p |
|---|---|---|---|---|---|
| *One cut per year* | | | | | |
| Main effect | Years (1) | 7.713 | 2 | 13.119 | <0.01 |
| | Cutting time (2) | 29.260 | 1 | 99.524 | <0.01 |
| Replicate | | 1.627 | 3 | 1.844 | n.s. |
| Interaction | 1 × 2 | 3.098 | 2 | 5.269 | <0.05 |
| Residual | | 4.407 | 15 | | |
| Total | | 46.105 | 23 | | |
| *2–4 cuts per year* | | | | | |
| Main effect | Years (1) | 61.119 | 2 | 26.141 | <0.01 |
| | Fertilization (2) | 19.670 | 3 | 5.609 | <0.01 |
| | Cutting frequency (3) | 36.765 | 2 | 15.725 | <0.01 |
| Replicate | | 2.001 | 3 | 1.752 | n.s. |
| Interaction | 1 × 2 | 5.828 | 6 | 1.203 | n.s. |
| | 1 × 3 | 56.238 | 4 | 12.027 | <0.01 |
| | 2 × 3 | 4.164 | 6 | 1.684 | n.s. |
| | 1 × 2 × 3 | 20.459 | 12 | 1.458 | n.s. |
| Residual | | 122.707 | 105 | | |
| Total | | 326.376 | 143 | | |

*SS* sum of squares, *DF* degrees of freedom, *n.s.* non-significant effect

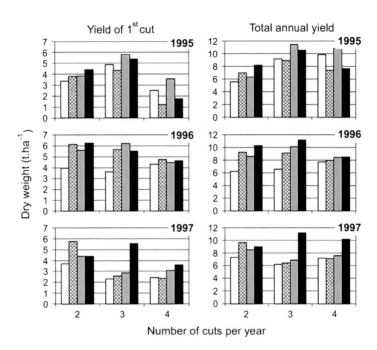

**Fig. 1.4** Yield obtained in the first cut and annual sum of yields in all successive (2–4) cuts, respectively, gained in the field manipulative experiment on a wet grassland dominated with *Phalaris arundinacea*. Note the different scale of vertical axis for the first cut and total annual yield, respectively. Values are based on four replicates within treatments. The legend indicates fertilization treatment. *Open columns*, control; *hatched columns*, PK addition; *grey columns*, NPK addition; *black columns*, 2NPK addition

the only rare exceptions were all in 1995 for the first cut and total annual yields for three and four cuts (which exceeded either the PK or PK and 2NPK applications, respectively). Comparison of annual yield among treatments also indicates that except for treatment with the highest fertilizer addition, only a small benefit resulted from increasing the cutting frequency from two to three cuts per year.

Among the treatments with $\geq 2$ cuts/year with the four fertilization treatments, significant effects for years, fertilization treatments and cutting frequency were found ($p \leq 0.01$). Years × cutting frequency was the only significant interaction ($p \leq 0.01$). Year explained the greatest part of the variation, followed by cutting frequency (Table 1.2). Yield was significantly lower ($p < 0.05$) in 1995 than in the two subsequent years. The plots mown three times a year gave significantly greater yields than plots mown twice or four times a year. Fertilization treatments did not differ between 0 NPK (control), PK and NPK, whereas 2NPK was significantly greater (Table 1.3).

The three-year long treatments brought about conspicuous changes in vegetation composition (Table 1.4). Species richness ranged from $n = 4$ (3 cuts/year, 2NPK) to $n = 17$ (4 cuts/year, no fertilization) and was generally highest in unfertilized treatments and lowest in treatments with the highest fertilizer addition (2NPK).

# 1 Biomass Production in Permanent Wet Grasslands Dominated with *Phalaris*...

**Table 1.3** Significant differences between levels of main factors (years, cutting frequency, fertilization treatments) in the aboveground biomass yield of stands dominated by *Phalaris arundinacea* in the field manipulative experiment, as determined by a three-factor analysis of variance (cf. Table 1.2)

| Years |   | Cutting frequency |   | Fertilization treatments |   |
|---|---|---|---|---|---|
| 1995 | a | 2x | b | 0 | a |
| 1996 | b | 3x | a | PK | a |
| 1997 | b | 4x | b | NPK | a |
|   |   |   |   | 2NPK | b |

Mean separations are based on Tukey post-hoc test at $\alpha = 0.05$; identical letters within the same column indicate homogenous groups at $p < 0.05$

**Table 1.4** Stand composition as indicated by species richness (expressed as the number of species), Simpson index and coverage proportion (%) of plant functional groups at the time of the first cut of the third year of field manipulative experiment with 0–4 cuts/year and four fertilizer treatments (0, PK, NPK, 2NPK) on a wet grassland dominated with *P. arundinacea*

| No of cuts/ year | Fert. trmt. | Species richness | Simpson index | Cover (%) PA | GR | C+J | V | D | BL |
|---|---|---|---|---|---|---|---|---|---|
| 0[a] | 0 | 12 | 5.32 | 13 | 9 | 6 | 0 | 47 | 25 |
| 1[b] | 0 | 8 | 1.48 | 12 | 0 | 81 | 0 | 4 | 0 |
| 1[c] | 0 | 8 | 2.04 | 66 | 2 | 21 | 0 | 11 | 0 |
| 2 | 0 | 14 | 4.37 | 28 | 9 | 36 | 0 | 24 | 3 |
| 2 | PK | 7 | 1.06 | 97 | 0 | 0 | 1 | 2 | 0 |
| 2 | NPK | 13 | 2.44 | 60 | 14 | 11 | 0 | 15 | 0 |
| 2 | 2NPK | 5 | 1.08 | 96 | 0 | 0 | 0 | 4 | 0 |
| 3 | 0 | 12 | 5.18 | 30 | 15 | 37 | 0 | 18 | 0 |
| 3 | PK | 12 | 1.34 | 86 | 3 | 5 | 0 | 6 | 0 |
| 3 | NPK | 12 | 2.36 | 60 | 15 | 20 | 0 | 5 | 0 |
| 3 | 2NPK | 4 | 1.34 | 85 | 15 | 0 | 0 | 0 | 0 |
| 4 | 0 | 17 | 3.60 | 19 | 44 | 2 | 0 | 35 | 0 |
| 4 | PK | 13 | 3.52 | 40 | 0 | 0 | 30 | 24 | 6 |
| 4 | NPK | 14 | 4.33 | 27 | 23 | 0 | 12 | 38 | 0 |
| 4 | 2NPK | 11 | 1.18 | 92 | 0 | 0 | 0 | 8 | 0 |

*PA* Phalaris arundinacea, *GR* other grasses, *C+J* Cyperaceae and Juncaceae, *V* Viciaceae, *D* Dicotyledons, *BL* bare land
[a]The plots were subjected to mulching in September
[b]Cut in June
[c]Cut in August

However, plots subjected to 4 cuts/year had a consistently higher species diversity and higher proportion of dicotyledons – including legumes – than plots cut less frequently. Simpson index values, integrating both the species richness and evenness, ranged from 1.08 (2 cuts/year, 2NPK fertilizer) to 5.32 (0 cuts/year, control fertilizer treatment) and generally copied the trend of the species richness. Application of fertilizer promoted the proportion of grasses, especially of

**Table 1.5** Mean ± standard deviation ($n=8$) of aboveground biomass (t.ha$^{-1}$), separated into yield and stubble, in the first and second cuts of wet grassland stands dominated by *Elytrigia repens*, *Alopecurus pratensis* or *Phalaris arundinacea*, subjected to long-term intensive application of pig sewage

|  | 1st cut | | 2nd cut | | |
|---|---|---|---|---|---|
| Dominant species | Yield | Stubble | Yield | Stubble | Total annual yield |
| *E. repens* | 5.17 ± 1.76 | 0.69 ± 0.51 | 5.08 ± 1.48 | 1.46 ± 0.62 | 10.25 |
| *A. pratensis* | 6.14 ± 0.84 | 0.41 ± 0.09 | 3.95 ± 0.79 | 1.50 ± 0.40 | 10.09 |
| *P. arundinacea* | 9.10 ± 1.69 | 0.41 ± 0.09 | 3.88 ± 1.06 | 1.00 ± 0.56 | 12.98 |

*P. arundinacea*, ranging as high as 97 % (2 cuts/year, PK fertilizer). On the other hand, unfertilized plots regardless of the number of cuts/year had consistently the lowest proportion of *P. arundinacea*, resulting in a greater proportion of sedges (*Carex* sp. div.) and dicotyledons.

### 1.4.4 Biomass Production of a Heavily Fertilized Wet Grassland Across a Moisture Gradient

All stands of the intensively fertilized wet grassland reached high mean values of aboveground biomass (Table 1.5). The highest mean yield of 9.1 t ha$^{-1}$ was reached by *P. arundinacea* stand in the first cut. In most cases, the first cut yielded significantly greater biomass than the second cut in stands dominated with *Alopecurus pratensis* and *P. arundinacea*, while mean yields of the *Elytrigia repens* stand were comparable in both cuts. The total annual mean yield was 13.0 t ha$^{-1}$ for the stand dominated with *P. arundinacea* and 10.1 and 10.3 t ha$^{-1}$ for the two other stand types (*A. pratensis* and *E. repens*, respectively). In contrast with mean yields of the first and second cuts, stubble biomass was significantly higher in the second cut, sometimes by as much as ~4× higher (*Alopecurus*). The greatest stubble variation, however, also occurred with the second cut for all dominant species and, in one instance, in the first cut (*Elytrigia*).

## 1.5 Concluding Remarks

### 1.5.1 Production

The results of this study document the production potential of wet grasslands dominated with *P. arundinacea* in conditions of Central Europe. The somewhat lower production of communities of wet grasslands (as compared with more intensely flushed floodplain biotopes) may reflect the differences in supply of main resources, i.e. nutrients and oxygen. If this is so, the yield would be more dependent on artificial nutrient supplement in wet grasslands on organic soils than in floodplain biotopes.

In one-cut regimes, the yield is determined mainly by timing of the harvest (Fig. 1.1, Table 1.2). Regarding cutting frequencies, three cuts per year appear to give highest annual yields in climatic conditions of Central Europe, provided adequate nutrition is ensured. Considering energy supplement necessary for each additional harvest, also the two-cut or even the one-cut regime, may be a suitable option for an extensive use of the grassland.

Both data on the biomass production of natural communities (Table 1.2) and results of the field experiment (Fig. 1.3, Table 1.3) document the extent of inter-annual variation in aboveground biomass. The key underlying factor is water availability during the vegetation season, which may be varied in biotopes suitable for the growth of *P. arundinacea*. Spring floods resulting from snow melt rather support the aboveground production of *P. arundinacea*, while summer floods can have a deleterious effect by damaging the stand. Flooding regime is a source of variation that cannot always be controlled by management and must then be considered a source of risk.

## *1.5.2 Plant Species Composition*

This study confirms that various types of management affect differently the plant species composition and representation of various plant groups, with extensive types of management giving more support to biodiversity. This corresponds to the general view that cutting of grasslands is a powerful means of affecting plant species composition. Both this and previous experimental studies have shown that several years of a certain cutting regime are enough to reverse a grassland degradation that proceeded for decades.

The results of the field manipulative experiment focused on the cutting frequency and fertilization options indicate that one- to three-year cut regime of wet grasslands dominated with *P. arundinacea* can be sustainable provided the nutrients removed from the ecosystem with the yield are replenished by suitable fertilization. Both nutrient depletion and nutrient enrichment can result in a change in plant species composition as soon as after three years.

Caution is necessary when making implications on the basis of short-term ecosystem studies for their long-term use or management. While extensive use probably imposes a fairly small risk on the ecosystem stability, a question arises about the effect of long-term intensive use of the wet grasslands. Some insight can be got on the basis of impacts of the long-term intensive management on the ecosystem of the wet meadows near Třeboň.

## 1.5.3 Possibilities of Sustainable Use

In conditions of the Czech Republic, Central Europe, stands of *P. arundinacea* were cut and used as fodder for cattle and horses until collectivization of agricultural management in the 1950s. The use of *P. arundinacea* stands has been limited since by sufficient offer of fodder of better quality on the one hand and lack of suitable mechanization for soft wetland soils. At the present time, it attracts attention as a potential energy crop. In comparison with Northern Europe, its use in Central Europe is limited by area of suitable land/habitats and existence of other plant species which give comparable or even greater yield. Nevertheless, the association of *P. arundinacea* with wet habitats favours its use in specific conditions of temporarily flooded areas, which cannot support regular production of terrestrial crops. The increasing interest in "soft" flood protection on the one hand and the increasing frequency of extreme meteorological events including floods, associated with the ongoing climate change, make *P. arundinacea* a suitable species in areas aimed primarily for soft flood control. Regular harvesting and subsequent use of biomass of *P. arundinacea* for bioenergy can therefore become an integral part of multifunctional use of temporarily flooded areas in cultural landscapes.

The selection of suitable management regime therefore depends on priorities. Extensive use with one or two cuts per year is regarded adequate for management of wet grasslands in temporarily flooded areas serving primarily for soft food control. When a trade-off is considered between production and energy supplement in the form of management, the two-cut treatment fertilized with P and K may be advantageous.

## References

Burvall, J. (1997). Influence of harvest time and soil types on fuel quality in reed canary grass (*Phalaris arundinacea* L.). *Biomass and Bioenergy, 12*, 149–154.

Buxton, D. R., Anderson, I. C., & Hallam, A. (1998). Intercropping sorghum into alfalfa and reed canarygrass to increase biomass yield. *Journal of Production Agriculture, 11*, 481–486.

Casler, M. D., Undersander, D. J., Frederick, C., Combs *Crop Science*, D. K., & Reed, J. D. (1998). An on-farm test of perennial forage grass varieties under management intensive grazing. *Journal of Production Agriculture*, 11, 92–99.

Casler, M. D., Phillips, M. M., & Krohn, A. L. (2009). DNA polymorphisms reveal geographic races of reed canarygrass. *Crop Science, 49*, 2139–2148.

Chekol, T., Vough, L. R., & Chaney, R. L. (2002). Plant-soil-contaminant specificity affects phytoremediation of organic contaminants. *International Journal of Phytoremediation, 4*, 17–26.

Cherney, J. H., Cherney, D. J. R., & Casler, M. D. (2003). Low intensity harvest management of reed canarygrass. *Agronomy Journal, 95*, 627–634.

Coulman, B. E. (1995). Bellevue reed canarygrass (*Phalaris arundinacea* L.). *Canadian Journal of Plant Science, 75*, 473–474.

Coulman, B. E., Woods, D. L., & Clark, K. W. (1977). Distribution within the plant, variation with maturity, and heritability of gramine and hordenine in reed canary grass. *Canadian Journal of Plant Science, 57*, 771–777.

Dierschke, H. (1994). *Pflanzensoziologie. Grundlagen und Methoden.* Stuttgart: Eugen Ulmer.

Dore, W. G., & McNeill, J. (1980). Grasses of Ontario. *Canada Department of Agriculture Monographs, 26*, 1–566.

Figiel, C. R., Collins, B., & Wein, G. (1995). Variation in survival and biomass of two wetland grasses at different nutrient and water levels over a six week period. *Bulletin of Torrey Botanical Club, 122*, 24–29.

Finel, l M. (2003). *The use of reed canary-grass (Phalaris arundinacea) as a short fibre raw material for the pulp and oaper industry.* Doctoral Thesis. Umeå: Swedish University of Agricultural Sciences.

Galatowitsch, S. M., Anderson, N. O., & Ascher, P. D. (1999). Invasiveness in wetland plants in temperate North America. *Wetlands, 19*, 733–755.

Groffman, P. M., Axelrod, E. A., Lemunyon, J. L., & Sullivan, W. M. (1991). Denitrification in grass and forest vegetated filter strips. *Journal of Environmental Quality, 20*, 671–674.

Gubanov, I. A., Kiseleva, K. B., Novikov, B. C., & Tihomirov, B. N. (1995). *Flora of vascular plants of central European Russia.* Moscow: Argus.

Hadders, G., & Olsson, R. (1997). Harvest of grass for combustion in late summer and in spring. *Biomass and Bioenergy, 12*, 171–175.

Hallam, A., Anderson, I. C., & Buxton, D. R. (2001). Comparative economic analysis of perennial, annual, and intercrops for biomass production. *Biomass and Bioenergy, 21*, 407–424.

Hellqvist, S., Finell, M., & Landstrom, S. (2003). Reed canary grass – observations of effects on crop stand and fibre quality caused by infestation of *Epicalamus phalaridis*. *Agriculture Food Science Finland, 12*, 49–56.

Hoveland, C. S. (1992). Grazing systems for humid regions. *Journal of Production Agriculture, 5*, 23–27.

Hrivnák, R., & Ujházy, K. (2003). The stands with the Phalaroides arundinacea dominance in the Ipeľ River catchment area (Slovakia and Hungary). *Acta Botanica Hungarica, 45*, 297–314.

Jones, T. A., Carlson, I. T., & Buxton, D. R. (1988). Persistence of reed canarygrass clones in binary mixture with alfalfa and birdsfoot trefoil. *Agronomy Journal, 80*, 967–970.

Kading, H., & Kreil, W. (1990). Optimum use and management of reed canary grass swards. *Archiv für Acker und Pflanzenbau und Bodenkunde – Archives of Agronomy and Soil Science, 34*, 489–495.

Květ, J., & Westlake, D. F. (1998). Primary production in wetlands. In D. F. Westlake, J. Květ, & A. Szczepánski (Eds.), *The production ecology of wetlands* (pp. 78–168). Cambridge: Cambridge University Press.

Květ, J., Jeník, J., & Soukupová, L. (Eds.). (2002). *Freshwater wetlands and their sustainable future: A case study of Třeboň basin biosphere reserve, Czech republic* (Man and the biosphere series 28). Paris: UNESCO and the Parthenon Publishing Group.

Lamb, J. F. S., Russelle, M. P., & Schmitt, M. A. (2005). Alfalfa and reed canarygrass response to midsummer manure application. *Crop Science, 45*, 2293–2300.

Lasat, M. M., Norvell, W. A., & Kochian, L. V. (1997). Potential for phytoextraction of Cs-137 from a contaminated soil. *Plant and Soil, 195*, 99–106.

Lavergne, S., & Molofsky, J. (2004). Reed canary grass (*Phalaris arundinacea*) as a biological model in the study of plant invasions. *Critical Reviews in Plant Science, 23*, 415–429.

Lewandowski, I., & Schmidt, U. (2006). Nitrogen, energy and land use efficiencies of miscanthus, reed canary grass and triticale as determined by the boundary line approach. *Agriculture, Ecosystems and Environment, 112*, 335–346.

Lewandowski, I., Scurlock, J. M. O., Lindvall, E., & Christou, M. (2003). The development and current status of perennial rhizomatous grasses as energy crops in the US and Europe. *Biomass and Bioenergy, 25*, 335–361.

Lyons, K. E. (1998). *Element stewardship abstract for Phalaris arundinacea L. reed canarygrass*. Arlington: The Nature Conservancy.
Merigliano, M. F., & Lesica, P. (1998). The native status of reed canary grass (*Phalaris arundinacea* L.) in the inland Northwest, USA. *Natural Areas Journal, 18*, 223–230.
Milner, C., & Hughes, R. E. (1968). *Methods for the measurement of primary production of grassland* (International Biological Programme handbook no. 6). Oxford: Blackwell Science Publisher.
Narasimhalu, P., McRae, K. B., & Kunelius, H. (1995). Hay composition, and intake and digestibility in sheep of newly introduced cultivars of timothy, tall fescue, and reed canary grass. *Animal Feed Science and Technology, 55*, 77–85.
Nilsson, D., & Hansson, P. A. (2001). Influence of various machinery combinations, fuel proportions and storage capacities on cost of co-handling of straw and reed canary grass to district heating plants. *Biomass and Bioenergy, 20*, 247–260.
Olsen, F. J., & Chong, S. K. (1991). Reclamation of acid coal refuse. *Landscape and Urban Planning, 20*, 309–313.
Ostrem, L. (1987). Studies on genetic variation in reed canary grass, *Phalaris arundinacea* L. I. Alkaloid type and concentration. *Hereditas, 107*, 235–248.
Ostrem, L. (1988). Studies on genetic variation in reed canary grass, *Phalaris arundinacea* L. II. Forage yield and quality. *Hereditas, 108*, 103–113.
Papatheofanous, M. G., Koullas, D. P., Koukios, E. G., Fuglsang, H., Schade, J. R., & Lofqvist, B. (1995). Biorefining of agricultural crops and residues: Effect of pilot-plant fractionation on properties of fibrous fractions. *Biomass and Bioenergy, 8*, 419–426.
Prach, K., Jeník, J., & Large, A. R. G. (Eds.). (1996). *Floodplain ecology and management. The Lužnice River in the Třeboň Biosphere Reserve, Central Europe*. Amsterdam: SPB Academic Publishing.
Prochnow, A., Heiermann, M., Plöchl, M., Linke, B., Idler, C., Amon, T., et al. (2009). Bioenergy from permanent grassland – A review: 1. Biogas. *Bioresource Technology, 100*, 4931–4944.
Riesterer, J. L., Undersander, D. J., Casler, M. D., & Combs, D. K. (2000). Forage yield of stockpiled perennial grasses in the upper midwest USA. *Agronomy Journal, 92*, 740–747.
Saijonkari-Pahkala, K. (2001). Non-wood plants as raw material for pulp and paper. *Agriculture and Food Science Finland, 10*, 1–101.
Samecka-Cymerman, A., & Kempers, A. J. (2001). Concentrations of heavy metals and plant nutrients in water, sediments and aquatic macrophytes of anthropogenic lakes (former open cut brown coal mines) differing in stage of acidification. *Science of Total Environment, 281*, 87–98.
Schmidt, M. A., Russelle, M. P., Randall, G. W., Sheaffer, C. C., Greub, L. J., & Clayton, P. D. (1999). Effect of rate, timing, and placement of liquid dairy manure on reed canarygrass yield. *Journal of Production Agriculture, 12*, 239–243.
Sheaffer, C. C., & Marten, G. C. (1992). Seeding patterns affect grass and Alfalfa yield in mixtures. *Journal of Production Agriculture, 5*, 328–332.
Sikora, F. J., Tong, Z., Behrends, L. L., Steinberg, S. L., & Coonrod, H. S. (1995). Ammonium removal in constructed wetlands with recirculating subsurface flow, removal rates and mechanisms. *Water Science and Technology, 32*, 193–202.
Vymazal, J. (1995). Constructed wetlands for wastewater treatment in the Czech Republic – state of the art. *Water Science and Technology, 32*, 357–364.
Vymazal, J. (2001). Constructed wetlands for wastewater treatment in the Czech Republic. *Water Science and Technology, 44*, 369–374.
Wittenberg, K. M., Duynisveld, G. W., & Tosi, H. R. (1992). Comparison of alkaloid content and nutritive values for tryptamine free and beta-carboline free cultivars of reed canary grass (*Phalaris arundinacea* L.). *Canadian Journal of Animal Science, 72*, 903–909.
Zhu, T., & Sikora, F. J. (1995). Ammonium and nitrate removal in vegetate and unvegetated gravelbed microcosm wetlands. *Water Science and Technology, 32*, 219–228.

# Chapter 2
# Greenhouse Gas Fluxes from Restored Agricultural Wetlands and Natural Wetlands, Northwestern Indiana

**Brianna Richards and Christopher B. Craft**

**Abstract** We measured gas fluxes and production efficiency rates of $CO_2$, $CH_4$, and $N_2O$ from natural and restored freshwater marshes in northwestern Indiana to evaluate the contribution of restored wetlands to regional greenhouse gas fluxes. Anaerobic soil incubations were used to determine production efficiencies, and static flux chamber measurements were used to measure fluxes during the growing season. Restored wetlands contained less soil organic carbon (1.5 % versus 6.3 %) than natural wetlands yet emitted comparable greenhouse gases in anaerobic incubations. Production efficiency rates, though, were significantly higher in restored wetlands. Mean growing season fluxes from static flux chamber measurements were 10.1 kg $CO_2$-C $ha^{-1}$ $day^{-1}$, $-0.2$ g $CH_4$-C $ha^{-1}$ $day^{-1}$, and 0.6 g $N_2O$-N $ha^{-1}$ $day^{-1}$ from natural wetlands and 3.8 kg $CO_2$-C $ha^{-1}$ $day^{-1}$, 0.1 g $CH_4$-C $ha^{-1}$ $day^{-1}$, and 0.4 g $N_2O$-N $ha^{-1}$ $day^{-1}$ from restored wetlands and did not differ among the two wetland types. We conclude that the ecological benefits gained from restoring wetlands in the glaciated interior plains outweigh the negative impact of their greenhouse gas contribution.

**Keywords** Gas flux • Methane • Restored wetlands • Natural marsh

## 2.1 Introduction

Freshwater wetlands are known sources of carbon dioxide ($CO_2$), methane ($CH_4$), and nitrous oxide ($N_2O$) (Bartlett and Harriss 1993), but their individual and cumulative contribution to global warming potential is poorly quantified (Bridgham et al. 2006). High variability exists in GHG fluxes between and within created and restored freshwater wetlands (Gleason et al. 2009; Sha et al. 2011), leading to difficulty in modeling landscape-scale fluxes. This difficulty is compounded by a

---

B. Richards • C.B. Craft (✉)
School of Public and Environmental Affairs, Indiana University, 702 North Walnut Grove, MSB II Room 408, 47405 Bloomington, IN, USA
e-mail: ccraft@indiana.edu

scarcity of data on GHG emissions from restored wetlands (Brinson and Eckles 2011).

We measured GHG ($CO_2$, $CH_4$, and $N_2O$) emissions from four restored wetlands and four natural wetlands using anaerobic incubations and static flux chambers to better understand their relative contribution to local and regional fluxes. We hypothesized that because restored wetlands often contain less soil organic matter (SOM) than natural wetlands (Bruland and Richardson 2005), they would exhibit lower fluxes of GHG. Furthermore, many restored wetlands have shorter hydroperiods and thus more aerobic soil conditions than natural sites (Knutsen and Euliss 2001) that may lead to lower $CH_4$ fluxes from restored sites.

## 2.2 Methods

### 2.2.1 Site Description

Four natural wetlands and four restored wetlands located in Newton County, northwestern Indiana, were selected for sampling (Fig. 2.1). The natural wetlands were located in Willow Slough Fish and Wildlife Area, a publicly owned area maintained by the Indiana Department of Natural Resources (DNR). Willow Slough, which encompasses 4,030 ha of land that was mostly used for agriculture in the past, was purchased by the state of Indiana between 1949 and 1951 (IN DNR 2011).

The restored wetlands were embedded in a restored prairie and oak-savanna landscape and were located in the 3,160 ha Kankakee Sands Efroymson Family Prairie Restoration owned by The Nature Conservancy (TNC). Prior to restoration, the land was a lake bed that was drained in the late 1800s for row crop agriculture (USFWS 1999). In 1999 and 2001, TNC filled in drainage ditches with sediment and regraded topography to restore wetlands as part of the Wetlands Reserve Program (C. O'Leary pers. comm.). The restored wetlands have been actively managed using a variety of methods including prescribed burns, woody and invasive species removal, and reintroduction of native species. Sites are burned approximately every three years. Undesirable plant species, such as *Equisetum hyemale* (L.), *Populus deltoides*, and *Salix nigra*, are controlled by mechanical removal and through the application of herbicides through backpack spraying, boom spraying, and basal bark treatments. Native grasses and forbs are introduced through annual seeding and plantings.

The natural wetlands are dominated by *Calamagrostis canadensis* (Michx.), *Boehmeria cylindrica* (L.), and *Scirpus cyperinus* (L.) (Hopple and Craft 2013). Dominant species in the restored wetlands include *Schoenoplectus pungens* (Vahl), *Scirpus cyperinus* (L.), *Leersia oryzoides* (L.), and *Solidago altissima* (L.). The natural wetland soils are predominantly in the Newton series (sandy, mixed, mesic typic humaquept) (NRCS 2011). Soils of the restored wetlands are in the Conrad

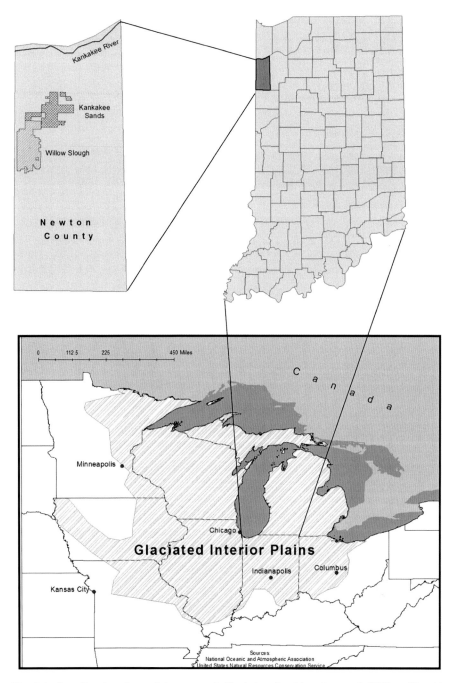

**Fig. 2.1** Sampling locations of the restored (Kankakee Sands) and natural (Willow Slough) wetlands. The sites are located in Newton County, IN, and are within the glaciated interior plains region

series (mixed, mesic typic psammaquent), which contains much less soil organic matter than the Newton series (NRCS 2011). All natural and restored wetlands sampled in this study are classified as depressional wetlands according to the HGM classification (Brinson 1993) receiving most of their water and nutrients from precipitation and are approximately 0.5 ha in size.

### 2.2.2 Greenhouse Gas Production Efficiency: Anaerobic Incubations

Five soil cores (8.5 cm diameter by 10 cm deep) were collected monthly from each wetland during the growing season from May 2011 through September 2011. Soils were transported to the lab on ice where they were homogenized and stored at 4 °C. For incubations, two replicates were prepared from each core for a total of 80 incubations per month. Incubations were conducted in glass bottles equipped with gray butyl septa. Forty mL of deoxygenated, deionized water were added to 25 g of field-moist soil at room temperature to create anaerobic conditions. Samples were then flushed with $N_2$. Five mL gas samples were withdrawn at intervals of 0.25, 1, 2, 3, and 4 h and injected into evacuated 5 mL glass vials with gray butyl septa. Five mL of $N_2$ was injected into the incubation bottles to replace the headspace. Gas samples were stored in the light at room temperature until they were analyzed.

### 2.2.3 Greenhouse Gas Measurements: Static Flux Chamber

A static flux chamber was used to measure fluxes of $CO_2$, $CH_4$, and $N_2O$ from permanent plots at two restored wetlands and two natural wetlands. Four aluminum chamber bases (50.8 cm × 50.8 cm × 3.8 cm in size) were installed in each wetland prior to the first sampling event and were held in place by 20 cm long spikes driven into the ground. Chamber bases were placed at randomly selected locations. The chamber (51 cm × 51 cm × 121 cm) was made from ¼" thick clear plexiglass and was large enough to include vegetation. During sampling, the chamber was shaded with an opaque tarp to regulate the internal chamber temperature. A small, battery-operated fan was attached to the inside of the chamber to ensure mixing. Septa were installed in the plexiglass at heights of 0.3 m and 0.9 m, and gas samples taken from each height were compared to confirm that the air in the chamber was well mixed.

Gases were sampled monthly from June 2011 to September 2011 between 8 am and 5 pm, and the order in which wetlands were sampled was varied each month. At 0, 5, 10, 20, and 30 min intervals, 5 mL of gas was withdrawn from the chamber and injected into evacuated 5 mL glass vials with gray butyl septa. Vials were stored in the light at room temperature and were analyzed as described below.

## 2.2.4 Laboratory Analyses

Gas samples were analyzed using a SRI 8610C gas chromatograph (GC) (SRI Instruments, Menlo Park, CA). Methane concentrations were determined using a flame ionization detector, and $CO_2$ concentrations were determined using a flame ionization detector coupled with a methanizer. Nitrous oxide concentrations were determined using an electron capture detector. 500 µL of each sample was injected, and $N_2$ was used as a carrier gas. The GC was calibrated using four standard concentrations, each for $CO_2$, $CH_4$, and $N_2O$ (Scotty Analyzed Gases balanced with $N_2$, Plumsteadville, PA). A blank and check standard were run after every ten samples.

## 2.2.5 Supporting Data

Surface water levels were measured monthly at the chamber base and soil core locations. Soil cores were collected from the top 10 cm. Soil moisture content and percent soil moisture of incubation soils (0–10 cm) were determined gravimetrically by weighing 3–5 g of soil at field-moist conditions and then reweighing after drying to a constant weight at 70 °C. Bulk density was calculated by dividing the weight of the dried core by the volume of the core (Blake and Hartge 1986). Porosity was calculated by the formula (Flint and Flint 2002):

1. $(Bulk\ Density / Particle\ Density)$

The particle density was assumed to be 2.65 (Singer and Munns 2006). Water-filled pore space (WFPS) was calculated using soil moisture and porosity data. Total nitrogen and organic carbon were measured using a Perkin-Elmer CHN Analyzer (Perkin-Elmer, Waltham, MA). An in-house soil standard (5.85 % C, 0.35 % N) had average recovery rates for organic carbon and total nitrogen of 96 % and 95 %.

## 2.2.6 Statistical Analyses

The data were analyzed using a three-way analysis of variance with wetland type, wetland number, and sampling date as the main effects variables (SAS 1996). The Kolmogorov-Smirnov and Shapiro-Wilk tests were used to test for normality. Chamber fluxes were determined from the change in concentration over time and were analyzed in SAS using repeated measure analysis of variance with wetland type and sampling date as the main effects variables (SAS 1996). Fluxes were natural log transformed prior to statistical analysis to achieve normality. Organic carbon, total nitrogen, bulk density, and WFPS were analyzed with a *t*-test to test for differences between natural and restored wetlands. Pearson correlation test was used to explore associations between supporting data, incubation emissions, and chamber fluxes. All tests of significance were performed at $\alpha = 0.05$

## 2.3 Results

### 2.3.1 Greenhouse Gas Measurements: Anaerobic Incubations

There were no differences in mean GHG emissions of anaerobic incubations from natural and restored wetland soils, and though differences occurred in some months, there were no consistent trends (Table 2.1). In June and September, mean $CO_2$ emissions from natural wetland soils (0.5 and 0.4 mg C kg$^{-1}$ h$^{-1}$) were significantly higher than from restored wetland soils (0.2 and 0.1 mg C kg$^{-1}$ h$^{-1}$). Conversely, mean $CO_2$ emissions in May and July were significantly higher from restored wetland soils (1.3 and 0.4 mg C kg$^{-1}$ h$^{-1}$) than from natural wetland soils (0.8 and 0.1 mg C kg$^{-1}$ h$^{-1}$). Nitrous oxide emissions also varied among months. Emissions in May and September were greater from natural wetland soils (0.7 and 0.2 µg N kg$^{-1}$ h$^{-1}$) than from restored wetland soils (0.1 and −0.02 µg N kg$^{-1}$ h$^{-1}$), whereas in July and August, $N_2O$ emissions from restored wetlands (1.0 and 0.2 µg N kg$^{-1}$ h$^{-1}$) were greater than from natural wetlands (0.1 and 0.1 µg N kg$^{-1}$ h$^{-1}$). Methane emissions did not vary between months, and natural and restored emissions did not differ in any month.

Restored wetlands had significantly higher GHG production efficiency rates than natural wetlands (Table 2.2). Carbon dioxide production efficiency rates were higher in restored wetlands in May, July, August, and September and were not significantly different in June. Methane production efficiency rates were higher in restored wetlands in June and August and were not significantly different in May, July, and September. Nitrous oxide production efficiencies were higher in restored wetlands in July and August and were not significantly different in May, June, and September.

### 2.3.2 Greenhouse Gas Measurements: Static Flux Chamber

Similar to the incubations of greenhouse gas production, wetland fluxes measured with the chamber did not differ between natural and restored wetlands (Fig. 2.2). No differences in $CO_2$ fluxes were observed between natural and restored wetlands in any month. Differences in $CH_4$ and $N_2O$ fluxes were observed in a few months, but there were no consistent trends. All fluxes had very high spatial and temporal variability. Mean $CO_2$ fluxes from natural and restored wetlands were 10.1 kg $CO_2$-C ha$^{-1}$ day$^{-1}$ and 3.6 kg $CO_2$-C ha$^{-1}$ day$^{-1}$, respectively. $CH_4$ flux means

**Table 2.1** Mean emissions (± standard errors) during the 2011 growing season of $N_2O$, $CO_2$, and $CH_4$ from restored and natural wetland soil using an anaerobic incubation laboratory method

| Month | Type | $N_2O$ emission ($\mu g\ N\ kg^{-1}\ h^{-1}$) | $CO_2$ emission ($mg\ C\ kg^{-1}\ h^{-1}$) | $CH_4$ emission ($\mu g\ C\ kg^{-1}\ h^{-1}$) |
|---|---|---|---|---|
| May | Restored | 0.13 ± 0.03* | 1.27 ± 0.25* | 0.17 ± 0.09 |
| | Natural | 0.73 ± 0.26* | 0.79 ± 0.10* | 0.09 ± 0.05 |
| June | Restored | 0.04 ± 0.03 | 0.18 ± 0.09* | 0.31 ± 0.18 |
| | Natural | 0.09 ± 0.06 | 0.47 ± 0.11* | 0.40 ± 0.15 |
| July | Restored | 1.03 ± 0.84* | 0.40 ± 0.05* | 0.07 ± 0.23 |
| | Natural | 0.08 ± 0.03* | 0.11 ± 0.08* | 0.09 ± 0.17 |
| Aug | Restored | 0.23 ± 0.03* | 0.33 ± 0.07 | 0.28 ± 0.08 |
| | Natural | 0.09 ± 0.01* | 0.25 ± 0.12 | 0.21 ± 0.09 |
| Sept | Restored | −0.02 ± 0.01* | 0.13 ± 0.01* | 0.00 ± 0.00 |
| | Natural | 0.19 ± 0.01* | 0.41 ± 0.03* | −0.06 ± 0.09 |
| **Mean** | **Restored** | **0.28 ± 0.19** | **0.37 ± 0.29** | **0.17 ± 0.07** |
| | **Natural** | **0.23 ± 0.13** | **0.39 ± 0.15** | **0.15 ± 0.08** |
| | GA FW forests[a] | 38–400 | 0.40–1.1 | 120–190 |
| | NC salt marshes[b] | | | |
| | Created | – | 0.01–0.03 | <0.001 |
| | Natural | – | 0.01–0.04 | <0.001 |

Emissions are reported per kilogram of soil on a dry weight basis. Asterisks represent significant differences ($p < 0.05$) between wetland types according to ANOVA analysis
[a]Marton et al. (2012)
[b]Cornell et al. (2007)

**Table 2.2** Mean production efficiency rates (±standard errors) during the 2011 growing season of $N_2O$, $CO_2$, and $CH_4$ from restored and natural wetland soil using an anaerobic incubation laboratory method.

| Month | Type | $N_2O$ emission (µg N kg$^{-1}$ h$^{-1}$) | $CO_2$ emission (mg C kg$^{-1}$ h$^{-1}$) | $CH_4$ emission (µg C kg$^{-1}$ h$^{-1}$) |
|---|---|---|---|---|
| May | Restored | 0.13 ± 0.05 | 1.01 ± 0.24* | 0.08 ± 0.07 |
|  | Natural | 0.18 ± 0.07 | 0.14 ± 0.03* | 0.02 ± 0.03 |
| June | Restored | 0.02 ± 0.01 | 0.18 ± 0.05 | 0.40 ± 0.16* |
|  | Natural | 0.01 ± 0.01 | 0.07 ± 0.03 | 0.06 ± 0.04* |
| July | Restored | 0.79 ± 0.28* | 0.38 ± 0.08* | 0.10 ± 0.17 |
|  | Natural | 0.01 ± 0.01* | 0.02 ± 0.01* | −0.03 ± 0.07 |
| Aug | Restored | 0.14 ± 0.05* | 0.24 ± 0.04* | 0.27 ± 0.11* |
|  | Natural | 0.03 ± 0.01* | 0.04 ± 0.01* | 0.05 ± 0.01* |
| Sept | Restored | −0.01 ± 0.06 | 0.11 ± 0.01* | 0.00 ± 0.00 |
|  | Natural | 0.04 ± 0.01 | 0.08 ± 0.01* | −0.02 ± 0.02 |
| **Mean** | **Restored** | **0.23 ± 0.09*** | **0.39 ± 0.10*** | **0.17 ± 0.11*** |
|  | **Natural** | **0.06 ± 0.02*** | **0.07 ± 0.02*** | **0.02 ± 0.03*** |
|  | GA tidal | 0.53–1.48 | 0.001–0.002 | 392–4,076 |
|  | FW forests[a] |  |  |  |
|  | NC salt marshes[b] |  |  |  |
|  | Natural | – | 0.70–1.11 | <0.001 |
|  | Created | – | 0.81–3.89 | <0.001 |

Rates are reported in carbon and nitrogen per kilogram of soil organic carbon basis. Asterisks represent significant differences ($p < 0.05$) between wetland types according to ANOVA analysis

[a]Marton et al. (2012)
[b]Cornell et al. (2007)

**Fig. 2.2** Monthly net ecosystem fluxes of (**a**) $CO_2$ (g-$CO_2$-C ha$^{-2}$ day$^{-1}$), (**b**) $CH_4$ (g-$CH_4$-C ha$^{-2}$ day$^{-1}$), and (**c**) $N_2O$ (g-$N_2O$-N ha$^{-2}$ day$^{-1}$) from natural and restored wetlands sampled from June to September 2011 using a static flux chamber. Different letters represent significant differences between wetland types according to ANOVA analysis. Fluxes were natural log transformed prior to statistical analysis. Error bars represent the untransformed standard error of the mean

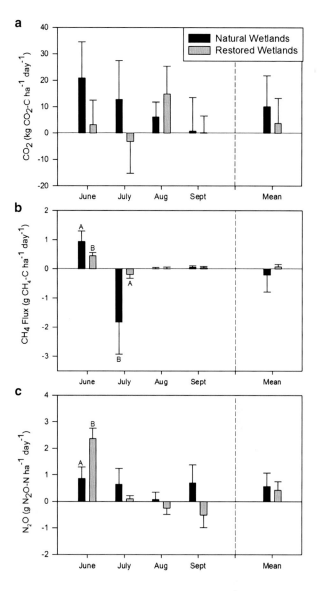

from natural and restored wetlands were very low, −0.2 and 0.1 g $CH_4$-C ha$^{-1}$ day$^{-1}$, respectively. Negative fluxes of $CH_4$ were observed in natural and restored wetlands in July. Growing season $N_2O$ fluxes from natural and restored wetlands were 0.6 g $N_2O$-N ha$^{-1}$ day$^{-1}$ and 0.4 g $N_2O$-N ha$^{-1}$ day$^{-1}$, respectively. Restored wetlands had a greater flux than natural wetlands in June.

Carbon dioxide was the largest contributor to global warming potential (in $CO_2$ equivalents) in restored (97.9 %) and natural (98.3 %) wetlands, and $CH_4$ had the smallest contribution, 0.1 % and 0 % in restored and natural wetlands, respectively

**Table 2.3** Growing season means (±standard errors) of water-filled pore space (WFPS), bulk density, percent organic carbon, and percent total nitrogen (equivalent mass to 15 cm) for restored and natural wetland soils

| | Bulk density (g cm$^{-3}$) | Organic C Mg ha$^{-1}$ | (%) | Total N Mg ha$^{-1}$ | (%) | Moisture (%) | Porosity (%) | WFPS (%) |
|---|---|---|---|---|---|---|---|---|
| Restored | 0.97 ± 0.02* | 21.2 | 1.5 ± 0.6* | 1.1 | 0.12 ± 0.02* | 24.7 ± 0.7* | 63.2 ± 0.9* | 24.7 ± 1.0* |
| Natural | 0.44 ± 0.02* | 41.9 | 6.3 ± 2.5* | 2.6 | 0.58 ± 0.02* | 41.2 ± 1.4* | 83.3 ± 0.9* | 18.1 ± 1.0* |
| ND restored wetlands[a] | .... | 49.8 | .... | 2.9 | .... | .... | .... | 20–80 |
| ND natural wetlands[b] | | | | | | | | |
| Deep marsh | .... | 54 | .... | .... | .... | 65 | .... | .... |
| Shallow marsh | .... | 54 | .... | .... | .... | 55 | .... | .... |
| Wet meadow | .... | 49 | .... | .... | .... | 45 | .... | .... |
| OH created wetlands[c] | .... | .... | 7.2 | .... | 0.3 | .... | .... | .... |
| Prairie pothole restored Wetlands[d] | .... | 50–100 | .... | .... | .... | .... | .... | .... |

Asterisks represent significant differences ($p < 0.05$) between wetland types according to ANOVA analysis. Values of mean organic carbon, mean total nitrogen, and mean water-filled pore space (WFPS) of natural and restored wetlands in northwestern Indiana are compared with other freshwater mineral soil wetlands

[a]Gleason et al. (2009) *Values were used from the grassland wetlands only
[b]Phillips and Beeri (2008)
[c]Hernandez and Mitsch (2006) *Values were used from the 2005 high marsh study only
[d]Badiou et al. (2011)

(Table 2.3). Nitrous oxide contributed 2.0 % of the global warming potential in restored wetlands and 1.7 % in natural wetlands.

### 2.3.3 Supporting Data

Although we observed no differences in GHG emissions between natural and restored wetlands, soil properties were dramatically different (Table 2.3). Bulk density in natural wetlands (0.44 g cm$^{-3}$) was half of that measured in restored wetlands (0.97 g cm$^{-3}$). Soil organic C and total N were four times greater in natural wetlands (6.33 % C, 0.58 % N) than in restored wetlands (1.45 % C, 0.12 % N). Soil organic C concentrations in the top 10 cm of soil were 21.8 Mg ha$^{-1}$ in restored wetlands and 41.6 Mg ha$^{-1}$ in natural wetlands. Porosity and percent moisture were also greater in natural wetlands (41 %, 83 %) than in restored wetlands (25 %, 63 %). Water-filled pore space (WFPS) was significantly higher in restored wetland soils (25 %) than in natural wetland soils (18 %).

## 2.4 Discussion

Greenhouse gas emissions and fluxes did not differ between natural and restored wetlands. Fluxes were comparable with prairie pothole wetlands in North Dakota, which share similar characteristics such as undergoing periodic burns and existing in a prairie mosaic. Natural (10.1 kg $CO_2$-C ha$^{-1}$ day$^{-1}$) and restored (3.8 kg $CO_2$-C ha$^{-1}$ day$^{-1}$) flux chamber measurements of $CO_2$ were comparable to fluxes from deep marsh (4.5 kg $CO_2$-C ha$^{-1}$ day$^{-1}$), shallow marsh (11.6 kg $CO_2$-C ha$^{-1}$ day$^{-1}$), and wet meadow (8.6 kg $CO_2$-C ha$^{-1}$ day$^{-1}$) vegetative zones in North Dakota prairie pothole wetlands (Phillips and Beeri 2008).

Methane fluxes also did not differ between natural and restored wetland soils, but unlike $CO_2$ emissions, levels were low compared with other studies. Methane fluxes for natural (−0.2 g $CH_4$-C ha$^{-1}$ day$^{-1}$) and restored (0.1 g $CH_4$-C ha$^{-1}$ day$^{-1}$) wetlands were five orders of magnitude lower than in North Dakota deep marshes (18,000 g $CH_4$-C ha$^{-1}$ day$^{-1}$), shallow marshes (7,000 g $CH_4$-C ha$^{-1}$ day$^{-1}$), and wet meadows (3,000 g $CH_4$-C ha$^{-1}$ day$^{-1}$) (Phillips and Beeri 2008). We attribute the low rates of $CH_4$ emissions in our natural and restored wetland to the predominantly aerobic soil conditions during sampling. Our wetlands dried down in late June and maintained a mean water-filled pore space (WFPS) of between 15 and 30 % throughout the entire growing season (Fig. 2.3). Gleason et al. (2009) observed net $CH_4$ consumption in restored grassland wetlands when WFPS dropped below 40 %. These findings suggest that, during the growing season, our wetlands did not maintain the anaerobic conditions required for net $CH_4$ production. Furthermore, the prolonged aerobic conditions may have suppressed

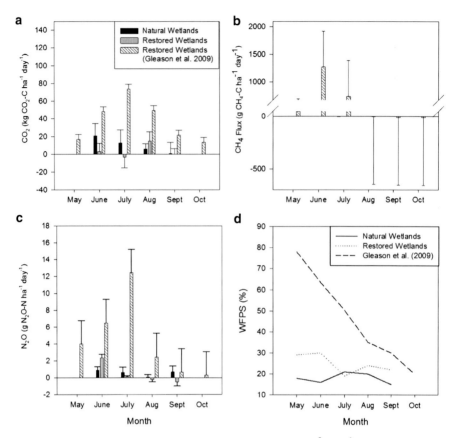

**Fig. 2.3** Monthly net ecosystem fluxes of (**a**) $CO_2$ (g-$CO_2$-C ha$^{-2}$ day$^{-1}$), (**b**) $CH_4$ (g-$CH_4$-C ha$^{-2}$ day$^{-1}$), and (**c**) $N_2O$ (g-$N_2O$-N ha$^{-2}$ day$^{-1}$) and (**d**) monthly water-filled pore space from the natural and restored wetlands in this study and the restored grassland wetlands in Gleason et al. (2009). Flux measurements were made during the 2011 growing season using a static flux chamber. No measurements for natural and restored wetlands were made in May and October. *Error bars* represent the untransformed standard error of the mean

the populations of methanogens in the soil (Shannon and White 1994), leading to low $CH_4$ production rates observed in our anaerobic laboratory incubations.

Similar to $CH_4$, $N_2O$ fluxes did not differ between natural and restored wetlands and were low compared to values reported in the literature. Nitrous oxide fluxes from our natural and restored wetlands measured using static chambers were one order of magnitude lower than natural and restored wetlands in North Dakota (Fig. 2.3c). However, once the WFPS of the restored grassland wetlands measured by Gleason et al (2009) dropped below 40 %, their mean flux (0.6 g $N_2O$-N ha$^{-1}$ day$^{-1}$) was comparable to growing season fluxes of our natural wetlands (0.6 g-$N_2O$-N ha$^{-1}$ day$^{-1}$) and restored wetlands (0.4 g-$N_2O$-N ha$^{-1}$ day$^{-1}$) (Fig. 2.3c, d), suggesting that the aerobic soil conditions limited net $N_2O$ production.

**Table 2.4** The relative global warming potential contributions of $CO_2$, $CH_4$, and $N_2O$ from natural and restored wetlands compared with other freshwater mineral soil wetlands

|  | $CO_2$ (%) | $CH_4$ (%) | $N_2O$ (%) |
|---|---|---|---|
| Restored | 97.9 | 0.1 | 2.0 |
| Natural | 98.3 | 0.0 | 1.7 |
| ND restored wetlands[a] | 90 | 9 | 1 |
| ND natural wetlands[b] |  |  |  |
| Deep marsh | 6.0 | 92.0 | 2.0 |
| Shallow marsh | 49.0 | 48.0 | 3.0 |
| Wet meadow | 83.6 | 0.4 | 16.0 |
| OH created wetlands[c] | 99.8 | 0.2 | <0.1 |

[a]Gleason et al. (2009). Values were used from the grassland wetlands only
[b]Phillips and Beeri (2008)
[c]Altor and Mitsch (2008) ($CO_2$ and $CH_4$ values), Hernandez and Mitsch (2006) ($N_2O$ values). Values were used from 2005 only

We observed no correlations between greenhouse gas fluxes and measured soil properties (organic carbon, total nitrogen, bulk density, and WFPS). This is likely because hydrology was the main limiting factor. Because the wetlands dried down so early in the growing season, the soil did not maintain anaerobic conditions ideal for greenhouse gas production. This led to a low WFPS. Other studies have observed that maximum $N_2O$ emissions occur when WFPS is between 40 % and 60 % (Davidson et al. 2000; Gleason et al. 2009), and maximum $CH_4$ emissions occur when WFPS approaches 100 % (Gleason et al. 2009). The WFPS of wetland soils in this study only exceeded 40 % in 3.5 % of the soil samples and never exceeded 60 %. This suggests that soil saturation is a key limiting factor in $N_2O$ and $CH_4$ production of our restored and natural wetlands.

Our study is among the few that compared greenhouse gas emission ($CO_2$, $CH_4$, and $N_2O$) between natural and restored freshwater mineral soil wetlands. Our chamber measurements found no differences in greenhouse gas fluxes between natural and restored wetlands. This finding is in conflict with Badiou et al. (2011) and Nahlik and Mitsch (2010). Badiou et al. (2011) found that restored wetlands in the prairie pothole region of Canada contributed more $CH_4$ and $N_2O$ than natural wetlands. Nahlik and Mitsch (2010) found the opposite trend and observed higher $CH_4$ fluxes in a natural Ohio wetland than in two nearby created wetlands. All three studies exhibited high spatial and temporal variability in greenhouse gas emissions, suggesting that more frequent measurements are needed to constrain the contribution of restored wetlands on a landscape scale.

Of the three GHGs, carbon dioxide contributed the most (98 %) to cumulative GHG emissions in our natural and restored wetlands (Table 2.4). This trend was also seen in created wetlands in Ohio (99.8 %) (Nahlik and Mitsch 2010), restored wetlands in North Dakota (90 %) (Gleason et al 2009), and, to a lesser extent, shallow marshes (49 %) and wet meadows (84 %) in North Dakota (Phillips and Beeri 2008). Methane was the lowest contributor to global warming potential in our natural (0 %) and restored (0.1 %) wetlands and was also the lowest contributor in the North Dakota wet meadows (0.4 %). Nitrous oxide was the lowest contributor in created marshes in Ohio (<0.1 %), restored wetlands in North Dakota (1 %), and deep (2 %) and shallow (3 %) marshes in North Dakota.

Using our flux data, we estimate that the restored wetlands contributed 0.66 Mg $CO_2$ equivalents ha$^{-1}$ during the growing season, and natural wetlands contributed 1.9 Mg $CO_2$ equivalents ha$^{-1}$. It should be noted that these values are likely lower than the true emissions as sampling began after late April, thus missing the beginning of the growing season.

Previous studies have demonstrated that the restored wetlands used in this study provide water quality and biodiversity benefits (Hopple and Craft 2013, Marton et al. 2014), and from our measurements, we conclude that the ecological benefits of the restored wetlands outweigh potential negative impacts due to GHG emissions. Because of the high spatial and temporal variability of fluxes observed, additional GHG measurements are needed to constrain greenhouse emissions of restored wetlands in the glaciated interior plains and elsewhere.

**Acknowledgments** We would like to thank Anya Hopple, Ellen Herbert, Laura Trice, Anne Altor, and Aida Haddad for their assistance with field sampling and laboratory analyses. We would also like to the thank The Nature Conservancy at Kankakee Sands and the Indiana Department of Natural Resources at Willow Slough Fish and Wildlife Area for access to their wetlands. This project was funded by the USDA Conservation Effects Assessment Project (CEAP) through Cooperative Agreement Number 68-7482-9-516 from the Great Lakes – Northern Forests Cooperative Ecosystems Studies Unit (CESU).

# References

Altor, A., & Mitsch, W. (2008). Pulsing hydrology, methane emissions and carbon dioxide fluxes in created marshes: a 2-year ecosystem study. *Wetlands, 28*(2), 423–438.

Badiou, P., McDougal, R., Pennock, D., & Clark, B. (2011). Greenhouse gas emissions and carbon sequestration potential in restored wetlands of the Canadian prairie pothole region. *Wetlands Ecology and Management, 19*, 237–256.

Bartlett, K. B., & Harriss, R. C. (1993). Review and assessment of methane emissions from wetlands. *Chemosphere, 26*, 261–320.

Blake, G. R., & Hartge, K. H. (1986). Bulk density. In A. Klute (Ed.), *Methods of soil analysis, Part I. Physical and mineralogical methods* (Agronomy monograph no. 9 2nd ed., pp. 363–375). Madison: American Society of Agronomy.

Brander, L. M., Florax, R., & Vermaat, J. E. (2006). The empirics of wetland valuation: A comprehensive summary and a meta-analysis of the literature. *Environmental and Resource Economics, 33*(2), 223.

Bridgham, S. D., Megonigal, J. P., Keller, J. K., Bliss, N. B., & Trettin, C. (2006). The carbon balance of North American wetlands. *Wetlands, 26*(4), 889–916.

Brinson, M. (1993). *A hydrogeomorphic classification for wetlands*. Wetlands Research Program TR-WRP-DE-4. US Army Waterways Exp. Station, Vicksburg, MS.

Brinson, M., & Eckles, S. (2011). U.S. Department of Agriculture conservation program and practice effects on wetland ecosystem services: A synthesis. *Ecological Applications, 21*(3), S116–S127.

Bruland, G., & Richardson, C. (2005). Spatial variability of soil properties in created, restored, and paired natural wetlands. *Soil Science Society of America Journal, 69*, 273–284.

Cornell, J., Craft, C., & Megonigal, P. (2007). Ecosystem gas exchange across a created salt marsh chronosequence. *Wetlands, 27*(2), 240–250.

Craft, C. B., & Casey, W. P. (2000). Sediment and nutrient accumulation in floodplain and depressional freshwater wetlands of Georgia, USA. *Wetlands, 20*(2), 323–332.

Craft, C. B., Broome, S., & Campbell, C. (2002). Fifteen years of vegetation and soil development after brackish-water marsh creation. *Restoration Ecology, 10*(2), 248–258.

Craft, C., Krull, K., & Graham, S. (2007). Ecological indicators of nutrient enrichment, freshwater wetlands, Midwestern United States (U.S.). *Ecological Indicators, 7*, 733–750.

Dahl, T. E. (1990). *Wetland losses in the United States: 1780's to 1980's*. Washington, DC: U.S. Fish and Wildlife Service.

Davidson, E. A., Keller, M., Erickson, H. E., Verchot, L. V., & Veldkamp, E. (2000). Testing a conceptual model of soil emissions of nitrous and nitric oxides. *BioScience, 50*, 667–680.

Euliss, N. H., Gleason, R. A., Olness, A., McDougal, R. L., Murkin, H. R., Robarts, R. D., Bourbonniere, R. A., & Warner, B. G. (2006). North American prairie wetlands are important nonforested land-based carbon storage sites. *Science of the Total Environment, 361*, 179–188.

Fennessy, S., & Craft, C. (2011). Agricultural conservation practices increase wetland ecosystem services in the Glaciated Interior Plains. *Ecological Applications, 21*(3), S49–S64.

Flint, L. E., & Flint, A. L. (2002). Porosity. In A. Klute (Ed.), *Methods of soil analysis, Part I. Physical and mineralogical methods* (Agronomy monograph no. 9 2nd ed., pp. 241–253). Madison: American Society of Agronomy.

Gleason, R. A., Tangen, B. A., Browne, B. A., & Euliss, N. H. (2009). Greenhouse gas flux from cropland and restored wetlands in the Prairie Pothole Region. *Soil Biology and Biochemistry, 41*, 2501–2507.

Hernandez, M. E., & Mitsch, W. J. (2006). Influence of hydrologic pulses, flooding frequency, and vegetation on nitrous oxide emissions from created riparian marshes. *Wetlands, 26*, 862–77.

Hopple, A., & Craft, C. (2013). Managed disturbance enhances biodiversity of restored wetlands in the agricultural Midwest. *Ecological Engineering, 61B*, 505–510.

Indiana Department of Natural Resources (IN DNR). Willow Slough. http://www.in.gov/dnr/fishwild/3080.htm. Accessed 15 Nov 2011.

Keller, M., Mitre, M. E., & Stallard, R. F. (1990). Consumption of atmospheric methane in soils of central Panama: Effects of agricultural development. *Global Biogeochemical Cycles, 4*, 21–27.

Knutsen, G. A., & Euliss, N. H., Jr. (2001). *Wetland restoration in the Prairie Pothole Region of North America: A literature review* (Biological science report 2001–0006). Reston: U.S. Geological Survey.

Marton, J., Herbert, E., & Craft, C. (2012). Effects of salinity on denitrification and greenhouse gas production from laboratory-incubated tidal forest soils. *Wetlands, 32*, 347–357.

Marton, J. M., Fennessy, M. S., & Craft, C. B. (2014). Functional differences between natural and restored wetlands in the Glaciated Interior Plains. *Journal of Environmental Quality, 43*, 409–417.

Mitsch, W. J., Day, J. W. Jr., Gilliam, J. W., Groffman, P. M., Hey, D. L., Randall, G. W., & Wang, N. (1999). *Reducing nutrient loads, especially nitrate–nitrogen, to surface water, ground water, and the Gulf of Mexico: Topic 5 report for the integrated assessment on hypoxia in the Gulf of Mexico*. NOAA Coastal Ocean Program Decision Analysis Series No. 19. NOAA Coastal Ocean Program, Silver Spring, MD, 111 pp.

Nahlik, A. M., & Mitsch, W. J. (2010). Methane emissions from created riverine wetlands. *Wetlands, 30*, 783–793.

Natural Resources Conservation Service. (2011). Web soil survey, United States Department of Agriculture. http://websoilsurvey.sc.egov.usda.gov/App/HomePage.htm. Accessed Nov 2011.

Nelson, D. W., & Sommers, L. E. (1996). Total carbon, organic carbon, and organic matter. In D. L. Sparks (Ed.), *Methods of soil analysis, Part 3: Chemical methods* (Soil Science Society of America, book series, Vol. 5, pp. 961–1010). Madison: Soil Science of Society of America.

Phillips, R., & Beeri, O. (2008). The role of hydropedologic vegetation zones in greenhouse gas emissions for agricultural wetland landscapes. *Catena, 72*, 386–394.

Poffenbarger, H. J., Needleman, B. A., & Megonigal, J. P. (2011). Salinity influence on methane emissions from tidal marshes. *Wetlands, 31*, 831–842.

Power, J. F., Koerner, P. T., Doran, J. W., & Wilhelm, W. W. (1998). Residual effects of crop residues on grain production and selected soil properties. *Soil Science Society of America Journal, 62*, 1393–1397.

SAS (Statistical Analysis Systems). (1996). *SAS user's guide*. Cary: SAS Institute.

Segers, R. (1998). Methane production and methane consumption: A review of processes underlying wetland methane fluxes. *Biogeochemistry, 41*, 23–51.

Sha, C., Mitsch, W., Mander, U., Lu, J., Batson, J., Zhang, L., & He, W. (2011). Methane emissions from freshwater riverine wetlands. *Ecological Engineering, 37*, 16–24.

Shannon, R., & White, J. (1994). A three-year study of controls on methane emissions from two Michigan peatlands. *Biogeochemistry, 27*(1), 35–60.

Singer, M. J., & Munns, D. N. (2006). *Soils: An introduction* (6th ed.). Upper Saddle River: Pearsons Prentice Hall.

Soil Survey Staff. (2010). Natural Resources Conservation Service, United States Department of Agriculture. Web Soil Survey, Newton County, Indiana. Available online at: http://websoilsurvey.nrcs.usda.gov/. Accessed 18 Oct 2011.

U. S. FWS. (1999). Chronology of Important Events on the Kankakee River. In *Grand Kankakee Marsh National Wildlife Refuge environmental assessment*. Fort Snelling, MN. Available online at: http://www.fws.gov/midwest/planning/GrandKankakee/. Accessed 4 Dec 2011

United States Department of Agriculture. (2008). *At a glance: Wetlands reserve program*. Natural Resource Conservation Service.

Zedler, J. (2003). Wetlands at your service: Reducing impacts of agriculture at the watershed scale. *Frontiers in Ecology and the Environment, 1*(2), 65–72.

# Chapter 3
# Assessment of Immobilisation and Biological Availability of Iron Phosphate Nanoparticle-Treated Metals in Wetland Sediments

**Herbert John Bavor and Batdelger Shinen**

**Abstract** A study was carried out to investigate the feasibility of use of iron phosphate (vivianite) nanoparticles for long-term stabilisation of heavy metals, in particular copper and zinc, in wetland sediments. The effectiveness of the treatment was examined by assessing nanoparticle sequestered metal leachability, bioavailability and speciation in sediments using USEPA standard methods: a toxicity characteristic leaching procedure (TCLP), a physiologically based extraction test (PBET) and a sequential extraction procedure (SEP). The impact of the nanoparticles and nanoparticle sequestered metal on bioactivity of plants and bacterial growth was also investigated. Inhibition of microbial growth was reduced in sediments treated with the nanoparticle preparation, and a bioaccumulating capacity was found for selected plants at high concentrations (1,000 ppm) of Cu and Zn, in the presence of nanoparticles. This finding suggests that metal bioavailability may have been reduced for microbial populations in nanoparticle-treated microcosms; however, the bioaccumulation finding suggests that select plant species could be used in combination with the vivianite as phytoremediation tools.

**Keywords** Bioavailability • Heavy metals • Immobilisation • Nanoparticles • Phytoremediation • Wetlands

## 3.1 Introduction

Heavy metals may be biologically available or may be stable in soils, and many are considered persistent bioaccumulative toxins. In order to mitigate metal toxicity, researchers have studied several remediation techniques, including mobilisation or immobilisation of the metals by physicochemical techniques (Raicevic et al. 2005; Hettiarachchi et al. 2000) and biological approaches (Marchiol et al. 2004; Marchiol et al. 2007) in addition to physical removal of sediments. More recently,

---

H.J. Bavor (✉) • B. Shinen
Centre for Water and Environmental Technology – Water Research Laboratory, University of Western Sydney – Hawkesbury, Locked Bag 1797, Penrith, NSW 2751, Australia
e-mail: j.bavor@uws.edu.au

nanotechnology-based methods have been reported in remediation of contaminated soils and sediments (He et al. 2006; Liu and Zhao 2007; Li et al. 2007; Fajardo et al. 2012). Long-term stability data for metals sequestered with metal oxide surfaces is scant. In general, information on management of heavy metals in wetland sediments and sequestration in plants is very limited (Vymazal et al. 2011). Further, due to significantly different properties of sediments and soils, technologies that work for soil may not be as effective when utilised in sediments (Mulligan et al. 2001).

We investigated the application of iron phosphate (vivianite) nanoparticles for immobilisation of heavy metals in wetland sediment. Vivianite nanoparticles were synthesised and used to treat metal-contaminated wetland sediment. Following treatment, the distribution and mobility of metals in sediments amended and not amended with the nanoparticles were determined and microbial activity of the sediment stabilised with iron phosphate nanoparticles and the impact of the nanoparticles and nanoparticle sequestered metal on bioactivity of plants were assessed.

## 3.2 Methods

To investigate the application of an iron-phosphate complex for metals immobilisation, vivianite nanoparticle synthesis was carried out by mixing iron sulphate and sodium phosphate under anoxic conditions using a technique adapted from Liu and Zhao (2007). Column experiments were then performed in the laboratory in order to examine nanoparticle sequestration of metal in wetland sediment loaded with metal then amended with a vivianite nanoparticle suspension. A flow diagram showing the sequence of wetland sediment spiking with metals, application of vivianite nanoparticle preparation (vNP) and subsequent extraction/leaching bioavailability assessment is shown in Fig. 3.1.

Metal leachability, bioavailability and speciation in the sediment were assessed using USEPA standard methods: a toxicity characteristic leaching procedure (TCLP) (USEPA 1992), a physiologically based extraction test (PBET) (Kelly et al. 2002; Nordwick 2008; Ruby et al. 1999) and a sequential extraction procedure (SEP) (Rauret 1998).

Metal concentrations in sediments and leachates were determined by atomic absorption spectrometry, and microbial populations were characterised using methods described in Roser et al. (1987).

The bioavailability of metals in wetland sediment treated with vivianite was assessed in the terrestrial plants, *Allium tuberosum* and *Helianthus annuus*. These species have been reported by several researchers to accumulate high concentrations of heavy metals such as lead, cadmium, copper and zinc and to have potential in phytoremediation applications (Soudek et al. 2010; Broadley et al. 2001). The species are also often found in wetland habitats in fringing communities.

# 3 Assessment of Immobilisation and Biological Availability of Iron Phosphate...

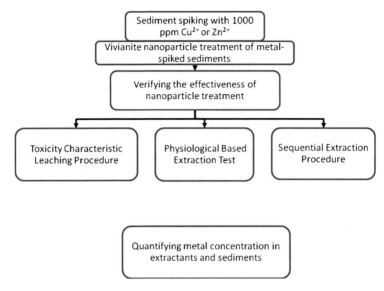

**Fig. 3.1** Flow chart showing sequence of wetland sediment spiking with metals, application of vivianite nanoparticle preparation and subsequent extraction/leaching. Sediments and leachate from control and treatment cores were then analysed, using leaching, extraction and in vivo techniques, to quantify metals

## 3.3 Results and Discussion

Using the TCLP procedure, metal immobilisation was assessed in sediments, with and without vNP amendment. Figure 3.2 shows metal concentrations in the leachates for the sediments spiked with both metals.

After applying vNP suspension (pH 5.6) to the sediment spiked with both metals, the leachate copper concentration lowered from 515.4 mg $L^{-1}$ to 119.8 mg $L^{-1}$ (77 % reduction), while the zinc concentration reduced from 272.5 mg $L^{-1}$ to 52.8 mg $L^{-1}$ (81 % reduction) in the leachates. The results indicated that the nanoparticle treatment is effective in reducing the leachability of metal in single metal-spiked sediment and also sediment spiked with both Cu and Zn.

The PBET assay, adapted from Kelly et al. (2002), evaluated the bioavailability of Cu and Zn in sediments treated with vNPs. In the untreated sediment 22.5 % of the Cu and 17.4 % of the Zn was bioavailable (Fig. 3.3). However, the bioavailability of Cu and Zn was reduced to 5 % and 0.6 %, respectively, when the sediments were amended with vNPs. The total reduction of bioavailability was 78 % for Cu and 97 % for Zn, indicating that the vivianite nanoparticles were effective in reducing bioavailability of copper and zinc in sediments.

Chemical speciation is an important determinant of metal mobility and binding characteristics. The SEP test was used to assess the effect of the nanoparticle treatment on Cu and Zn speciation in the sediment. The availability of each metal speciation in sediment treated and not treated with vNP was investigated using five

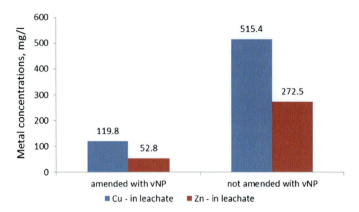

**Fig. 3.2** Comparison of copper and zinc concentrations in TCLP leachates from sediment spiked with both elements, 1,000 ppm Cu and 1,000 ppm Zn, to concentrations in the leachates from sediment with 1,000 ppm Cu and 1,000 ppm Zn content but amended with 10 mM vivianite nanoparticle suspension

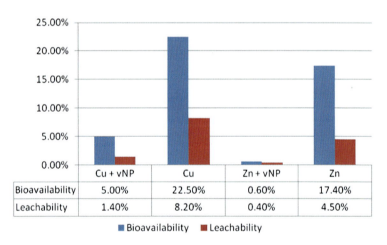

**Fig. 3.3** Leachability (using TCLP) and bioavailability (using PBET) characteristics of copper and zinc in sediments spiked with 1,000 ppm Cu and Zn, amended and not amended with 10 mM vivianite nanoparticle suspension

steps in a sequential extraction procedure. Figure 3.4 shows the percentage of the copper species in each fraction of sediment, amended and not amended with 10 mM vNP suspension. The copper species was determined in five fractions as the following: (1) exchangeable (EX), (2) carbonate bound (CB), (3) oxides bound (OX), (4) organic matter bound (OM) and (5) residual (RS). In metal-spiked and non-amended sediments, 11 % of copper was bound to sediment in the EX, 60.3 % in the CB, 18.3 % in the OX, 7.3 % in the OM and 3.1 % in the RS forms. In the sediment amended with vNP, the EX and CB bound fractions reduced from 11 % to 6.5 % and from 60.3 % to 52.6 %, respectively, while the OX, OM and RS fractions

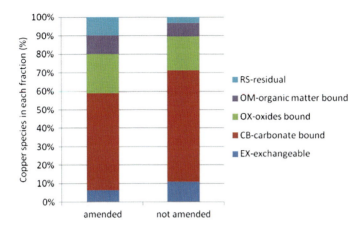

**Fig. 3.4** Sequential extraction (SEP) of copper into five operationally defined fractions in sediment spiked with 1,000 ppm Cu and Zn, amended and not amended with 10 mM vivianite nanoparticle suspension

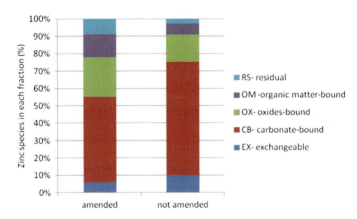

**Fig. 3.5** Sequential extraction of zinc into five operationally defined fractions in sediment spiked with 1,000 ppm Cu and Zn, amended and not amended with 10 mM vivianite nanoparticle suspension

increased from 18.3 % to 21.1 %, from 7.3 to 9.9 % and from 3.1 % to 9.9 %, respectively. The results shown in Fig. 3.4 indicate the copper species moved from the EX and CB to the OX, OM and RS phases when the sediment was amended with 10 mM vivianite nanoparticle suspension. After nanoparticle treatment, the copper concentration in the RS sediment increased by almost 3 times (from 0.6 mg kg$^{-1}$ to 1.7 mg kg$^{-1}$) compared to the non-amended sediments.

Similarly, in 1,000 ppm Cu- and Zn-spiked sediment, Fig. 3.5 shows the percentage of the zinc species in each of five fractions of sediments amended and not amended with 10 mM vNP suspension.

In non-amended sediments, 10 % of the zinc was bound to sediment in the EX, 65.3 % in the CB, 15.8 % in the OX, 6.2 % in the OM and 2.7 % in the RS forms. In sediments amended with 10 mM vNP suspension, the EX and CB bound fractions reduced from 10 % to 5.9 % and from 65.3 % to 49.2 %, respectively, while the OX, OM and RS fractions increased from 15.8 % to 22.7 %, from 6.2 to 13.3 % and from 2.7 % to 8.9 %, respectively. Zinc species shifted from the EX and CB fractions to the less mobile OX, OM and RS phases when the sediment was amended with the vivianite nanoparticles. After nanoparticle treatment, the zinc concentration in the RS sediment increased by more than 3 times from 2.7 mg kg$^{-1}$ to 8.9 mg kg$^{-1}$ compared to non-amended sediment.

The results of batch experiments to quantify the copper and zinc uptake in *H. annuus* and *A. tuberosum* species grown in a mixture of wetland sediment and peat moss loaded with Cu or Zn or both Cu and Zn with concentration of (1) 100, (2) 250, (3) 500 and (4) 1,000 ppm, respectively, showed a significant difference in the uptake and metal distribution within plants, with and without nanoparticle treatment. Figure 3.6 shows metal distribution in *A. tuberosum* trials. At lower metal spiking concentrations of 100 to 200 ppm, the root compartment accumulated higher concentrations of both Cu and Zn.

In experiments using *H. annuus*, much higher metal spiking concentrations were used, reflecting the higher tolerance to Cu and Zn by the plant (Fig. 3.7). The metals were found to be concentrated in the root fraction of the plant, similar to the response shown by *A. tuberosum*.

The results for the uptake partitioning are similar to those reported for the uptake of copper and zinc concentrations, which were found to be higher in the root- and sediment-associated compartments than in the leaves and stems of *Phalaris*

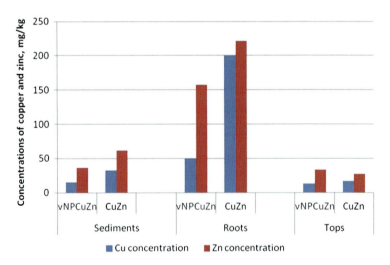

**Fig. 3.6** Copper and zinc distribution in *Allium tuberosum* growing on mixture of sediment and peat moss, with ratio 1:1, loaded with both metals, 100 ppm Cu and 100 ppm Zn, then amended and not amended with 10 mM vivianite nanoparticle suspension

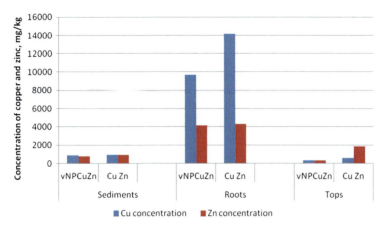

**Fig. 3.7** Copper and zinc distribution in *Helianthus annuus* growing on mixture of sediment and peat moss, with ratio 1:1, loaded with both metals, 1,000 ppm Cu and 1,000 ppm Zn, then amended and not amended with 10 mM vivianite nanoparticle suspension

*arundinacea* and *Phragmites australis* when these species were used to remove heavy metals from constructed wetland treating municipal sewage (Vymazal and Krasa 2005).

## 3.4 Conclusions

Previous researchers have synthesised iron phosphate nanoparticles and studied stabilisation of heavy metals in the shorter-term, 1–3 month periods. This study investigated the interaction of nanoparticles with biological systems and the stabilising heavy metals in wetland sediments in simulated "long-term" studies using a toxicity characteristic leaching procedure (TCLP), a physiologically based extraction test (PBET) and a sequential extraction procedure (SEP).

Experimental data indicated that stabilised vivianite nanoparticles can effectively decrease the leachability and bioimpact of heavy metals. In considering the treatment mechanism of vivianite nanoparticle application, it may be hypothesised that stable metal complexes have formed, causing a shift in metal speciation from more soluble fractions in the sediment to less soluble, more stable *residual* fractions, reducing metal mobility. The nanoparticle treatment was found to stimulate microbial populations in wetland sediments spiked with copper and/or zinc. The study found that metal bioavailability may be reduced for microbial populations. However, bioaccumulation results suggest that select plant species could be used with vivianite as phytoremediation tools.

# References

Broadley, M. R., Willey, N. J., Wilkins, J. C., Baker, A. J. M., Mead, A., & White, P. J. (2001). Phylogenetic variation in heavy metal accumulation in angiosperms. *New Phytologist, 152*(1), 9–27.

Fajardo, C., Ortíz, L. T., Rodríguez-Membibre, M. L., Nande, M., Lobo, M. C., & Martín, M. (2012). Assessing the impact of zero-valent iron (ZVI) nanotechnology on soil microbial structure and functionality: A molecular approach. *Chemosphere, 86*, 802–808.

He, F., Zhao, D., Liu, J., & Roberts, C. B. (2006). Stabilization of Fe – Pd nanoparticles with sodium carboxymethyl cellulose for enhanced transport and dechlorination of trichloroethylene in soil and groundwater. *Industrial and Engineering Chemistry Research, 46*, 29–34.

Hettiarachchi, G. M., Pierzynski, G. M., & Ransom, M. D. (2000). In situ stabilisation of soil lead using phosphorus and manganese oxide. *Environmental Science and Technology, 34*(21), 4614–4619.

Kelly, M. E., Brauning, S. E., Schoof, R. A., & Ruby, M. V. (2002). *Assessing oral bioavailability of metals in oil* (124p.). Columbus: Battelle Press.

Li, X., Brown, D. G., & Zhang, W. (2007). Stabilization of biosolids with nanoscale zero-valent iron (nZVI). *Journal of Nanoparticle Research, 9*, 233–243.

Liu, R., & Zhao, D. (2007). Reducing leachability and bioaccessibility of lead in soils using a new class of stabilized iron phosphate nanoparticles. *Water Research, 41*, 2491–2502.

Marchiol, L., Assolari, S., Sacco, P., & Zerbi, G. (2004). Phytoextraction of heavy metals by canola (*Brassica napus*) and radish (*Raphanus sativus*) grown on multicontaminated soil. *Environmental Pollution, 132*, 21–27.

Marchiol, L., Fellet, G., Perosa, D., & Zerbi, G. (2007). Removal of trace metals by *Sorghum bicolor* and *Helianthus annuus* in a site polluted by industrial wastes: A field experience. *Plant Physiology and Biochemistry, 45*, 379–387.

Mulligan, C. N., Yong, R. N., & Gibbs, B. F. (2001). Remediation technologies for metal-contaminated soils and groundwater: An evaluation. *Engineering Geology, 60*, 193–207.

Nordwick, S. (2008). Linking waterfowl with contaminant speciation in riparian soils. Mine Waste Technology Program. US-EPA Report: EPA/600/R-08/060

Raicevic, S., Kaludjerovic-Radoicic, T., & Zouboulis, A. (2005). In situ stabilisation of toxic metals in polluted soils using phosphates: Theoretical prediction and experimental verification. *Journal of Hazardous Materials, 117*(1), 41–53.

Rauret, G. (1998). Extraction procedures for the determination of heavy metals in contaminated soil and sediment. *Talanta, 46*, 449–455.

Roser, D. J., McKersie, S. A., Fisher, P. F., Breen, P. F., & Bavor, H. J. (1987). Sewage treatment using aquatic plants and artificial wetlands. *Water, 14*, 20–24.

Ruby, M. V., Schoof, R., Brattin, W., Goldade, M., Post, G., Harnois, M., Mosby, D. E., Casteel, S. W., Berti, W., Carpenter, M., Edwards, D., Cragin, D., & Chappell, W. (1999). Advances in evaluating the oral bioavailability of inorganics in soil for use in human health risk assessment. *Environmental Science and Technology, 33*, 3697–3705.

Soudek, P., Petrová, Š., Benešová, D., & Vaněk, T. (2010). Phytoextraction of toxic metals by sunflower and corn plants. *Journal of Food Agriculture and Environment, 8*(3–4), 383–390.

USEPA. (1992). US EPA Method 1311. Environmental Protection Agency, Washington, DC, USA.

Vymazal, J., & Krása, P. (2005). Heavy metals budget for constructed wetland treating municipal sewage. In J. Vymazal (Ed.), *Natural and constructed wetlands: Nutrients, metals and management* (pp. 135–142). Leiden: Backhuys Publishers.

Vymazal, J., Kröpfelová, L., Švehla, J., & Němcová, J. (2011). Heavy metals in *Phalaris arundinacea* growing in a constructed wetland treating municipal sewage. *International Journal of Environmental Analytical Chemistry, 91*(7–8), 753–767.

# Chapter 4
# Spatial Variability in Sedimentation, Carbon Sequestration, and Nutrient Accumulation in an Alluvial Floodplain Forest

Jacob M. Bannister, Ellen R. Herbert, and Christopher B. Craft

**Abstract** We measured soil properties, vertical accretion, and nutrient (organic C, N, and P) accumulation across a range of habitats to evaluate spatial variability of soil properties and processes of alluvial floodplain wetlands of the Altamaha River, Georgia, USA. The habitats vary in elevation and distance from the river channel, creating differences in the depth and duration of inundation. Habitats closer to the river had lower bulk density and higher total P than habitats further removed. $^{137}$Cs and $^{210}$Pb accretion rates were also greater at sites closer to the channel. Mineral sediment deposition and nutrient accumulation were greater in sloughs closer to the channel and lower in elevation relative to other habitats. We found distance to be a significant predictor of mineral soil properties across the floodplain. Bulk density increased whereas TP and silt content decreased with distance from the river channel. $^{137}$Cs accretion, P accumulation, and mineral sediment deposition also decreased with distance from the main channel. Elevation was not a significant predictor of soil properties or processes measured. Long-term (100 year) sediment accumulation rates based on $^{210}$Pb were significantly higher than 50-year rate of sedimentation based on $^{137}$Cs, perhaps as the result of greater land clearing for agriculture and lack of best management practices in the southeastern USA prior to 1950. Distance from the main channel is the driving force behind the spatial variability of soil properties and processes measured; however, slough habitats closest to the channel and lowest in elevation relative to other habitats maintain distinct vegetation patterns and are hotspots for N, P, and sediment accumulation. Characterization of soil properties and processes of alluvial floodplain forests and other wetlands should take into consideration microtopographic and spatial variation across the wetland.

**Keywords** Carbon sequestration • Alluvial floodplain forest • Wetland soils • Soil accretion

---

J.M. Bannister • E.R. Herbert • C.B. Craft (✉)
School of Public and Environmental Affairs, Indiana University, 702 N. Walnut Grove Ave., MSB II Room 408, Bloomington, IN 47405, USA
e-mail: ccraft@indiana.edu

## 4.1 Introduction

Freshwater wetland soils are important sinks for sediment, organic carbon (C), nitrogen (N), and phosphorus (P) (Brinson et al. 1981a, b; Burke 1975; Craft and Casey 2000; Hopkinson 1992; Martin and Hartman 1987; Mausbach and Richardson 1994; Naiman and Décamps 1997), and a wetland's ability to trap sediment and accumulate nutrients depends on its connectivity to other systems (Mitsch and Goselink 2000). Wetlands such as riparian and alluvial floodplain forests are hydrologically open systems, with larger catchment areas and higher anthropogenic inputs (e.g., nutrients and sediment) than hydrologically closed systems, such as depressional wetlands (Craft and Casey 2000), and as a result, alluvial floodplain wetlands act as nutrient and sediment sinks, reducing sediment deposition and eutrophication downstream in receiving water bodies (Noe and Hupp 2005).

The alluvial floodplain forest floor is a heterogeneous mix of hummocks, hollows, and other microtopographic features that are defined by differences in hydroperiod and soil type (Wharton 1978). The small-scale differences in elevation between microtopographic sites have important implications for ecosystem structure and function (Courtwright and Findlay 2011). Microsites are associated with the maintenance of species diversity in both upland and wetland forest ecosystems (Ehrenfeld 1994). Microtopography influences the frequency and duration of flooding (Rheinhardt 1992) and creates variation in biogeochemical cycling (Darke and Walbridge 2000). Microtopographic relief results in adjacent areas of aerobic and anaerobic soil, which facilitate nutrient cycling (Bridgham et al. 2001; Mosier et al. 2009) including carbon sequestration and N cycling and removal (Wolf et al. 2011). Higher rates of litterfall production and phosphorus and cation (K, Mg) circulation have been found on lower, wetter sites in the floodplain relative to higher, drier sites (Schilling and Lockaby 2005). Studies of alluvial floodplain forest soils to date have focused on the effects of topographic heterogeneity and microtopography on plant communities, net primary production (NPP), and nutrient cycling (Ehrenfeld 1994; Schilling and Lockaby 2005, 2006; Courtwright and Findlay 2011; Wolf et al. 2011), but none have investigated relationships between microtopography and sediment deposition, carbon sequestration, or nutrient (N and P) accumulation.

Using feldspar marker layers, Noe and Hupp (2005) found that sediment and nutrient (C, N, P) accumulation was greater in hydraulically connected floodplain wetlands than in floodplains (Virginia-Maryland-Delaware, USA) where the river was channelized and levees emplaced. Kroes et al. (2007) investigated sediment accretion using the same methods in one of the same rivers (Pokomoke MD). They reported no difference in accretion (mm/year) across the floodplain. Percent organic matter increased with distance from the river, suggesting that mineral sediment deposition decreases with distance from the channel (Kroes et al. 2007). In tidal marshes, sedimentation and nutrient accumulation tend to be greater adjacent to the tidal creek as compared to interior areas (Craft 2007; Loomis and Craft 2010; Craft

2012). Also in tidal marshes, studies indicate that sedimentation is a function of depth and duration of tidal inundation (Morris et al. 2002) and distance from the creek (Temmerman et al. 2005). However, direct measurements are lacking for nontidal alluvial forested wetlands.

We measured soil accretion and nutrient and sediment accumulation across four microtopographic habitats spanning the gradient from the river to upland, in an alluvial floodplain forest along the Altamaha River, Georgia, USA. The four habitats vary in elevation and distance from the river and presumably differ in depth, duration, and frequency of inundation. We hypothesize that sediment deposition and phosphorus accumulation will decrease with distance from the main channel and with elevation in the floodplain. We also hypothesize that carbon sequestration and nitrogen accumulation will be greatest in the areas of lowest elevation.

## 4.2 Methods

### 4.2.1 Site Description

The study site is located on the Altamaha River in Griffin Ridge Wildlife Management Area, Long Co., southeastern Georgia (31°68′ N, 81°82′ W) (Fig. 4.1). The Altamaha is a red water river high in sediment and mineral nutrients derived from the oxidized clay soils of the piedmont, which are reddish in color (Conner et al. 2007). The Altamaha drains nearly 23 million hectares of the piedmont and coastal plain, with an average discharge of 166.3 $m^3 \ s^{-1}$ and sediment yield of 57.1 $kg \ ha^{-1} year^{-1}$ (Schilling and Lockaby 2006) (Altamaha River #02226000 1939-2009; USGS 2010).

Four habitats were sampled along a cross-sectional gradient of the floodplain: levee, slough, floodplain, and upper slough. Each habitat differs in elevation and distance from the main river channel and hence in the degree of inundation and alluvial inputs from floodwaters (Fig. 4.2). Vegetation is dominated by *Platanus occidentalis* L. (sycamore) and *Betula nigra* L. (river birch) along the levee; *Nyssa ogeche* Bartr. ex Marsh. (Ogeechee gum), *Nyssa aquatic* L. (water tupelo), *Taxodium distichum* L. (bald cypress), and *Taxodium ascendens* Brongn. (pond cypress) in the sloughs; and *Fraxinus pennsylvanica* Marsh. (green ash), *Acer rubrum* L. (red maple), and *Quercus spp.* (oak) in floodplain locations (Craft pers. obs.). Soils consist of Tawcaw series (somewhat poorly drained, fine, kaolinitic, thermic Fluvaquentic Dystrudepts) in the higher-lying areas (e.g., levee and floodplain) with a Chastain series (poorly drained, fine, mixed, semiactive, acid, thermic Fluvaquentic Endoaquepts) in the sloughs and lower-lying positions (USDA 2010). Sample collection sites were located in areas close to the river and within the floodplain. However, Griffin Ridge, a dry, elevated sand ridge that

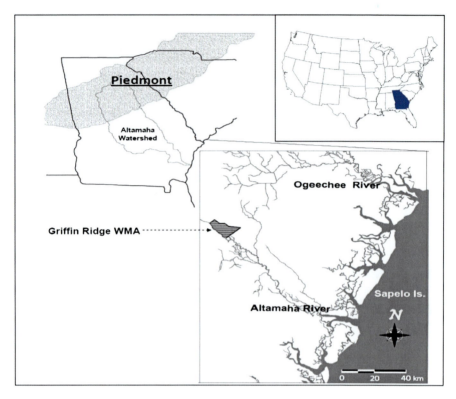

**Fig. 4.1** Map of Griffin Ridge Wildlife Management Area in relation to the Piedmont Region and Georgia Coast

straddles the floodplain, was excluded from sampling due to soils and vegetation not characteristic of wetland areas.

## 4.2.2 Soil Sampling and Analysis

Soil cores, 8.5 cm diameter by 60 cm deep, were collected from each habitat. Two cores each were collected from the slough and upper slough habitats, while three cores each were collected from the levee and floodplain habitats (Fig. 4.2.). Cores were sectioned in the field with the top 30 cm separated into 2-cm increments and the bottom 30 cm into 5-cm increments. Each increment was air dried, weighed for bulk density, and then ground and sieved through a 2-mm mesh screen. A subsample was oven-dried at 85 °C, for determination of moisture content, organic carbon, total nitrogen, and total phosphorus. Bulk density was calculated from the dry weight per unit volume for each depth increment after correcting for moisture content with the oven-dried subsample (Blake and Hartge 1986). Organic C and

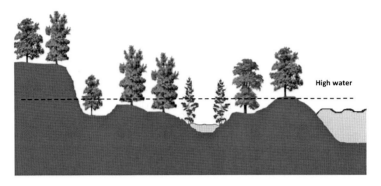

**Fig. 4.2** Schematic of the four landscape positions sampled relative to typical spring peak floodwater levels

N were determined using a PerkinElmer 2400 CHN analyzer (PerkinElmer Corporation, Waltham, MA). Analysis of an in-house standard (% $C = 6.1$, % $N = 0.365$) returned average recovery rates of 103 % and 102 % for C and N, respectively. Total P was determined by colorimetric analysis following digestion with $HNO_3$–$HClO_4$ (Sommers and Nelson 1972). Analysis of estuarine sediment (NIST #1646a) returned an average recovery rate of 88 % for total P. Percent sand, silt, clay, and organic matter (0–10 cm) were determined by the hydrometer method (Gee and Bauder 1986).

Soil accretion was determined using ground and sieved soil increments packed into 50 mm-diameter by 9 mm-deep petri dishes which were analyzed for $^{137}Cs$ and $^{210}Pb$ by gamma analysis of the 661.62 keV and 46.5 keV photopeaks, respectively (Craft and Casey 2000). The $^{137}Cs$ maximum in each core corresponds with the peak fallout from ground thermonuclear weapons testing, which occurred in 1964. $^{210}Pb$ is a naturally occurring radionuclide (half-life of 22 years) that can be used to estimate soil accretion during the past 100–150 years (Craft and Casey 2000). The constant activity model was used to determine accretion using the distribution of excess $^{210}Pb$ (total $^{210}Pb$ minus background $^{210}Pb$) (Oldfield and Appleby 1984; Schelske et al. 1988). Accumulation of organic C, N, and P was calculated using the $^{137}Cs$ and $^{210}Pb$ accretion rates, bulk density, and nutrient concentrations (C, N, and P) down to the increment containing the peak $^{137}Cs$ and excess $^{210}Pb$ activities, respectively, for each core.

## 4.2.3 GIS/LiDAR Distance and Elevation Calculation

Elevations of the four habitats were estimated through interactive examination of a LiDAR dataset of the area (Photo Science and Fugro EarthData, Inc. 2010). A 10 m-resolution surface model was generated by using NOAA's Digital Coast Data Access Viewer to filter the average, last return points from the 2010 Coastal Georgia Topographic LiDAR dataset for the study area (NOAA 2011). The dataset was queried using ArcGIS 10 Desktop software (ESRI 2011) for 10 m areas along the transect to estimate the elevations (relative to NAVD88) for the primary topographic features noted in the field (levee, sloughs, and floodplain).

## 4.2.4 Statistical Analysis

Bulk soil properties, accretion, and accumulation were analyzed using a one-way analysis of variance with habitat as the main effects variable. Student's t-test was used to test for differences in $^{137}$Cs versus $^{210}$Pb accretion and accumulation. Tests of normality and homoscedasticity revealed that the data were normally distributed and estimate a common variance. Regression analysis was used to test the effects of distance from main channel and elevation on soil properties, accretion, and accumulation (SAS 1996). All tests of significance were conducted at $\alpha = 0.05$.

## 4.3 Results

Bulk soil properties (0–30 cm) varied among habitats across the Altamaha River floodplain. Bulk density was greater at sites furthest from the river, in the upper floodplain and upper slough (1.2 g cm$^{-3}$), and lowest in the slough (0.75 g cm$^{-3}$) ($p < 0.02$, $F = 7.5$) (Table 4.1). There was no difference in percent organic carbon or total nitrogen among habitats. Total P, however, was higher in the slough (530 μg/g) than in the upper slough and floodplain (120 μg g$^{-1}$) ($p < 0.05$, $F = 7.1$) (Table 4.1). Soils (0–10 cm) were composed predominantly of sand particles (63–82 %) with lesser amounts of silt (10–24 %), clay (3–7 %), and organic matter (5–10 %) (Fig. 4.3). There was no significant difference in particle size fractions among habitats.

$^{137}$Cs profiles revealed interpretable peaks, corresponding to 1964, of the maximum $^{137}$Cs fallout from aboveground nuclear weapons tests. $^{210}$Pb also exhibited interpretable total and excess $^{210}$Pb concentrations, decreasing with depth corresponding to its exponential decay with time (Fig. 4.4). $^{137}$Cs soil accretion ranged from 0.2 to 3.3 mm year$^{-1}$ and was greater in the slough than in the floodplain and upper slough ($p < 0.01$, $F = 12.2$) (Fig. 4.5). $^{137}$Cs organic C sequestration ranged from 7 to 61 g m$^{-2}$ year$^{-1}$ (Fig. 4.5). Total N and P

# 4 Spatial Variability in Sedimentation, Carbon Sequestration, and Nutrient...

**Table 4.1** Mean bulk density, organic carbon, total nitrogen, and phosphorus (0–30 cm) (±1 S.E) across the floodplain of the Altamaha River. Different letters represent significant differences ($p < 0.05$) according to ANOVA analysis

| Landscape position | Bulk density (g/cm$^3$) | Carbon (%) | Nitrogen (%) | Total P (ug/g) |
|---|---|---|---|---|
| Upper slough | 1.2 ± 0.06 a | 1.02 ± 0.31 | 0.07 ± 0.02 | 120 ± 21 b |
| Floodplain | 1.2 ± 0.07 a | 2.35 ± 0.48 | 0.18 ± 0.03 | 240 ± 25 b |
| Slough | 0.8 ± 0.04 b | 2.33 ± 0.21 | 0.20 ± 0.01 | 530 ± 17 a |
| Levee | 0.9 ± 0.04 a, b | 1.83 ± 0.27 | 0.16 ± 0.02 | 370 ± 22 a, b |

**Fig. 4.3** Particle-size distribution and organic matter content (0–10 cm) of soils across the floodplain of the Altamaha River

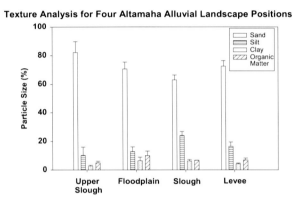

accumulation ranged from 0.4 to 5.4 g N m$^{-2}$ year$^{-1}$ and 0.04 to 1.15 g P m$^{-2}$ year$^{-1}$. As with accretion, N ($p < 0.05$, $F = 6.4$) and P ($p < 0.01$, $F = 20.9$) accumulation were significantly greater in the slough than the upper slough and floodplain (P only). Sediment deposition on the levee and in the slough also was greater relative to habitats (floodplain, upper slough) far removed from the main channel ($p < 0.01$, $F = 12.2$) (Fig. 4.6).

Like $^{137}$Cs, $^{210}$Pb accretion was significantly greater in the slough ($p < 0.05$, $F = 5$) than in the upper slough and the levee as well. C sequestration ranged from 34 to 68 g C m$^{-2}$ year$^{-1}$, while N and P accumulation ranged from 2.7 to 7.8 g N m$^{-2}$ year$^{-1}$ and 0.41 to 1.5 g P m$^{-2}$ year$^{-1}$. The slough also exhibited greater P accumulation than other habitats ($p < 0.05$, $F = 7.2$). Likewise, N accumulation was greater in the slough than the levee and upper slough habitats ($p < 0.05$, $F = 6.5$). No differences were found in $^{210}$Pb sediment deposition (1,170–2,760 g m$^{-2}$ year$^{-1}$) across the floodplain.

Mean rates of $^{210}$Pb soil accretion (2.5 ± 0.6 mm year$^{-1}$) were consistently but not significantly higher than accretion based on $^{137}$Cs (1.6 ± 0.6 mm/year). However, $^{210}$Pb accumulation of mineral sediment (1,870 ± 300 g m$^{-2}$ year$^{-1}$) was significantly greater than sedimentation rates based on $^{137}$Cs (1,030 ± 250 g m$^{-2}$ year$^{-1}$) ($p < 0.05$). Like accretion, $^{210}$Pb accumulation of organic C, N, and P was consistently higher than rates based on $^{137}$Cs but not significantly so.

**Fig. 4.4** Representative $^{137}$Cs and $^{210}$Pb profiles for 1 L levee site

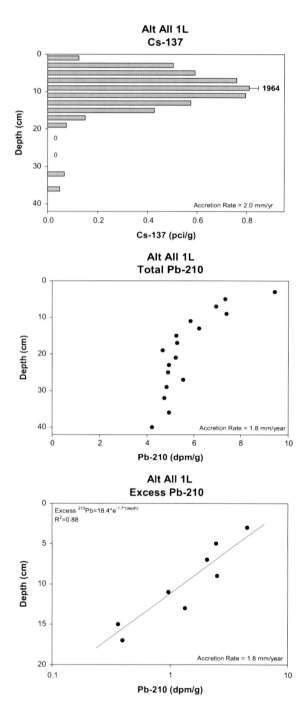

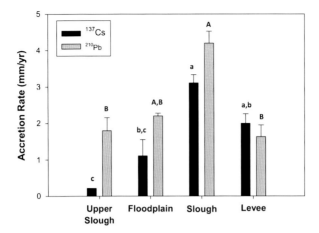

**Fig. 4.5** Mean $^{137}$Cs- and $^{210}$Pb-based soil accretion ($\pm 1$ S.E.) across the floodplain of the Altamaha River. Different letters indicate significant differences ($p < 0.05$) according to ANOVA analysis

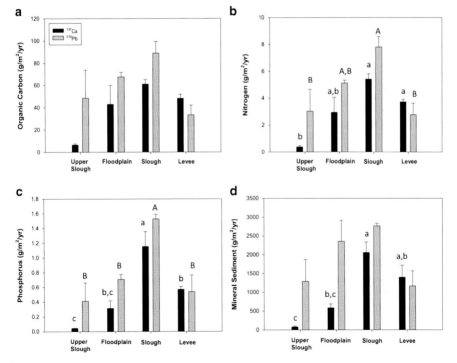

**Fig. 4.6** Mean (**a**) carbon sequestration and accumulation of (**b**) nitrogen, (**c**) phosphorus, and (**d**) mineral sediment ($\pm 1$ S.E.) across the floodplain of the Altamaha River. Different letters indicate significant differences ($p < 0.05$) according to ANOVA analysis

Multiple regression analysis revealed that soil properties and processes associated with mineral fraction were significantly related to distance from the main channel. Bulk density increased with distance from the main channel ($r^2 = 0.78$, $p = 0.002$), whereas silt content ($r^2 = 0.43$, $p = 0.05$), $^{137}$Cs-based soil accretion ($r^2 = 0.68, p = 0.006$), P accumulation ($r^2 = 0.69, p = 0.006$), and mineral sediment deposition ($r^2 = 0.75, p = 0.002$) decreased with distance from the channel, though $^{210}$Pb accretion and accumulation rates were not predicted by distance. We found no relationship between soil elevation and surface soil properties or processes.

## 4.4 Discussion

### 4.4.1 Spatial Patterns

Microtopography is known to affect wetland plant productivity and diversity and biogeochemical processes, yet little research has investigated its effect on sedimentation and nutrient accumulation processes. It has been shown that small variations in elevation can affect wetland structure and function, such as nutrient flow and redox conditions (Courtwright and Findlay 2011), diversity of plant communities (Ehrenfeld 1994), and biogeochemical cycling (Rheinhardt 1992). Surprisingly, in our study, we observed no relationship between elevation and soil properties or processes at our study site. This may be attributed in part to the small change in elevation relative to distance among our four habitats ($\approx$1 m.).

Shilling and Lockaby (2005) investigated vegetation and soil properties within floodplains on the Satilla and Altamaha rivers, GA, and found lower bulk density at lower, wetter habitats relative to higher, drier habitats. We also found lower bulk density in the slough than in higher, drier habitats, but we observed no relationship between bulk density and elevation in our study. In the Shilling and Lockaby (2005) study, bulk density ranged from 0.9 g cm$^{-3}$ in wetter habitats to 1.3 g cm$^{-3}$ in intermediate habitats, as compared to bulk density of 0.8 g cm$^{-3}$ in the slough and 1.2 g cm$^{-3}$ in the floodplain measured in our study. Shilling and Lockaby (2005) also found that habitats intermediate between their wet and dry sites had significantly more percent soil organic C, but they observed no difference in soil N among their habitats. We observed no difference in percent organic C or N among our habitats. Using a high-precision elevation model, Drouin et al. (2011) also found that alluvial wetlands in the St. Lawrence River basin, Vermont, much like ours, that are separated from the channel by a natural levee with a parallel slough exhibit no significant difference in percent organic C produced by differences in hydroperiod in sloughs (1.5–1.75 %) versus higher elevation habitats (0.75–1.0 %). Soil type is another factor that creates variation in microtopography, producing areas differing in water-holding capacity and mineral composition which lead to distinct plant communities (Wharton 1978). Our floodplain forest exhibits little difference in soil texture (sand, silt, clay, and OM) among habitats, which may further explain a lack of significant differences in C and N accumulation between habitats, as the accumulation of these elements is highly associated with variation in soil organic matter (Brunet and Astin 1997; Noe and Hupp 2005).

Connectivity to the river channel is also known to affect soil processes, notably sediment and P deposition. For example, Noe and Hupp (2005), measuring accretion over feldspar clay marker horizons, found that floodplains that were more connected to a water source had greater amounts of mineral sediment and P accumulation. We also found that habitats closer to the main channel had higher sediment deposition and P accumulation than those far removed from the channel (Fig. 4.5). Noe and Hupp (2005) also found that more connected systems had higher rates of organic C and N accumulation, but we did not observe this in our study. Inorganic P is often sorbed to mineral sediments, and it has been shown that P concentrations and the rates of P accumulation are strongly correlated with rates of sediment deposition and that floodplains capturing more sediment will capture more P, with higher P concentrations and accumulation rates close to the main channel and decreasing with distance from the channel (Noe and Hupp 2005; Kroes et al. 2007). We observed this in our study as well, as the positions closest to the main river channel had higher concentrations and accumulation of P and mineral sediment that decreased with distance from the main channel. Distance was a significant predictor of mineral sediment processes, but not for organic C or N, which are driven more by in situ organic matter production and deposition rather than sediment transport.

## 4.4.2 Temporal Patterns

$^{210}$Pb sediment accumulation was significantly higher than rates based on $^{137}$Cs, which suggests greater sediment deposition in the floodplain in the past 100 years than in the past 50 years. A similar pattern was observed in Georgia floodplain wetlands that drain to the Gulf of Mexico, which was attributed to historical changes in land use and demographic trends in the watershed, and the more recent implementation of BMPs reducing erosion and sediment runoff (Craft and Casey 2000). Georgia and the southeastern states saw the greatest erosive land use (ELU) during the period from 1860 to 1920 (Trimble 1974). It was commonplace in agriculture at that time to grow crops in an area until the land was deemed unproductive, then abandon it, and leave it bare (Trimble 1974). In the time after the Civil War, there was an increase in tenant farmers who were also more prone to poor land-use practices than landowners. This severe abuse of the land promoted rapid soil loss that filled streams and severely altered water flows in the area. Streams topped their banks changing bottomland fields to bottomland swamps, pushing cultivation to upland areas, and in the process creating more erosion (Trimble 1974).

In the twentieth century, ELU decreased greatly over all of the Southern Piedmont, where the headwaters of the Altamaha River are located, and was low

**Table 4.2** Comparison of sediment and nutrient (organic carbon, nitrogen, and phosphorus) accumulation (g m$^{-2}$ year$^{-1}$) in floodplain wetlands

|  | Sediment | Organic C | N | P |
|---|---|---|---|---|
| IL[a] | – | – | – | 3.4 |
| NC[b] | 39–305 | 37 | – | 0.1–0.28 |
| FL[c] | – | – | – | 3.2 |
| AR[d] | 600–800 | – | – | – |
| WI[e] | 2,000 | – | 12.8 | 2.6 |
| GA[f] | 710 (120–1,300) | 63 (18–107) | 4.7 (1.4–8) | 0.44 (0.12–0.75) |
| VA[h*] | – | 74–212 | 4.2–13.4 | 0.44–4.13 |
| **This study** | **1465 (90–2760)** | **50 (7–90)** | **4.0 (0.4–7.8)** | **0.7 (0.04–1.5)** |

[a]Mitsch et al. (1979)
[b]Kuenzler et al. (1980)
[c]Brown (1978)
[d]Kleiss (1996)
[e]Johnston et al. (1984)
[f]Craft and Casey (2000)
[h]Noe and Hupp (2005) *used feldspar clay pads

everywhere by 1967. The greatest decreases of ELU in the area occurred in regions including the Piedmont areas of Georgia (Trimble 1974). During this period, there was a large decrease in harvested cropland across the Southern Piedmont. The abandonment of land deemed unsuitable for crops, the effects of the boll weevil making cotton cultivation difficult and expensive, a drop in cotton prices, modern technology allowing for fewer acres of corn and oats necessary, and government controls on crop acreage all contributed to the decrease in farm acreage (Trimble 1974). This was also paired with an increase in land conservation as abandoned agricultural fields were allowed, or promoted, to convert back into forest. Practices such as crop rotation, contour plowing, and terracing were also widely implemented, and these new practices contributed to a large decrease in erosion across the Piedmont area (Trimble 1974). These land-use changes along with a shift in agricultural practices reduced the sediment load into streams and wetlands, accounting for greater sediment input in the years before their implementation.

Our limited sample size ($n = 10$ cores across the floodplain) limits the conclusions we can draw. However, our findings of differences in mineral soil properties (and lack of differences in organic soil properties) are supported by other studies (Schilling and Lockaby 2005; Noe and Hupp 2005; Franklin et al. 2009; Drouin et al. 2011). Furthermore, our measured rates of C sequestration and N, P, and sediment accumulation are within the range of other published studies of nutrient accumulation and sediment deposition in floodplain wetlands (Table 4.2).

In conclusion, distance from the main river channel was a significant predictor of mineral soil properties. Bulk density increased with distance from the channel, and processes such as P accumulation and sediment deposition decreased with distance. Elevation was not a strong predictor of soil properties or processes across the floodplain. Sloughs, especially those nearest to the main channel, which are the lowest sites within our floodplain, were hotspots of sediment and nutrient accumulation as evidenced by greater soil accretion, as well as greater N and P concentrations and accumulation, and mineral sediment deposition. In recent years, there has been increasing emphasis on quantifying soil properties and processes at a landscape scale, especially C stocks, C sequestration, and nutrient removal. This study adds to a body of literature which suggests a high degree of small-scale heterogeneity across alluvial landscapes, leading to spatial variability in soil properties and processes. Thus, scaling measurements from a single habitat across a heterogeneous alluvial floodplain wetland may over- or underestimate nutrient accumulation and retention.

In order to accurately inventory soil C and nutrient stocks, studies of soil properties and processes of alluvial floodplains and other wetlands should account for spatial variability, especially distance, across the wetland. Similarly, it is important to recognize temporal variability in sedimentary process (e.g., the measured difference in long-term $^{210}$Pb versus 50-year $^{137}$Cs rates) that relates to land-use change when attempting to project soil process like C sequestration and nutrient accumulation by these ecosystems in the future.

**Acknowledgments** We thank Nathan Knowles for his help in sample collection, Sarah Sutton for her help with sample analysis and input into the methods section, Jeff Ehman for his help acquiring the GIS/LiDAR elevation and distance datasets, and Anya Hopple and Anne Altor for edits to early versions of the manuscript. This research was supported by the US Department of Energy through grant #TUL-563-07/08 and the National Science Foundation grant #OCE-9982133 to the Georgia Coastal Ecosystems Long-Term Ecological Research Program. This is contribution number 1037 from the University of Georgia Marine Institute.

# References

Blake, G. R., & Hartge, K. H. (1986). Bulk density. In A. Klute (Ed.), *Methods of soil analysis. Part 1. Physical and mineralogical methods* (Agronomy monograph 9 2nd ed., pp. 363–375). Madison: ASA and SSSA.

Bridgham, S., Johnston, C. A., Schubauer-Berigan, J. P., & Weishampel, P. (2001). Phosphorus sorption dynamics in soils and coupling with surface and porewater in riverine wetlands. *Soil Science Society of America Journal, 65*, 577–588.

Brinson, M. M., Swift, R. C., Plantico, C., & Barclay, J. S. (1981a). Riparian ecosystems: Their ecology and status. U. S. Fish and Wildlife Service, Biological Services Program, Washington, DC, USA. FWS/OBS-81/17.

Brinson, M. M., Lugo, A. E., & Brown, S. (1981b). Primary productivity, decomposition and consumer activity in fresh water wetlands. *Annual Review of Ecological Systems, 12*, 123–161.

Brown, S. L. (1978). A comparison of cypress ecosystems in the landscape of Florida. PhD dissertation. Gainesville, FL, USA: University of Florida.

Brunet, R. C., & Astin, K. B. (1997). Spatio-temporal variations in sediment nutrient levels: The River Adour. *Landscape Ecology, 12*, 171–184.

Burke, W. (1975). Fertilizer and other chemical losses in drainage water from blanket bog. *Irish Journal of Agricultural Research, 14*, 163–178.

Conner, W. H., Doyle, T. W., & Krauss, K. R. (2007). *Ecology of tidal forested wetlands of the Southeastern United States*. New York: Springer.

Courtwright, J., & Findlay, S. E. G. (2011). Effects of microtopography on hydrology, physicochemistry, and vegetation in a tidal swamp of the Hudson River. *Wetlands, 31*, 239–249.

Craft, C. B. (2007). Freshwater input structures soil properties, vertical accretion and nutrient accumulation of Georgia and U.S. tidal marshes. *Limnology and Oceanography, 52*(3), 1220–1230.

Craft, C. B. (2012). Tidal freshwater forest accretion does not keep pace with sea level rise. *Global Change Biology, 18*, 3615–3623.

Craft, C. B., & Casey, W. P. (2000). Sediment and nutrient accumulation in floodplain and depressional freshwater wetlands of Georgia, USA. *Wetlands, 20*, 323–332.

Darke, A. K., & Walbridge, M. R. (2000). Al and Fe biogeochemistry in a floodplain forest: Implications for P retention. *Biogeochemistry, 51*, 1–32.

Drouin, A., Saint-Laurent, D., Lavoie, L., & Oullet, C. (2011). High-precision elevation model to evaluate the spatial distribution of soil organic carbon in active floodplains. *Wetlands, 31*, 1151–1164.

Ehrenfeld, J. G. (1994). Microtopography and vegetation in Atlantic white cedar swamps: The effects of natural disturbances. *Canadian Journal of Botany, 73*, 474–484.

ESRI (Environmental Systems Resource Institute). (2011). *ArcMap 9.3*. Redlands: ESRI.

Franklin, S. B., Kupfer, J. A., Pezeshki, S. R., Gentry, R., & Smith, R. D. (2009). Complex effects of channelization and levee construction on western Tennessee floodplain forest function. *Wetlands, 29*(2), 451–464.

Gee, G. W., & Bauder, J. W. (1986). Particle-size analysis. In A. Klute (Ed.), *Methods of soil analysis, part 1* (pp. 383–411). Madison: American Society of Agronomy.

Hopkinson, C. S. (1992). A comparison of ecosystem dynamics in fresh water wetlands. *Estuaries, 15*, 549–562.

Johnston, C. A., Bubenzer, G. D., Lee, G. B., Madison, F. W., & McHenry, J. R. (1984). Nutrient trapping by sediment deposition in a seasonally flooded lakeside wetland. *Journal of Environmental Quality, 13*, 283–290.

Kleiss, B. A. (1996). Sediment retention in a bottomland hardwood wetland in eastern Arkansas. *Wetlands, 16*, 321–333.

Kroes, D. E., Hupp, C. R., & Noe, G. B. (2007). Sediment, nutrient, and vegetation trends along the tidal, forested Pocomoke River, Maryland. In W. H. Conner, T. W. Doyle, & K. W. Krauss (Eds.), *Ecology of tidal freshwater forested wetlands of the southeastern United States* (pp. 113–137). New York: Springer.

Kuenzler, E. J., Mulholland, P. J., Yarbro, L. A., Smock, L. A. (1980). *Distributions and budgets of carbon, phosphorus, iron and manganese in a floodplain swamp ecosystem*. Water Resources Research Institute of the University of North Carolina, Raleigh, NC, USA. Report no. 157.

Loomis, M. J., & Craft, C. B. (2010). Carbon sequestration and nutrient (nitrogen, phosphorus) accumulation in river-dominated tidal marshes, Georgia, USA. *Soil Science Society of America Journal, 74*(3), 1028–1036.

Martin, D. B., & Hartman, W. A. (1987). Correlations between selected trace elements and organic matter and texture in sediments of northern prairie wetlands. *Journal of the Association of Official Analytical Chemists, 70*, 916–919.

Mausbach, M. J., & Richardson, J. L. (1994). Biogeochemical processes in hydric soil formation. *Current Topics and Wetland Biogeochemistry, 1*, 68–127.

Mitsch, W. J., & Goselink, J. G. (2000). The value of wetlands: Importance of scale and landscape setting. *Ecological Economics, 35*, 25–33.

Mitsch, W. J., Dorge, C. L., & Wiemhoff, J. R. (1979). Ecosystem dynamics and a phosphorus budget of an alluvial cypress swamp in southern Illinois. *Ecology, 60*, 1116–1124.

Morris, J. T., Sundareshwar, P. V., Nietch, C. T., Kjerfve, B., & Cahoon, D. R. (2002). Response of coastal wetlands to rising sea level. *Ecological Applications, 83*, 2869–2877.

Mosier, K., Ahn, C., & Noe, G. (2009). The influence of microtopography on soil nutrients in created mitigation wetlands. *Restoration Ecology, 17*, 641–651.

Naiman, R. J., & Décamps, H. (1997). Ecology of interfaces: Riparian zones. *Annual Review of Ecology and Systematics, 28*, 621–658.

National Oceanic and Atmospheric Administration's Digital Coast Data Access Viewer. http://www.csc.noaa.gov/dataviewer. 18 Nov 2011.

Noe, G. B., & Hupp, C. R. (2005). Carbon, nitrogen, and phosphorus accumulation in floodplains of Atlantic Coastal Plain Rivers, USA. *Ecological Applications, 15*(4), 1178–1190.

Oldfield, F., & Appleby, P. G. (1984). Empirical testing of $^{210}$Pb models for dating lake sediments. In E. Y. Haworth & J. W. G. Lund (Eds.), *Lake sediments and environmental history* (pp. 93–124). Minneapolis: University of Minnesota Press.

Photo Science and Fugro EarthData, Inc. (2010). Coastal Georgia Elevation Project Lidar Data, 1:2,400 scale, digital LiDAR data.

Rheinhardt, R. (1992). A multivariate analysis of vegetation patterns in tidal freshwater swamps of Lower Chesapeake Bay, USA. *Bulletin of the Torrey Botanical Club, 119*, 192–207.

SAS Institute. (1996). *SAS user's guide: Statistics*. Cary: SAS Institute, Inc.

Schelske, C. L., Robbins, J. A., Gardner, W. D., Conley, D. J., & Bourbonniere, R. A. (1988). Sediment record of biogeochemical responses to anthropogenic perturbations of nutrient cycles in Lake Ontario. *Canadian Journal of Fisheries and Aquatic Sciences, 45*, 1291–1303.

Schilling, E. B., & Lockaby, B. G. (2005). Microsite influences on productivity and nutrient circulation within two southeastern floodplain forests. *Soil Sciences Society of America Journal, 69*, 1185–1195.

Schilling, E. B., & Lockaby, B. G. (2006). Relationships between productivity and nutrient circulation within two contrasting southeastern U.S. floodplain forests. *Wetlands, 26*, 181–192.

Sommers, L. E., & Nelson, D. W. (1972). Determination of total phosphorus in soils: A rapid perchloric acid digestion procedure. *Soil Science Society of America Journal, 36*, 902–904.

Temmerman, S., Bouma, T. J., Govers, G., & Lauwaet, D. (2005). Flow paths of water and sediment in a tidal marsh: Relations with marsh development stage and tidal inundation height. *Estuaries, 28*, 338–352.

Trimble, S. W. (1974). Man-induced soil erosion on the Southern Piedmont 1700–1970. Soil Conservation Society of America. Milwaukee, WI: University of Wisconsin, Department of Geography.

USDA (U.S. Department of Agriculture). (2010). Natural Resources Conservation Service, United States Department of Agriculture. Web Soil Survey. http://websoilsurvey.usda.gov/. Last accessed 12 July 2010.

USGS (U.S. Geological Survey). (2010). Georgia real-time stream flow data. [Online] http://waterdata.usgs.gov/ga/nwis/current?type = flowgroup_key = basin_cd. Last accessed 12July2010. USGS, Reston, VA.

Wharton, C. H. (1978). Natural environments of Georgia. Geologic and Water Resources Division and Resource Planning Section, Office of Planning and Research, Georgia Department of Natural Resources, Atlanta, GA.

Wolf, K. L., Ahn, C. W., & Noe, G. B. (2011). Microtopography enhances nitrogen cycling and removal in created mitigation wetlands. *Ecological Engineering, 37*, 1398–1406.

# Chapter 5
# Natural and Restored Wetland Buffers in Reducing Sediment and Nutrient Export from Forested Catchments: Finnish Experiences

Mika Nieminen, Annu Kaila, Markku Koskinen, Sakari Sarkkola, Hannu Fritze, Eeva-Stiina Tuittila, Hannu Nousiainen, Harri Koivusalo, Ari Laurén, Hannu Ilvesniemi, Harri Vasander, and Tapani Sallantaus

**Abstract** One of the water quality management practices in forested catchments is to construct wetland buffers between managed areas and recipient water courses. Wetland buffers can be constructed simply by routing runoff from forested areas to natural peatlands and wetlands, or by rewetting lower sections of drained peatlands by filling in or blocking the drainage ditches. The use of natural and restored wetland buffers for reducing nutrient and sediment export from forested catchments, particularly catchments dominated by forestry-drained peatlands, has been studied actively in Finland during the last 15 years. The studies have shown highly variable retention capacity for wetland buffers with different site characteristics and under different environmental conditions. In favorable conditions, high amounts of sediments and adhered mineral elements may be deposited within peat and surface vegetation of the buffer. Dissolved nutrients can be retained biologically into plant

M. Nieminen (✉) · A. Kaila · S. Sarkkola · H. Fritze · H. Nousiainen · H. Ilvesniemi
Finnish Forest Research Institute, P. O. Box 18, FI-01301 Vantaa, Finland
e-mail: mika.nieminen@metla.fi

M. Koskinen · H. Vasander
Department of Forest Sciences, University of Helsinki, P. O. Box 27, FI-00014 Helsinki, Finland

E.-S. Tuittila
School of Forest Sciences, University of Eastern Finland, P. O. Box 111, FI-80101 Joensuu, Finland

H. Koivusalo
Department of Civil and Environmental Engineering, Aalto University School of Science and Technology, P. O. Box 15200, FI-00076 Aalto, Finland

A. Laurén
Finnish Forest Research Institute, P. O. Box 68, FI-80101 Joensuu, Finland

T. Sallantaus
Natural Environment Center, Biodiversity Unit, Finnish Environment Institute, P. O. Box 140, FI-00251 Helsinki, Finland

and microbial biomasses and chemically into peat. In contrast, nitrogen can also be lost into the atmosphere in gaseous form. In this literature review, we summarize the results of the experiments established on natural and restored wetland buffers in forested catchments in Finland to clarify the different processes and factors controlling their nutrient and sediment retention capacity. We also discuss the limitations and possible negative consequences of using wetland buffers for managing water quality in forested catchments.

**Keywords** Drained peatland • Forestry • Restoration • Retention capacity • Wetland buffer

## 5.1 Introduction

Nutrient losses from forested catchments are generally low (Kortelainen and Saukkonen 1998; Mattsson et al. 2003), but after forest harvesting (Nieminen 2004), fertilization (Saura et al. 1995), and ditching operations (Joensuu et al. 2002; Nieminen et al. 2010), export of nutrients and sediments may increase. Harvesting of tree stands grown on drained peatlands was shown to increase nitrogen (N) export by over 4 kg ha$^{-1}$ year$^{-1}$ (Uusivuori et al. 2008), and ditch drainage of peatlands and wetlands increased the sediment loading up to several thousands of kilograms per hectare (Hynninen and Sepponen 1983; Ahtiainen and Huttunen 1999). Similarly, forest fertilizations with nitrogen and phosphorus (P) may cause an excess leaching of several kilograms per hectare during the first few years after application (Nieminen and Ahti 1993; Saura et al. 1995). Forestry is typically practiced in headwater catchments where other human influence is insignificant; in these areas forestry is locally the main source of nutrients and sediments in water courses. In order to prevent nutrient and sediment leaching to downstream water courses, the current water protection guidelines in Finland propose that runoff from forested catchments is conveyed to receiving surface waters through wetland buffer areas (Metsätalouden ympäristöopas 2004). Buffer wetlands can be created by simply conducting the discharge waters from forested catchments to pristine mires, or occasionally to paludified mineral soils. However, because most peatlands and wetlands in Finland have been drained, a common practice is to restore sections of drained peatlands by filling in or blocking the drainage ditches. Buffer area size may vary considerably from a few meter-wide buffer zone-type constructions (Liljaniemi et al. 2003) to over hundred meters-long natural mires or restored sections of drained peatlands (Vikman et al. 2010). If only productive forestry land is available for the construction of the buffer, small areas are preferred, and the area then rarely exceeds 1.0–1.5 ha.

In addition to water quality management in forestry areas, wetland buffers or constructed wetlands have been applied to reduce sediment and nutrient loads from peat mining areas and to improve the quality of municipal waste water (e.g., Ronkanen and Kløve 2007, 2008), as well as to reduce loads from agricultural fields (Braskerud 2002) and urban areas (Birch et al. 2004). The quality of runoff from these different land use areas and discharge from waste water treatment plants is different and typically worse compared with runoff from managed forest areas. The retention efficiency in terms of relation between input and output loads tends to be higher for runoff with high pollutant concentrations than runoff with low pollution levels. The functioning of the wetlands for different purposes shares the same mechanisms, and the lessons learned from forest studies have wider implications to all buffer wetlands.

Different studies have shown highly variable nutrient retention efficiency for different wetland buffer areas in managed forest areas. For example, the efficiency of wetland buffers in reducing P load has varied from complete 100 % retention (Kubin et al. 2000) through partial P removal (Silvan et al. 2005a, b; Väänänen et al. 2008) to even increased leaching of phosphate (Sallantaus et al. 1998; Liljaniemi et al. 2003; Vasander et al. 2003). Similarly, the retention of ammonium ($NH_4$-N) in six wetland buffers receiving runoff from upstream ditch-network maintenance areas ranged from clearly negative to complete 100 % retention capacity (Hynninen et al. 2011b). The retention of suspended solids by wetland buffers in seven ditch-network maintenance areas also showed high variation from slightly negative to over 80 % retention capacity (Nieminen et al. 2005a). The varying conditions of the buffer zone areas studied so far, such as their size and shape, vegetation composition and density, soil nutrient retention capacity, management history, life and construction method, environmental and meteorological conditions during the study period, as well as the varying length of the study period, complicate the detection of the common nominators for their nutrient and sediment retention efficiency. In this literature review, we summarize the factors controlling nutrient and sediment retention in wetland buffers used in forested catchments in order to be able to improve their functionality and retention capacity in operational forestry. We also discuss the limitations and possible drawbacks of using wetland buffers in managing water quality in forested catchments.

## 5.2 Nutrient Retention Efficiency: Contributing Factors

### 5.2.1 Buffer Size and Shape

A number of studies indicate that the key factor explaining the nutrient retention efficiency of a wetland buffer is its size, more precisely the size of the buffer relative to the size of the whole upstream catchment area. Nieminen et al. (2005a) showed efficient suspended solid (SS) reduction capacity for the wetland buffers covering >1 % of the catchments area, but no reduction in through-flow SS concentrations for the buffers covering <0.1 %. They conclude the reduction of water flow velocity to be a key factor in the reduction of SS via increasing the time for particles to settle down. Further, as larger buffers (relative to catchment area)

**Fig. 5.1** The over 300 m long Kallioneva buffer is highly efficient in retaining the sediments and nutrients discharging from the upstream forested catchment (Photo: Martti Vuollekoski)

slow down water flow velocity more than small buffers, SS removal increases with the relative size.

The relative size of wetland buffers also explains much of the dissolved nutrient retention. The large size itself is a contributing factor because the vegetation and soil sinks are correspondingly larger, which results in lower relative nutrient load and lower probability of saturation of these sinks. The large relative size also enables longer water residence time and thus a longer contact time between the chemical and biological nutrient sinks and nutrient-rich through-flow waters (Fig. 5.1). In the very small buffer areas, the nutrient retention is poor, particularly if the flow is channelized to form continuous flow channels across the buffer area. In such channels, flow velocity is high and contact time between vegetation and soil sinks and through-flow water nutrients short; both of these factors are disadvantageous for high nutrient retention capacity (Väänänen et al. 2006). Thus, the study by Liljaniemi et al. (2003) showed negligible nutrient retention for the 2–8 m wide buffer strips, and they concluded that wider buffer areas with extensive overland flow areas are needed to efficiently control diffuse pollution from forested areas. In an artificial N addition experiment by Vikman et al. (2010) on six wetland buffers with the relative buffer size between 0.1 and 4.9 %, the correlation between $NO_3$-N retention (% of added) and relative buffer size was 0.75 ($p = 0.008$), but only 0.42 (n.s.) for $NH_4$-N. Their results actually indicated that the buffer length may be an even more important factor for buffer N retention efficiency than buffer size. The correlations between $NH_4$-N and $NO_3$-N retention and buffer length were 0.65 ($p = 0.03$) and 0.92 ($p < 0.001$), respectively. The effect of buffer length was interpreted to be because the probability of the formation of continuous flow channels across buffer area is lower for long buffers than short and wide buffers of the same size. The best-performing ($NO_3$-N retention 93.3–99.9 %) three wetland buffers in the study by Vikman et al. (2010) were all >100 m long and with no visible flow channels, while the 30 m long Asusuo buffer with a continuous flow channel across the buffer area had a $NO_3$-N retention capacity of <16 %. As the P retention capacity of the Asusuo buffer was also poor (Väänänen et al. 2008), Hynninen et al. (2010) question the rationale of constructing such small and short buffers.

## 5.2.2 Nutrient and Hydrological Loading

Although the size and shape of the buffer are important in sediment and nutrient retention, the results indicate that other factors also exist. The hydrological loading entering the buffer area and its temporal variability are considered to be one of the key factors (Väänänen et al. 2008). During high-flow episodes, the water residence time is short and the short contact time between nutrient sinks and through-flow water nutrients, as well as the formation of flow channels across the buffer, decreases the retention efficiency (Fig. 5.2). In an artificial nutrient addition experiment by Vikman et al. (2010), the correlation between $NO_3$-N retention and hydrological loading during 5 days after starting the N addition was $-0.73$ ($p < 0.010$), and $-0.42$ (n.s.) for $NH_4$-N. The study by Hynninen et al. (2011c), where they investigated the efficiency of buffer areas to reduce the ammonium export originating from ditch-network maintenance areas, also indicated that runoff during the study duration is a significant factor explaining the nutrient retention efficiency of buffer areas. As annual runoff increases toward northern latitudes, Hynninen et al. (2010) pointed out that larger buffer areas are needed in northern Finland to achieve similar retention efficiency as in southern Finland.

Although the hydrological loading and the buffer length were significant in explaining the $NH_4$-N retention originating from ditch-network maintenance areas in the study by Hynninen et al. (2011b), the contribution of these factors was minor compared with the strong influence of $NH_4$-N loading, i.e., the higher the $NH_4$-N loading into the buffer area was, the better the retention efficiency was. While the buffer length and the hydrological loading each explained about 5 % of the variation in ammonium retention, the rate of $NH_4$-N loading into the buffer areas was responsible for about 68 % of the variation. Hynninen et al. (2011b) argued that their results likely overestimate the effect of $NH_4$-N loading on retention efficiency and underestimate the contributions of other factors, such as buffer size, as the largest buffer areas with potentially high retention efficiency would probably have been able to retain much more ammonium from a higher loading than the buffers actually received after the ditch-network maintenance. Nevertheless, their results showed that the extent and pattern of nutrient loading may be a significant factor explaining the nutrient retention efficiency by wetland buffers. Also, Nieminen et al. (2005a) showed a strong positive correlation between

**Fig. 5.2** Channelization of water flow, particularly in small wetland buffers during high snowmelt flow periods, significantly decreases the nutrient retention capacity (Photo: Anu Hynninen)

sediment retention efficiency by wetland buffers and inflow water sediment concentrations.

The efficiency of buffer areas to retain nutrients and sediments is generally expressed as a load reduction percentage from the load input to the buffer. As significant reduction is not likely to occur from the inflow water with already low nutrient concentrations close to background levels of forested areas, it is not surprising that particularly poor retention efficiencies were reported when the performance of wetland buffers was assessed under such conditions (Vasander et al. 2003; Nieminen et al. 2005b). It is also to be noted that the very high retention efficiencies that are often reported after artificial nutrient additions (Silvan et al. 2005a; Väänänen et al. 2008; Vikman et al. 2010) may partly be explained by the fact that the high transient and steady loadings of artificial additions are retained more efficiently than the sporadically increased and long-lasting loadings typical for forested catchments after management options, such as harvesting (Nieminen 2004), ditch-network maintenance (Joensuu et al. 2002), and fertilization (Nieminen and Ahti 1993). In a nutrient addition experiment by Vikman et al. (2010), large and long buffers were able to retain almost all of the 25 kg of ammonium added during four days, but the model simulations by Hynninen et al. (2011b) indicated that only about half of the ammonium would be retained by similar long buffers from an equal annual loading caused by ditch-network maintenance.

Collectively, the previous results from forested catchments indicate that the wetland buffers have little effect on nutrient transport when the loadings are already near the background levels of forested catchments and that the pattern and duration of loading is a significant factor explaining nutrient retention efficiency under increased loadings. It is also to be noted here that the saturation of nutrient sinks in vegetation and soil due to chronic high loadings, as often occur in agricultural catchments and waste water treatment wetlands, may not be a common problem in forested catchments, where increased nutrient loadings due to forest management (harvesting, fertilization, ditch drainage) only occur 2–3 times during the tree rotation (80–100 years) and where the overall nutrient input into buffer areas is significantly lower.

### 5.2.3 Vegetation and Soil Processes

It is generally accepted that dense vegetation is important in nutrient retention, not only through nutrient accumulation in plant biomass but also because a dense vegetation cover forms a hydraulically rough surface and slows down the water flow velocity through the buffer area. A common argument supporting the importance of soil is the high cation exchange capacity (CEC) of peat in wetland buffers that enables a considerable potential to the retention $NH_4$-N, while the low P adsorption capacity of peat (Nieminen and Jarva 1996) may not enable much chemical retention of phosphate. Although vegetation has been found superior over microbes in compaction of N (Silvan et al. 2005b), the microbial communities are likely to thrive under high N inflow into buffer areas (Silvan et al. 2002; Hynninen et al. 2011c). A significant amount of N can be immobilized through an increase in the size and N concentrations of the microbial biomass. Despite these arguments, there are very few studies that attempted to quantify the roles of vegetation and soil processes in nutrient retention in wetland buffers in forested catchments.

Especially the role of different plant species in nutrient retention is weakly known. One aspect in the role of vegetation to be accounted for is the changing vegetation composition due to restoration succession promoted by increased water level in restored buffers. Species turnover might be further impacted by increased availability of nutrients that give competition advantage to opportunistic species such as cotton grass (Silvan et al. 2004).

The lack of information on the roles of vegetation and soil processes in nutrient retention may be due to experimental difficulties. The study by Silvan (2003) quantified the retention of P in soil and vegetation using the samples collected before and after an artificial P addition. The retention of P in peat was estimated to be 43 % of the added P, but then only the volumetric concentrations (mg cm$^{-3}$) indicated any P retention, while the gravimetric P concentrations (mg g$^{-1}$) remained unchanged. This reveals a possible experimental error in this type of approach; as the soil and vegetation samples collected before and after nutrient loading cannot be from exactly the same position, the different characteristics between the pre- and post-load sampling positions introduce error in the results. Thus, the higher volumetric concentrations in the post-addition samples in the study by Silvan (2003) may simply be because the samples were from more humified sampling positions than the pre-addition samples. Some of the error involved in pre- and post-load sampling could be decreased by increasing the number of sampling positions, but as the number already was high, e.g., in Silvan (2003), this would make the estimation very laborious, with no possibility to foresee any improvements in estimation accuracy. An additional problem with this type of approach is that, as the N and P stores in peat and vegetation may be large already before increased N and P loading or artificial addition, the changes in stores will easily remain too small for reliable estimation.

Experimental difficulties may also explain some of the very high variation in P and N retention estimates between the different studies. While Silvan (2003) estimated the retention of added N and P in vegetation biomass to be 70 and 25 %, respectively, Huttunen et al. (1996) in their study on a wetland buffer below a peat extraction area found more release than accumulation of P and only a slight retention of N (4 % of N input). The variation in nutrient retention estimates between the different studies may also be because soil and vegetation may not be permanent nutrient sinks, and while some of the nutrients are released after first being retained, the length of the study period affects the retention estimates. Some fraction of the nutrients are released from vegetation during the senescence and decay of the litter in the end and after the growing season, and also the nutrients retained in labile forms in the soil may be released when the nutrient concentrations in soil solution return to the levels before the increased nutrient loading. The only permanent nutrient "sink" in wetland buffers can actually be argued to be the loss of $N_2$ and $N_2O$ into the atmosphere. If the nutrients retained in the vegetation and soil of wetland buffers eventually end up as structural components of the constantly accumulating dead biomass, i.e., peat, then nearly permanent retention in terms of unforeseen future is also possible.

The use of labeled isotopes could be a more powerful tool in the estimation of N and P retention in plant biomass and peat than the comparison of pre- and post-loading samples, but only the study by Väänänen et al. (2006) has utilized the labeled isotope approach in a forested catchment. They estimated that the recovery of added $^{32}P$ in a natural mesotrophic fen used as a wetland buffer was 16 % of added P, of which 92 % was in surface peat and 3 % in vascular plants and mosses. They interpreted that the low overall recovery and low accumulation in vegetation were because the addition experiment was realized in early spring, when snowmelt in the upslope areas still contributed to the high hydrological loading and plant photosynthesis and P assimilation had not yet recovered after a winter period. Thus, besides the extent of nutrient loading (see Sect. 5.2.2), the timing of loading may be a significant factor behind wetland buffer retention efficiency. In areas with a distinct winter period with snow accumulation-melting cycles and ground freezing, significant retention is improbable, when the highest loadings occur during the snowmelt periods with sparse vegetation cover.

A major problem in using labeled P in retention studies is that only short-term experiments are possible, as the degradation of $^{32}P$ to $^{31}P$ with a half-life of 14.3 days rapidly lowers the radioactivity level below reliable detection limits.

The problem in using $^{15}$N isotope in N retention studies is its price; one kg of $^{15}$N labeled ammonium nitrate costs still several thousands of euros in 2014. Thus, only laboratory-scale estimations are so far feasible.

Collectively, the previous results from wetland buffers indicate that roles of vegetation and soil processes in nutrient retention are difficult to quantify reliably and that it is possible that the highly variable results between the different studies are, at least partly, due to these experimental difficulties. Indirect information on the roles of vegetation and soil in nutrient retention may also be achieved by using vegetation and soil factors (e.g., bottom or field layer vegetation coverage (%), soil CEC (mmol kg$^{-1}$), soil phosphate adsorption capacity) as explanatory variables in the experimental models explaining the buffer retention efficiency. However, their effects are easily hidden behind the factors that are more significant for the retention capacity, such as the buffer size and the nutrient and hydrological loading. This was also the case in the study by Hynninen et al. (2011b), where the ammonium retention efficiency of wetland buffers was modeled using buffer size, buffer length, the coverage of buffer bottom and field layer vegetation, tree stand volume, soil bulk density, soil CEC, hydrological loading, and ammonium loading as explanatory factors. Only ammonium loading, buffer length, and hydrological loading were significant in explaining the ammonium retention efficiency by six wetland buffers.

### 5.2.4 Other Factors

One likely factor behind sediment retention in wetland buffers is also the type of sediment particles. Light organic particles and fine-textured mineral soil particles are probably retained less efficiently than heavy and high-density mineral particles, but the effect of this factor on sediment retention in wetland buffers has not been studied. Water protection constructions based on the sedimentation of SS, such as sedimentation ponds, have been shown to be inefficient in reducing the transport of light organic particles and fine-textured mineral soil particles (Joensuu et al. 1999). As the increase in fine-textured mineral soil particles, in particular, may be substantial after ditching operations and ditching-induced sediment transport is regarded as the most harmful water quality impact of forestry in Finland (Joensuu et al. 2002), the contribution of wetland buffers to the reduction of fine-textured sediment transport is an important future research subject.

Also the age of the buffer has been shown to be a factor behind nutrient retention efficiency, particularly the retention of phosphate (Vasander et al. 2003). The wetlands recently restored for use as buffer areas may release more phosphate than accumulate it, probably because the redox-sensitive phosphate compounds in peat are released along with filling in or blocking the ditches and consequent rewetting and water table rising. If the restoration also involves harvesting of the tree stand from the buffer area, the harvest residues left on site also form a possible source of phosphate and other nutrients from the buffer area into receiving water courses. However, although enhanced export of P would occur during peatland restoration for use as a buffer area, and perhaps a few years after restoration (Vasander et al. 2003), Liljaniemi et al. (2003) and Nieminen et al. (2005a) pointed out that all wetland buffers are likely to turn into nutrient-accumulating systems in the long term. As the enhanced release of P due to rewetting may only last for 2–3 years (Vasander et al. 2003), relatively newly restored wetland buffers may already be efficient in retaining the nutrients released from the upstream forest area as a consequence of forest management operations. It should also be noted that only some of the peat soils appear to contain significant amounts of redox-sensitive P (Kaila et al. 2012), but the current level of understanding does not support the identification of sites with high risk for increased P release upon rewetting. As pointed out earlier, the aging of buffer areas in forested catchments is unlikely to result in decreased nutrient retention capacity (i.e., saturation of nutrient sinks), as may be true for the buffers in agricultural areas and the waste water treatment buffers, because the nutrient loadings into wetland buffers in forested areas are significantly lower.

## 5.3 Limitations and Possible Drawbacks

Blocking or filling in the ditches in the area planned to be used as a buffer results in water table rising not only in the buffer area itself but also in the upstream area. The size of the affected area and the rate of water table rising depend on the local land topography, soil depths, and soil hydraulic properties. In a sloping land, the rewetted area above the buffer area may be just a few meters or tens of meters long, but in the very flat lowlands, the rewetted area may extend to several hundreds of meters from the buffer area. This causes a major limitation in the use of wetland buffers in operational forestry. In the coastal area of western Finland, in particular, the lands are flat and constructing wetland buffers there could mean significant water table rising and decrease in the vitality and growth of trees in the large productive forestland areas above the buffer area. Thus, even if the use of wetland buffers is currently recommended as the most efficient means of decreasing diffuse pollution in forested catchments, their use in operational forestry is restricted to areas where sloping land facilitates the construction of the buffer without severely disturbing tree growth in the upstream productive forestland. In the very flat areas, typically consisting of drained peatland forests, use of the recently developed peak runoff control method (Marttila and Klöve 2010) is recommended to decrease ditch erosion and the export of suspended sediments and adhered mineral elements. Instead, the use of sedimentation ponds as the only water quality protection method should be avoided due to their very limited capacity for decreasing sediment export (Joensuu et al. 1999).

Another major limitation in the use of buffer areas arises from the need to conserve endangered wetland site types. Use of these sites as buffer areas may induce unwanted changes in the plant species composition. According to Hynninen et al. (2011a), grasses, sedges, as well as herbs are generally favored in wetland

5 Natural and Restored Wetland Buffers in Reducing Sediment and Nutrient...

**Fig. 5.3** The vegetation in the Hirsikangas during the time of buffer construction (Photo: Jorma Issakainen)

buffers (Figs. 5.3 and 5.4), and the changes are more apparent in the upstream parts of buffers than the lower parts. Also, the vegetation growing in the lawn-level surfaces changes more than the hummock vegetation. To protect endangered mire site types, such sites should be left aside from buffer use to avoid significant changes in vegetation composition (Hynninen et al. 2011a).

It should be noted that wetland buffers may be efficient in mostly reducing inorganic nutrient export, while much of the nutrient export from forestland occurs in organic forms. For example, while the increased export of inorganic N following harvesting of drained spruce mires was only a few hundred grams per hectare, the increase in dissolved organic N export was several kilograms per hectare (Nieminen 2004). Forest clear-felling may also result in substantially enhanced dissolved organic carbon (DOC) export (Nieminen 2004), but because organic soils, including wetland buffers, typically act as sources rather than sinks of DOC, wetland buffers may not be used as efficient means for decreasing DOC export. The efficiency of wetland buffers in reducing DOC and organic nutrient transport has not been assessed in forested catchments, but the studies from peat extraction areas showed that wetland buffers in those areas were not efficient in reducing dissolved carbon export (Klöve et al. 2012). The fate of DOC and organic nutrients in wetland buffers in forested catchments still needs to be assessed, but the hypothesis is that their retention may not be particularly efficient.

The use of buffer areas to filter high loads of N could change the dynamics of N cycling, including the production of greenhouse gas $N_2O$. However, the study by Hynninen et al. (2011c) indicated low emissions even when the artificial N addition

**Fig. 5.4** The vegetation in the Hirsikangas buffer 11 years after buffer construction (Photo: Anu Hynninen)

of 150 kg of $NH_4NO_3$ per buffer area increased the N loading to considerably higher levels than is likely to occur under actual conditions in managed forested catchments. The emissions (0.15 kg $N_2O$ ha$^{-1}$) were substantially lower than the annual emissions reported for drained minerotrophic peatland forests by Martikainen et al. (1995) (0.4–1.4 kg $N_2O$ year$^{-1}$), or for peat soils drained for agricultural purposes (Regina et al. 2004); (0.3–19 kg $N_2O$ ha$^{-1}$ year$^{-1}$). Saari et al. (2013) also showed low $N_2O$ emissions for a wetland buffer receiving water flows from a drained peatland forest, and they concluded the low inflow N concentrations to be the main reason for low emissions.

The restoration of drained peatlands for use as wetland buffers has also raised a concern of increased methane emissions. However, Juottonen et al. (2012) showed negligible methane emission from the wetland buffers rewetted >10 years earlier compared with corresponding natural wetland buffers. The emission from natural wetland buffers was similar as reported for the natural peatlands that are not used as buffer areas. The analysis of methanotrophic and methanogenic populations by Juottonen et al. (2012) indicated that, rather than enhanced methane oxidation, the reason behind low methane emissions was that the time after rewetting was still too short for the restoration of methanogen populations. The methanogenic populations in the three restored wetland buffers differed significantly from the populations in three natural wetland buffers sharing very identical populations.

## 5.4 Summary

We summarize the results of the experiments established on wetland buffers in forested catchments in Finland as follows.

The over 100 m long buffer areas with a relative size of over 1 % from the upstream catchment are generally highly efficient in reducing sediment and dissolved nutrient transport, while the short and small buffer zone-type constructions covering <0.1–0.2 % may be inefficient in retaining the nutrients released from upstream forest areas as a consequence of forest management operations.

High hydrological loadings decrease the buffer nutrient retention efficiency, and in areas with a distinct winter period, such as Finland, buffer areas may not be particularly efficient, if the highest nutrient loadings occur during high snowmelt flow periods with still sparse vegetation cover. As runoff increases toward northern latitudes, larger buffers are needed in northern Finland to achieve similar retention capacity as in southern Finland.

When the nutrient loadings into buffer areas are already low and near the background levels of forested catchments, wetland buffers may have little effect on nutrient transport. The timing, pattern, and duration of nutrient loading are significant factors explaining the nutrient retention efficiency under the increased loadings caused by forest management operations.

The roles of vegetation and soil processes in nutrient retention in wetland buffers are difficult to quantify reliably. The highly variable retention estimates for soil and vegetation between the different studies may, at least partly, be because of these experimental difficulties.

The major limitation in the use of wetland buffers in operational forestry is that their use is restricted to areas where sloping land enables the construction of the buffer without leading to the rising of water table in the upstream productive forestland. Another major limitation is that wetland buffers may not be particularly efficient in decreasing the transport of DOC and dissolved organic nutrients. It should also be noted that as the vegetation in natural mires used as wetland buffers is likely to undergo significant changes, endangered mire site types in their pristine state should not be used as buffer areas.

Even if the area of wetland buffers increased significantly from the present state, the emissions of greenhouse gases $N_2O$ and $CH_4$ from wetland buffers are unlikely to increase to levels causing problems from the viewpoint of global warming. The low $CH_4$ emissions from restored wetland buffers, even after >10 years after restoration, make them promising candidates for buffer areas to take advantage of their nutrient retention capacity without simultaneously causing high methane fluxes.

Future research should clarify the contribution of sediment type (organic vs. mineral, fine textured vs. coarse) to sediment retention efficiency by wetland buffers, as well as the retention of DOC and dissolved organic nutrients in different types of wetland buffers. An important future research topic is also to provide tools to identify the sites that are likely to release redox-sensitive P into drainage waters, when drained peat soils are restored and rewetted for use as a buffer area.

## References

Ahtiainen, M., & Huttunen, P. (1999). Long-term effects of forestry managements on water quality and loading in brooks. *Boreal Environment Research, 4*, 101–114.

Birch, G. F., Matthai, C., Fazeli, M. S., & Suh, J. Y. (2004). Efficiency of a constructed wetland in removing contaminants from stormwater. *Wetlands, 24*, 459–466.

Braskerud, B. C. (2002). Factors affecting phosphorus retention in small constructed wetlands treating agricultural non-point source pollution. *Ecological Engineering, 19*, 41–61.

Huttunen, A., Heikkinen, K., & Ihme, R. (1996). Nutrient retention in the vegetation of an overland flow treatment system in northern Finland. *Aquatic Botany, 55*, 61–73.

Hynninen, P., & Sepponen, P. (1983). Erään suoalueen ojituksen vaikutus purovesien laatuun Kiiminkijoen alueella, Pohjois-Suomessa. The effect of drainage on the quality of brook waters in the Kiiminkijoki river basin, Northern Finland. *Silva Fennica, 17*, 23–43.

Hynninen, A., Saari, P., Nieminen, M., & Alm, J. (2010). Pintavalutus metsätaloustoimien valumavesien puhdistamisessa – kirjallisuustarkastelu. Use of peatland buffer areas for water purification in forested catchments – A review. *Suo – Mires and Peat, 61*(3–4), 77–85.

Hynninen, A., Hamberg, L., Nousiainen, H., Korpela, L., & Nieminen, M. (2011a). Vegetation composition dynamics in peatlands used as buffer areas in forested catchments in southern and central Finland. *Plant Ecology, 212*, 1803–1818.

Hynninen, A., Sarkkola, S., Lauren, A., Koivusalo, H., & Nieminen, M. (2011b). Capacity of natural and restored peatland buffer areas in reducing ammonium export originating from ditch network maintenance areas in peatlands drained for forestry. *Boreal Environment Research, 16*, 430–444.

Hynninen, A., Fritze, H., Sarkkola, S., Kitunen, V., Nousiainen, H., Silvan, N., et al. (2011c). $N_2O$ fluxes from peatland buffer areas after high N loadings in five forested catchments in Finland. *Wetlands, 31*(6), 1067–1077.

Joensuu, S., Ahti, E., & Vuollekoski, M. (1999). The effects of peatland forest ditch maintenance on suspended solids in runoff. *Boreal Environment Research, 4*, 343–355.

Joensuu, S., Ahti, E., & Vuollekoski, M. (2002). Effects of ditch network maintenance on the chemistry of run-off water from peatland forests. *Scandinavian Journal of Forest Research, 17*, 238–247.

Juottonen, H., Vikman, A., Nieminen, M., Tuomivirta, T., Tuittila, E.-S., Nousiainen, H., et al. (2012). Methane-cycling microbial communities and methane emissions in natural and restored peatlands. *Applied and Environmental Microbiology, 78*(17), 6386–6389.

Kaila, A., Asam, Z., Sarkkola, S., Liwen, X., Laurén, A., & Nieminen, M. (2012). The effect of water table rising on nutrient and dissolved organic carbon (DOC) release from restored peatland forest. In T. Magnusson (Ed.), *The 14th international peat congress, peatlands in balance, Stockholm, Sweden, June 3–8, 2012. Book of abstracts* (p. 338). International Peat Society.

Klöve, B., Saukkoriipi, J., Tuukkanen, T., Heiderscheidt, E., Heikkinen, K., Marttila, H., et al. (2012). Turvetuotannon vesistökuormituksen ennakointi ja uudet hallintamenetelmät. *Suomen Ympäristö, 35*, 1–31.

Kortelainen, P., & Saukkonen, S. (1998). Leaching of nutrients, organic carbon and iron from Finnish forestry land. *Water, Air, and Soil Pollution, 105*, 239–250.

Kubin, E., Ylitolonen, A., Välitalo, J., & Eskelinen, J. (2000). Prevention of nutrient leaching from a forest regeneration area using overland flow fields. In M. Haigh & J. Křeček (Eds.),

Environmental reconstruction in headwater areas (pp. 161–169). Dordrecht: Kluwer Academic Publishers.
Liljaniemi, P., Vuori, K.-M., Tossavainen, T., Kotanen, J., Haapanen, M., Lepistö, A., et al. (2003). Effectiveness of constructed overland flow areas in decreasing diffuse pollution from forest drainages. Environmental Management, 32, 602–613. doi:10.1007/s00267-003-2927-4.
Martikainen, P., Nykänen, H., Alm, J., & Silvola, J. (1995). Change in fluxes of carbon dioxide, methane and nitrous oxide due to forest drainage of mire sites of different trophy. Plant and Soil, 168–169, 571–577.
Marttila, H., & Klöve, B. (2010). Managing runoff, water quality and erosion in peatland forestry by peak runoff control. Ecological Engineering, 135(2), 900–911.
Mattsson, T., Finér, L., Kortelainen, P., & Sallantaus, T. (2003). Brook water quality and background leaching from unmanaged forested catchments in Finland. Water, Air, and Soil Pollution, 147, 275–297.
Metsätalouden ympäristöopas. (2004). 159 p. Metsähallitus.
Nieminen, M. (2004). Export of dissolved organic carbon, nitrogen and phosphorus following clear-cutting of three Norway spruce forests growing on drained peatlands in southern Finland. Silva Fennica, 38, 123–132.
Nieminen, M., & Ahti, E. (1993). Talvilannoituksen vaikutus ravinteiden huuhtoutumiseen karulta suolta. Leaching of nutrients from an ombrotrophic peatland area after fertilizer application on snow. Folia Forestalia, 814, 1–22.
Nieminen, M., & Jarva, M. (1996). Phosphorus adsorption by peat from drained mires in southern Finland. Scandinavian Journal of Forest Research, 11, 321–326.
Nieminen, M., Ahti, E., Nousiainen, H., Joensuu, S., & Vuollekoski, M. (2005a). Capacity of riparian buffer zones to reduce sediment concentrations in discharge from peatlands drained for forestry. Silva Fennica, 39, 331–339.
Nieminen, M., Ahti, E., Nousiainen, H., Joensuu, S., & Vuollekoski, M. (2005b). Does the use of riparian buffer zones in forest drainage sites to reduce the transport of solids simultaneously increase the export of solutes? Boreal Environment Research, 10, 191–201.
Nieminen, M., Ahti, E., Koivusalo, H., Mattsson, T., Sarkkola, S., & Laurén, A. (2010). Export of suspended solids and dissolved elements from peatland areas after ditch network maintenance in south-central Finland. Silva Fennica, 44(1), 39–49.
Regina, K., Syväsalo, E., Hannukkala, A., & Esala, M. (2004). Fluxes of $N_2O$ from farmed soils in Finland. European Journal of Soil Science, 55, 591–599.
Ronkanen, A.-K., & Kløve, B. (2007). Use of stabile isotopes and tracers to detect preferential flow patterns in a peatland treating municipal wastewater. Journal of Hydrology, 347(3–4), 418–429.
Ronkanen, A.-K., & Kløve, B. (2008). Hydraulics and flow modelling of water treatment wetlands constructed on peatlands in Northern Finland. Water Research, 42, 3826–3836.
Saari, P., Saarnio, S., Heinonen, J., & Alm, J. (2013). Emissions and dynamics of $N_2O$ in a buffer wetland receiving water flows from a forested peatland. Boreal Environment Research, 18, 164–180.
Sallantaus, T., Vasander, H., & Laine, J. (1998). Metsätalouden vesistöhaittojen torjuminen ojitetuista soista muodostettujen puskurivyöhykkeiden avulla. Prevention of detrimental impacts of forestry operations on water bodies using buffer zones created from drained peatlands. Suo, 49, 125–133.
Saura, M., Sallantaus, T., Bilaletdin, Ä., & Frisk, T. (1995). Metsälannoitteiden huuhtoutuminen Kalliojärven valuma-alueelta. In S. Saukkonen & K. Kenttämies (Eds.), Metsätalouden vesistövaikutukset ja niiden torjunta. METVE-projektin loppuraportti (pp. 87–104). Suomen ympäristökeskus 1995/2.
Silvan, N. (2003). Nutrient retention in a restored peatland buffer. Helsingin yliopiston metsäekologian laitoksen julkaisuja 32 (44 p.). Yliopistopaino, Helsinki.

Silvan, N., Regina, K., Kitunen, V., Vasander, H., & Laine, J. (2002). Gaseous nitrogen loss from a restored peatland buffer zone. *Soil Biology and Biochemistry, 34*, 721–728.

Silvan, N., Tuittila, E.-S., Vasander, H., & Laine, J. (2004). *Eriophorum vaginatum* plays a major role in nutrient retention in boreal peatlands. *Annales Botanici Fennici, 41*, 189–199.

Silvan, N., Sallataus, T., Vasander, H., & Laine, J. (2005a). Hydraulic nutrient transport in a restored peatland buffer. *Boreal Environment Research, 10*, 203–210.

Silvan, N., Tuittila, E.-S., Kitunen, V., Vasander, H., & Laine, J. (2005b). Nitrate uptake by vegetation controls $N_2O$ fluxes from peatlands. *Soil Biology and Biochemistry, 37*, 1519–1526.

Uusivuori, J., Kallio, M., & Salminen, O. (2008). Vaihtoehtolaskelmat Kansallisen metsäohjelman 2015 valmistelua varten. *Working papers of the Finnish Forest Research Institute,* 75, 104 p.

Väänänen, R., Nieminen, M., Vuollekoski, M., & Ilvesniemi, H. (2006). Retention of phosphorus in soil and vegetation of a buffer zone area during snowmelt peak flow in southern Finland. *Water, Air, and Soil Pollution, 177*, 103–118.

Väänänen, R., Nieminen, M., Vuollekoski, M., Nousiainen, H., Sallantaus, T., Tuittila, E.-S., et al. (2008). Retention of phosphorus in peatland buffer zones at six forested catchments in southern Finland. *Silva Fennica, 42*, 211–231.

Vasander, H., Tuittila, E.-S., Lode, E., Lundin, L., Ilomets, M., Sallantaus, T., et al. (2003). Status and restoration of peatlands in northern Europe. *Wetland Ecology and Management, 11*, 51–63.

Vikman, A., Sarkkola, S., Koivusalo, H., Sallantaus, T., Laine, J., Silvan, N., et al. (2010). Nitrogen retention by peatland buffer areas at six forested catchments in southern and central Finland. *Hydrobiologia, 641*(1), 171–183.

# Chapter 6
# Do Reflectance Spectra of Different Plant Stands in Wetland Indicate Species Properties?

Katja Klančnik, Igor Zelnik, Primož Gnezda, and Alenka Gaberščik

**Abstract** This contribution discusses the relationships between reflectance spectra obtained by field spectroscopy and properties of the leaves of the species that form a stand and the relation between reflectance spectra and stand characteristics. We thus investigate the reliability of conclusions made at the species levels on the basis of the reflectance spectra of the stands. We studied monospecific and mixed stands that thrive in habitats along a hydrological gradient in the intermittent Lake Cerknica. The reflectance spectra differed significantly at the stand and leaf levels; however, although the shape of the reflectance spectra of a monospecific stand with *Phalaris arundinacea* was similar to the shape of the leaf spectra, this was not the case for mixed stands. The leaf morphological and biochemical properties that explain most of the variability of the spectra differed for graminoids and different dicotyledons. This study shows that based on the reflectance spectra, the species properties for monospecific stands can be deduced, while for mixed stands, such deductions can be misleading.

**Keywords** Macrophytes • Ecosystem structure • Hydrological gradient • *Phalaris arundinacea*

## 6.1 Introduction

Ecosystem structure and function depend on multiple environmental factors that affect habitats, species properties and their distribution (Ustin 2010). The key factor is the amount of incoming radiation and its fate in the plant community. The majority of light is absorbed by different plant organs, while some of the light can either penetrate through the stand or is reflected from the plant surface. Thus, only a small proportion of the incoming solar radiation reaches the stand floor. The interactions between the radiation and the plant communities are very complex, due

K. Klančnik • I. Zelnik • P. Gnezda • A. Gaberščik (✉)
Department of Biology, Biotechnical Faculty, University of Ljubljana, Večna pot 111, SI-1000 Ljubljana, Slovenia
e-mail: alenka.gaberscik@bf.uni-lj.si

to stand diversity, species architecture and leaf structural properties (Larcher 2003). Leaves that thrive in specific environments have specific traits that optimise their capture of solar energy and prevent damage due to excessive and/or harmful photons (Gurevitch et al. 2002). This fine-tuning is made possible through special adaptations of leaves at the morphological, anatomical, biochemical and functional levels (Robe and Griffiths 2000; Boeger and Poulson 2003; Šraj-Kržič and Gaberščik 2005; Klančnik et al. 2012).

Light that is reflected from plant leaves can provide a basis for an understanding of the photosynthetic performance and energy balance of plants (Vogelmann 1993). It also provides information on leaf biochemistry (Levizou et al. 2005; Castro and Sanchez-Azofeifa 2008) and nutrient and water status (Baltzer and Thomas 2005; Asner and Martin 2008) and can serve as a tool for stress detection (Gitelson et al. 2002); in some cases, this also allows species classification (i.e. through their spectral signatures) (Castro-Esau et al. 2006). Similarly, light that is reflected from plants can indicate the condition of a stand (Asner 1998; Ullah et al. 2012).

Different indices that are based on species and/or stand reflectance spectra have been developed to determine the properties of different plant species and plant functional groups (Levizou et al. 2005). However, without detailed knowledge of the basic parameters that define the spectral signatures at the species level, reflectance spectra might not provide reliable information (Milton et al. 2009).

In comparison to measurements of leaf optical properties, which are time-consuming, remote sensing allows for surveying and monitoring of relatively large areas, as well as comparisons of data across time and space (Ollinger 2010). Therefore, one of the main reasons for detailed research and a need to understand leaf optical properties is the establishment of libraries of species spectral signatures, along with species leaf properties (Chandrasekharan 2005).

Remote sensing includes two types of spectroscopy: 'field spectroscopy', which is based on measurements within or close to a stand, and 'imaging spectroscopy', which is the detection of the spectra from a distance (e.g. from aircrafts or satellites). In comparison to remote sensing, field spectroscopy is technically less demanding and less influenced by atmospheric conditions (Gao et al. 2009).

In the present study, we aimed to define the properties of stands and leaves in the intermittent Lake Cerknica affecting the reflectance spectra that can be obtained by field spectroscopy and to compare the reflectance spectra at the stand level to that at the leaf level. We also examined how reliable conclusions at the species level can be on the basis of the reflectance spectra of a stand.

## 6.2 Materials and Methods

### 6.2.1 Site Description

The intermittent Cerknica Lake appears at the bottom of the karst Cerknica Polje (38 km$^2$). Due to abundant precipitation in spring and autumn, the polje changes into a shallow lake of 20–25 km$^2$ in size. On average, the floods last for 260 days a year, and the dry period usually starts in late spring (Kranjc 2003). The result of this intermittence of Cerknica Lake is the zonation of the plant communities along a hydrological gradient that depends on the duration and extent of the flooding.

### 6.2.2 Field Spectroscopy and Stand Properties

For the purpose of the present study, we selected plant stands at 23 locations along the hydrological gradient (Table 6.1). The selected stands were homogenous, as either monospecific or mixed species. We performed two to four sets of 20 scans per stand in the vegetative period. These measurements of reflectance between 280 nm and 887 nm were carried out using a portable spectrometer (Jaz Modular Optical Sensing Suite; Ocean Optics, Inc., Dunedin, FL, USA). Prior to the leaf reflectance measurements, a white reference panel (Spectralon®, Labsphere, North Sutton, USA) was used to calibrate the spectrometer to 100 % reflectance. The reflectance spectra were then calculated as the ratios of the sample data to the white reference under the same illumination. The scans were recorded between 10:00 h and 14:00 h. The detector was positioned 90 cm above the stands, at a constant angle that was adjusted according to the position of the sun. At each sampling plot, the properties of plant stands were determined as the number of species, species abundance, total plant and specific species cover (%), height of the stand and species properties (i.e. plant phenological phases, vitality, leaf angle). The species abundance was estimated according to the Braun-Blanquet method (Braun-Blanquet 1964). The amount of photosynthetically active radiation and the air temperature and relative humidity were also measured.

### 6.2.3 Measurements at the Leaf Level

The reflectance spectra of the leaves were measured on the day of sampling with the above-mentioned portable spectrometer. The individual leaves were positioned under an integrating sphere (ISP-30-6-R; Ocean Optics, Inc., FL, USA) connected to the spectrometer via an optical fibre (QP600-1-SR-BX; Ocean Optics, Inc., Dunedin, FL, USA). During the illumination of the leaf with an ultraviolet-visible-near infrared (UV-VIS-NIR) light source (DH-2000, Ocean Optics, Inc., FL, USA),

**Table 6.1** Plant species composition and abundance (in brackets) at selected locations during the growing season

| Location | Month of measurement | RDA code[a] | Plant species composition (abundance[b]) |
|---|---|---|---|
| 1 | May | 1 | *Euphorbia lucida* (3), *Phalaris arundinacea* (3), *Carex elata* (3) |
|  | June | 2 | *E. lucida* (4), *P. arundinacea* (2), *C. elata* (2) |
|  | Aug | 3 | *E. lucida* (5), *C. elata* (2), *P. arundinacea* (2) |
|  | Sept | 4 | *P. arundinacea* (4), *E. lucida* (3), *C. elata* (2) |
| 2 | May, June, Aug, Sept | 5-8 | *P. arundinacea* (5) |
| 3 | Aug | 9 | *Myosotis scorpioides* agg. (5), *Mentha aquatica* (3), *Teucrium scordium* (2) |
| 4 | Sept | 10 | *T. scordium* (4), *M. aquatica* (3), *M. scorpioides* agg. (2), *Agrostis sp.* (2) |
| 5 | May | 11 | *Gratiola officinalis* (5), *Plantago altissima* (2) |
|  | Aug | 12 | *G. officinalis* (4), *P. altissima* (3) |
|  | Sept | 13 | *G. officinalis* (4), *P. altissima* (2), *C. elata* (2) |
| 6 | May | 14 | *Senecio paludosus* (4), *Polygonum amphibium* (3) |
|  | Aug, 4 | 15-16 | *S. paludosus* (5), *P. amphibium* (2) |
| 7 | June | 17 | *Phragmites australis* (5) |
| 8 | June | 18 | *Molinia caerulea* (5), *P. altissima* (2) |
| 9 | June | 19 | *Deschampsia cespitosa* (5), *P. altissima* (2) |
| 10 | June | 20 | *C. elata* (5) |
| 11 | May | 21 | Apiaceae (5), *M. scorpioides* agg. (2), *M. aquatica* (2) |
|  | Aug | 22 | *M. scorpioides* agg. (3), Apiaceae (2), *M. aquatica* (2), *T. scordium* (2) |
|  | Sept | 23 | Apiaceae (3), *M. aquatica* (2), *T. scordium* (2), *M. scorpioides* agg. (2) |
| 12 | May | 24 | *P. altissima* (4), *Carex panicea* (3), *Molinia caerulea* (2) |
|  | Aug | 25 | *P. altissima* (4), *C. panicea* (3), *M. caerulea* (2) |
|  | Sept | 26 | *P. altissima* (3), *M. caerulea* (2), *C. panicea* (2), *M. aquatica* (2) |
| 13 | May | 27 | *M. aquatica* (3), *Rorippa amphibia* (3), *P. arundinacea* (2) |
|  | Aug, Sept | 28-29 | *M. aquatica* (4), *R. amphibia* (2) |
| 14 | May | 30 | *R. amphibia* (3), *P. amphibium* (2), *M. aquatica* (2) |
|  | Aug, Sept | 31-32 | *P. amphibium* (4), *R. amphibia* (3), *M. scorpioides* agg. (2), *M. aquatica* (2) |
| 15 | May | 33 | *C. elata* (5), *P. altissima* (2), *G. officinalis* (2) |
|  | Aug | 34 | *C. elata* (4), *G. officinalis* (3), *P. altissima* (2), *L. salicaria* (2) |
|  | Sept | 35 | *C. elata* (4), *G. officinalis* (3), *P. altissima* (2) |
| 16 | May | 36 | *E. lucida* (5), *P. altissima* (3) |
|  | June | 37 | *E. lucida* (4), *P. altissima* (3) |

(continued)

6 Do Reflectance Spectra of Different Plant Stands in Wetland Indicate...

**Table 6.1** (continued)

| Location | Month of measurement | RDA code[a] | Plant species composition (abundance[b]) |
|---|---|---|---|
|  | Aug, Sept | 38-39 | *E. lucida* (5) |
| 17 | May | 40 | *C. panicea* (4), *P. altissima* (3), *M. caerulea* (2), *Succisa pratensis* (2) |
|  | June | 41 | *C. panicea* (3), *P. altissima* (2), *M. caerulea* (2) |
|  | Aug | 42 | *P. altissima* (4), *C. panicea* (3), *M. caerulea* (3) |
|  | Sept | 43 | *P. altissima* (4), *C. panicea* (2), *M. caerulea* (2) |
| 18 | May, Sept | 44-45 | *P. amphibium* (5) |
|  | Aug | 46 | *P. amphibium* (5), *R. amphibia* (2) |
| 19 | May, Aug, Sept | 47-49 | *P. amphibium* (5) |
| 20 | May, Aug, Sept | 50-52 | *P. amphibium* (5) |
| 21 | May | 53 | *G. officinalis* (4), *P. altissima* (3) |
|  | Aug | 54 | *P. altissima* (4), *G. officinalis* (3), *C. panicea* (2) |
|  | Sept | 55 | *G. officinalis* (4), *P. altissima* (3) |
| 22 | May | 56 | *Schoenus nigricans* (5), *Centaurea jacea* agg. (2) |
|  | June | 57 | *S. nigricans* (4), *P. altissima* (3), *C. jacea* agg. (2) |
|  | Aug, Sept | 58-59 | *S. nigricans* (4), *P. altissima* (3), *C. jacea* agg. (2), *M. caerulea* (2), *C. panicea* (2) |
| 23 | May, Aug | 60-54 | *Salix rosmarinifolia* (5) |
|  | Sept | 55 | *S. rosmarinifolia* (5), *M. caerulea* (2) |

[a]RDA code in Figure 5
[b]Abundance according to Braun-Blanquet (1964)

the total adaxial reflectance spectra of the leaves were recorded between 280 nm and 887 nm, with a resolution of approximately 0.3 nm.

For the same leaves, the following morphological, anatomical and biochemical properties were determined: specific leaf area; thickness of the leaf, cuticle, epidermis and mesophyll; density and length of the leaf stomata, trichome and prickle hairs (silicified trichome in graminoids); contents of chlorophyll $a$, chlorophyll $b$, carotenoids and anthocyanins; and amount of UV-B (280–320 nm) and UV-A (320–400 nm) absorbing compounds. These analyses followed the procedures and methods as described and cited previously (Klančnik et al. 2012, 2013a).

## 6.2.4 Statistical Analysis

Measurements of the reflectance spectra are given as the means across 5-nm intervals. The significances of the differences between reflectance spectra were assessed by Kruskal-Wallis tests with Bonferroni correction. Detrended correspondence analysis was used for exploratory data analysis, using the CANOCO 4.5

program package. The gradient length was <3 S.D., and therefore, redundancy analysis (RDA) was used to determine the possible effects of explanatory variables (i.e. leaf traits, stand properties) on the reflectance spectra variability (ter Braak and Šmilauer 2002). Each variable was entered separately into the analysis, and the significance of its gross effects was assessed using Monte Carlo tests with 999 permutations. To avoid possible collinearity between explanatory variables, forward selection was used. Nonsignificant variables ($p > 0.05$) were excluded from the further analysis.

## 6.3 Results

### 6.3.1 Reflectance Spectra at Leaf and Stand Levels

Comparisons of the reflectance spectra differed among the stands and leaves. We compared the reflectance spectra of monospecific and mixed stands of *Phalaris arundinacea* and reflectance measurements on the leaves (Fig. 6.1). Three main differences were observed: (1) leaves reflected significantly more light than stands; (2) variability of the reflectance in different colour bands was more pronounced for leaves, with the least variability observed for mixed stands; and (3) the greatest differences were obtained in the UV, green and NIR ranges.

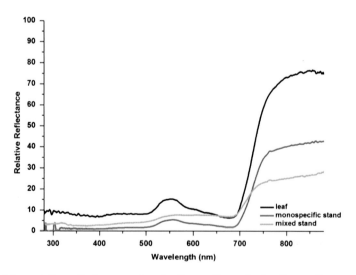

**Fig. 6.1** Mean relative reflectance spectra of a *P. arundinacea* leaf, a *P. arundinacea* monospecific stand and a mixed stand where *P. arundinacea* covered 25 % of the sampling plot (Data are means over 5-nm intervals ($n = 10$))

## 6.3.2 Leaf Reflectance Spectra and Leaf Traits

RDA was performed to define the parameters that explained most of the variability of the reflectance spectra, taking into account the different datasets. In the first run, data on the biochemical and anatomical leaf traits and the corresponding leaf reflectance spectra were used. In this case, the thickness of the upper epidermis explained as much as 17 % of the variability of the reflectance spectra; the trichome density, 16 %; the amount of carotenoids and the length of the prickle hairs, 8 % each; and the specific leaf area, an additional 7 % (Fig. 6.2). The length of the prickle hairs was negatively related to the reflectance, while the density of the trichome showed a positive relationship. The species studied were distributed along the full gradient of visible wavelengths, which showed differences in reflectance and formed optical groups, with the exception of specimens of *Myosotis scorpioides* agg., which were scattered throughout the whole plot. The graminoids *Carex elata*, *Molinia caerulea* and *Phragmites australis* formed a single group, while the single dicotyledonous species were located distinctly apart (Fig. 6.2).

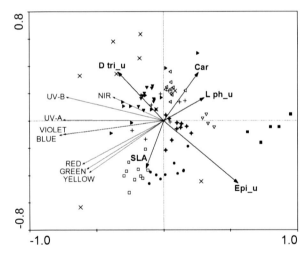

**Fig. 6.2** Redundancy analysis ordination diagram showing the strength of the associations between the significant leaf traits ($p < 0.05$) and the regions of the leaf reflectance spectra. Plant species: *filled circles*, samples of *P. arundinacea*; *open squares*, *Gratiola officinalis*; *filled squares*, *Polygonum amphibium*; *filled upside-down triangles*, *C. elata*; *open upside-down triangles*, *Euphorbia lucida*; *filled right-pointing triangles*, *M. caerulea*; *open left-pointing triangles*, *P. australis*; pluses (+), *Deschampsia cespitosa*; crosses (×), *M. scorpioides* agg.; thick pluses (+), *Senecio paludosus*. **D tri_u** mean trichome density on the adaxial leaf surface, **L ph_u** mean prickle-hair length on the adaxial leaf surface, **Epi_u** epidermis thickness on the adaxial leaf surface, **SLA** specific leaf area, **Car** carotenoids content per leaf area

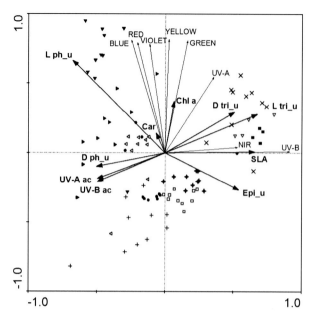

**Fig. 6.3** Redundancy analysis ordination diagram showing the strength of the associations between the significant morphological and biochemical leaf traits ($p < 0.05$) and the regions of the monospecific stand reflectance spectra. Plant species: *filled circles*, samples of *P. arundinacea*; *open squares*, *Gratiola officinalis*; *filled squares*, *Polygonum amphibium*; *filled upside-down triangles*, *C. elata*; *open upside-down triangles*, *Euphorbia lucida*; *filled right-pointing triangles*, *M. caerulea*; *open left-pointing triangles*, *P. australis*; pluses (+), *Deschampsia cespitosa*; crosses (×), *M. scorpioides* agg.; thick pluses (+), *Senecio paludosus*. **D tri_u** trichome density on the adaxial leaf surface, **L tri_u** mean trichome length on the adaxial leaf surface, **D ph_u** prickle-hair density on the adaxial leaf surface, **L ph_u** prickle-hair length on the adaxial leaf surface, **Epi_u** epidermis thickness on the adaxial leaf surface, **SLA** specific leaf area, **Chl a** chlorophyll *a* content per leaf area, **Car** carotenoids content per leaf area, **UV-A ac** UV-A absorbing compounds per leaf area, **UV-B ac** UV-B absorbing compounds per leaf area

## 6.3.3 Monospecific Stand Reflectance and Leaf Traits

In the second RDA, we examined relationships between the reflectance spectra of monospecific stands and the biochemical and anatomical leaf traits of the species that formed these stands. The amount of total explained variance was 76 %, which was even higher than in the first RDA. The length of the prickle hairs of the upper epidermis and the density of the trichome explained 32 % and 18 % of the spectra variability, respectively; the UV-A absorbing compounds and chlorophyll *a*, 6 % each; and other significant parameters, 1–3 % each. As shown in Fig. 6.3, the graminoids reflected more in the UV range, while the reflectance in the visible range was very variable.

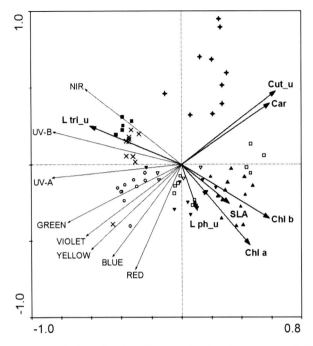

**Fig. 6.4** Redundancy analysis ordination diagram showing the strength of the associations between the significant morphological and biochemical leaf traits ($p < 0.05$) and the regions of the mixed stands reflectance spectra (prevailing species covers 50 %). Prevailing plant species: *open circles*, samples of *Mentha aquatica*; *open squares*, *Gratiola officinalis*; *filled squares*, *Polygonum amphibium*; *filled upside-down triangles*, *C. elata*; *open upside-down triangles*, *Euphorbia lucida*; *filled triangles*, *Plantago altissima*; *crosses* (×), *M. scorpioides* agg.; *thick pluses* (+), *Senecio paludosus*. **L tri_u** trichome length on the adaxial leaf surface, **L ph_u** prickle-hair length on the adaxial leaf surface, **Cut_u** cuticle thickness on the adaxial leaf surface, **SLA** specific leaf area, **Chl a** chlorophyll *a* content per leaf area, **Chl b** chlorophyll *b* content per leaf area, **Car** carotenoids content per leaf area

## 6.3.4  Mixed Stand Reflectance and Leaf Traits

In the next step, we related the reflectance spectra of the mixed stands to the biochemical and anatomical leaf traits of the species that covered half of the sampling area of the plot. In this case, the majority of species that formed the stands were dicotyledons (except *C. elata*), and therefore, the outcomes were somewhat different. With the species traits, a total of 74 % of the variability of the reflectance spectra was explained. Chlorophyll *a* explained 31 %, the thickness of the cuticle 26 %, and other parameters exerted little influence on spectra variability (up to 5 % each). The thickness of the cuticle was negatively related to all ranges of the spectra (Fig. 6.4).

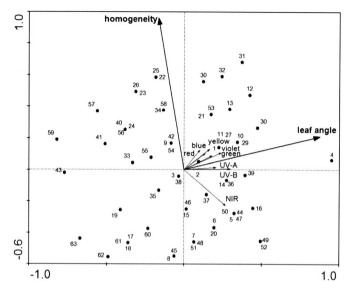

**Fig. 6.5** Redundancy analysis ordination diagram showing the strength of the associations between the significant stand properties ($p < 0.05$) and the regions of the stand reflectance spectra. The detailed species compositions of the stands represented by the numbers are given in Table 6.1

## 6.3.5 Stand Reflectance and Stand Properties

The last RDA was performed, taking into account the stand reflectance spectra and properties of the stand. Only two variables had significant effects on the stand reflectance: leaf angle and stand homogeneity. Together, these explained 14 % of the variability (Fig. 6.5). The leaf angle was positively related to all parts of the spectra. The distribution of the stands within a plot showed that the same plots had different distributions at different times of the season.

Field spectroscopy enables determination of the properties of the analysed object. In the past decade, this method has significantly enhanced the understanding of the interactions between matter and energy at the levels of plant leaves and stands (Gamon 2006). Some studies have concentrated on the reflection in narrow wave bands, with proposals of various vegetation indices, although these have usually been tested with only a few different species (Sims and Gamon 2002). Many field spectroscopy studies have aimed to define the species composition and species properties as, for example, levels of the chlorophylls, carotenoids and anthocyanins (Gamon et al. 1990; Gitelson et al. 2009). The present study has shown that such conclusions might not be always reliable.

To establish the relationships between reflectance spectra and species traits, we studied different stand types, with different species compositions and different species properties. The measured reflectance spectra differed significantly at the leaf and stand levels, as related to the leaf and stand properties. When the reflectance spectra of stands with different abundance of *P. arundinacea* were compared, this showed that the reflectance curves of monospecific stands of *P. arundinacea* have similar shapes to those of the *P. arundinacea* leaves, while mixed stands reflected less radiation along the whole spectra. The most pronounced differences were observed in the UV, green and NIR ranges. This was apparently related to the more complex architecture of the mixed stands, in comparison to the monospecific stands (Schulze et al. 2005).

The development of individual plant-leaf properties depends on the species genotype and site conditions, while the structure of a stand mainly depends on the species that constitute the stand, and especially on their growth forms (Larcher 2003). For Lake Cerknica, the specific water regime creates an environmental gradient (Martinčič and Leskovar 2003) that supports a variety of different communities with the different species that were included in the present study. We applied RDA to explain the variability of the spectra with the properties of the species that formed the stands. The data show that the reflectance spectra of monospecific stands can be explained by species properties, while different properties are indicative of different species or optical groups. The majority of the significant parameters to the monospecific stand reflectance were largely expected. The exceptions were for the contents of chlorophyll *a* and the carotenoids, where the relationships with reflectance were positive, although they explained minor parts of the spectra variance (i.e. 12 %). In some species and/or for some stands, the structural parameters were more important than the biochemical parameters (Klančnik et al. 2012, 2013a). It is generally accepted that the leaf surface relief greatly influences the surface reflection of light, while the structure of the mesophyll affects light penetration. The limited role of biochemical parameters in the reflectance spectra in some species/stands was therefore a consequence of structures on the leaf surface, such as the waxy cuticle, trichomes or prickle hairs, which dissipate the radiation and reduce its penetration into the mesophyll (Baldini et al. 1997; Holmes and Keiller 2002). Different trichomes are present in many plant species, as they are cost-effective due to their multiple functions, i.e. the prevention of water loss and protection against excessive radiation (Ehleringer 1980; Woodman and Fernandes 1991). The reflectance in the UV range is usually very low (<10 %) (Yoshimura et al. 2010; Qi et al. 2002; Holmes and Keiller 2002), due to the absorption of UV photons by phenolic substances, which usually accumulate in the upper leaf layers and mainly in the epidermis (Pfündel et al. 2007). However, it has also been reported that, in some cases, the increased reflectance is a consequence of silica structures (prickle hairs and cuticle) at the leaf surface (Klančnik et al. 2013a). Silica is a key structural element in graminoids, where it substitutes for carbon as a structural element, and enhances their strength, while preventing lodging and shading of leaves (Schoelynck et al. 2010; Schaller et al. 2012). Therefore, silica should be taken into account when studying reflectance of this plant group.

In the analyses of the mixed stands, mainly dicotyledonous species were included, and their biochemical properties were revealed as more important than their structural properties, together explaining 38 % of the variability of the reflectance spectra. As expected, chlorophylls *a* and *b*, which intercept the light inside the leaf, correlated negatively with the visible parts of the spectra (Klančnik et al. 2012, 2013b). Surprisingly, the cuticle thickness correlated negatively with the entire spectra, even though many studies have shown that wax on leaf surfaces effectively reflects radiation (Holmes and Keiller 2002; Klančnik et al. 2012). This unexpected relationship potentially arose because the accompanying species in the stands contributed to the shape of the reflectance spectra and masked the role of the studied species in the light reflectance. The data obtained indicate that the reflectance of the monospecific stands can be explained by the species properties, while in mixed stands, the data might be misleading, even in the case of a very abundant species. In addition, the architecture of the stand can also contribute to the shape of the spectra. With the RDA where the stand reflectance was related to the stand properties, this revealed that the leaf angle and the stand homogeneity significantly affect the stand reflectance, as has also been shown in previous studies (Ganapol et al. 1999; Rautiainen et al. 2008).

## 6.4 Conclusions

We can conclude that (1) the complexity of a stand negatively affects the amount of light that is reflected; (2) in monospecific stands, the reflectance can be explained by the leaf properties of the species that constitutes the stand, although the key properties differed among the various species; (3) this is not very likely for mixed stands, including those with species that occur at high abundance; (4) plant architecture might also have an important role in explaining the reflectance spectra variability; and (5) any interpretation of the results of field spectroscopy needs detailed knowledge of the structural and biochemical properties at the stand and species levels.

**Acknowledgements** The work was supported by the Ministry of Education, Science and Sport, Republic of Slovenia, through the programmes 'Biology of Plants' (P1-0212) and 'Young Researchers' (33135).

## References

Asner, G. P. (1998). Biophysical and biochemical sources of variability in canopy reflectance. *Remote Sensing of Environment, 64*, 234–253.
Asner, G. P., & Martin, R. E. (2008). Spectral and chemical analysis of tropical forests: Scaling from leaf to canopy levels. *Remote Sensing of Environment, 112*, 3958–3970.
Baldini, E., Facini, O., Nerozzi, F., Rossi, F., & Rotondi, A. (1997). Leaf characteristics and optical properties of different wood species. *Trees, 12*, 73–81.
Baltzer, J. L., & Thomas, S. C. (2005). Leaf optical responses to light and soil nutrient availability in temperate deciduous trees. *American Journal of Botany, 92*, 214–223.

Boeger, M. R., & Poulson, M. (2003). Morphological adaptations and photosynthetic rates of amphibious *Veronica anagallis-aquatica* L. (Scrophulariaceae) under different flow regimes. *Aquatic Botany, 75*, 123–135.

Braun-Blanquet, J. (1964). *Pflanzensoziologie. Grunzüge der Vegetationskunde*. Wien: Springer.

Castro, K. L., & Sanchez-Azofeifa, G. A. (2008). Changes in spectral properties, chlorophyll content and internal mesophyll structure of senescing *Populus balsamifera* and *Populus tremuloides* leaves. *Sensors, 81*, 51–69.

Castro-Esau, K. L., Sanchez-Azofeifa, G. A., Rivard, B., Wright, S. J., & Quesada, M. (2006). Variability in leaf optical properties of Mesoamerican trees and the potential for species classification. *American Journal of Botany, 93*, 517–530.

Chandrasekharan, R. (2005). Optical properties of leaves. http://www.docstoc.com/docs/49236849/Optical-properties-of-leaves. Accessed 4 Dec 2013.

Ehleringer, J. (1980). Leaf morphology and reflectance in relation to water and temperature stress. In N. C. Turner & P. J. Kramer (Eds.), *Adaptation of plants to water and high temperature stress* (pp. 295–308). New York: Wiley.

Gamon, J. A. (2006). Spectral Network SpecNet—what is it and why do we need it? *Remote Sensing of Environment, 103*, 227–235.

Gamon, J. A., Field, C. B., Bilger, W., Björkman, Ö., Fredeen, A. L., & Peñuelas, J. (1990). Remote sensing of the xanthophyll cycle and chlorophyll fluorescence in sunflower leaves and canopies. *Oecologia, 85*, 1–7.

Ganapol, B. D., Johnson, L. F., Hlavka, C. A., Peterson, D. L., & Bond, B. (1999). LCM2: A coupled leaf/canopy radiative transfer model. *Remote Sensing of Environment, 70*, 153–166.

Gao, B. C., Montes, M. J., Davis, C. O., & Goetz, A. F. H. (2009). Atmospheric correction algorithms for hyperspectral remote sensing data of land and ocean. *Remote Sensing of Environment, 113*, 17–24.

Gitelson, A. A., Zur, Y., Chivkunova, O. B., & Merzlyak, M. N. (2002). Assessing carotenoid content in plant leaves with reflectance spectroscopy. *Photochemistry and Photobiology, 75*, 272–281.

Gitelson, A. A., Chivkunova, O. B., & Merzlyak, M. N. (2009). Nondestructive estimation of anthocyanins and chlorophylls in anthocyanic leaves. *American Journal of Botany, 96*, 1861–1868.

Gurevitch, J., Scheiner, S. M., & Fox, G. (2002). *The ecology of plants*. Sunderland: Sinauer Associates.

Holmes, M. G., & Keiller, D. R. (2002). Effects of pubescence and waxes on the reflectance of leaves in the ultraviolet and photosynthetic wavebands: A comparison of a range of species. *Plant Cell Environment, 25*, 85–93.

Klančnik, K., Mlinar, M., & Gaberščik, A. (2012). Heterophylly results in a variety of "spectral signatures" in aquatic plant species. *Aquatic Botany, 98*, 20–26.

Klančnik, K., Vogel-Mikuš, K., & Gaberščik, A. (2013a). Silicified structures affect leaf optical properties in grasses and sedge. *Journal of Photochemistry and Photobiology B*. doi:10.1016/j.jphotobiol.2013.10.011.

Klančnik, K., Pančić, M., & Gaberščik, A. (2013b). Leaf optical properties in amphibious plant species are affected by multiple leaf traits. *Hydrobiologia*. doi:10.1007/s10750-013-1646-y.

Kranjc, A. (2003). Hidrološke značilnosti (Hydrological Characteristics). In A. Gaberščik (Ed.), *Jezero, ki izginja – Monografija o Cerkniškem jezeru (The Vanishing Lake – Monograph on Lake Cerknica)* (pp. 26–37). Ljubljana: Društvo ekologov Slovenije, [in Slovenian].

Larcher, W. (2003). *Physiological plant ecology, ecophysiology and stress physiology of functional groups* (4th ed.). Heidelberg: Springer.

Levizou, E., Drilias, P., Psaras, G. K., & Manetas, Y. (2005). Nondestructive assessment of leaf chemistry and physiology through spectral reflectance measurements may be misleading when changes in trichome density co-occur. *New Phytologist, 165*, 463–472.

Martinčič, A., & Leskovar, I. (2003). Vegetacija (Vegetation). In A. Gaberščik (Ed.), *Jezero, ki izginja – Monografija o Cerkniškem jezeru (The Vanishing Lake – Monograph on Lake Cerknica)* (pp. 81–95). Ljubljana: Društvo ekologov Slovenije, [in Slovenian].

Milton, E. J., Schaepman, M. E., Anderson, K., Kneubühler, M., & Fox, N. (2009). Progress in field spectroscopy. *Remote Sensing of Environment, 113*, 92–109.

Ollinger, S. V. (2010). Sources of variability in canopy reflectance and the convergent properties of plants. *New Phytologist, 189*, 375–394.

Pfündel, E. E., Agati, G., & Cerovic, Z. G. (2007). Optical properties of plant surfaces. In M. Riederer & C. Müller (Eds.), *Biology of the plant cuticle* (Annual plant reviews, Vol. 23, pp. 216–249). London: Blackwell.

Qi, Y. D., Bai, S., Vogelmann, T. C., Heisler, G. M., & Qin, J. (2002). Methodology for comprehensive evaluation of UV-B tolerance in trees. In: J. R. Slusser, J. R. Herman, & W. Gao (Eds.), *Proceedings of SPIE: Ultraviolet ground- and space-based measurements, models, and effects*, San Diego.

Rautiainen, M., Mõtus, M., Stenberg, P., & Ervasti, S. (2008). Crown envelope shape measurements and models. *Silva Fennici, 42*, 19–33.

Robe, W. E., & Griffiths, H. (2000). Physiological and photosynthetic plasticity in the amphibious, freshwater plant, *Litorella uniflora*, during the transition from aquatic to dry terrestrial environments. *Plant Cell Environment, 23*, 1041–1054.

Schaller, J., Brackhage, C., & Dudel, G. (2012). Silicon availability changes structural carbon ratio and phenol content of grasses. *Environmental and Experimental Botany, 77*, 283–287.

Schoelynck, J., Bal, K., Backx, H., Okruszko, T., Meire, P., & Struyf, E. (2010). Silica uptake in aquatic and wetland macrophytes: A strategic choice between silica, lignin and cellulose? *New Phytologist, 168*, 385–391.

Schulze, E. D., Beck, E., & Müller-Hohenstein, K. (2005). *Plant ecology*. Heidelberg: Springer.

Sims, D. A., & Gamon, J. A. (2002). Relationships between leaf pigment content and spectral reflectance across a wide range of species, leaf structures and developmental stages. *Remote Sensing of Environment, 812*, 337–354.

Šraj-Kržič, N., & Gaberščik, A. (2005). Photochemical efficiency of amphibious plants in an intermittent lake. *Aquatic Botany, 83*, 281–288.

ter Braak, C. J. F., & Šmilauer, P. (2002). *CANOCO reference manual and CanoDraw for Windows user's guide software for canonical community ordination (version 4.5)* (Microcomputer power). Ithaca: Biometris.

Ullah, S., Schlerf, M., Skidmore, A. K., & Hecker, C. (2012). Identifying plant species using mid-wave infrared 2.5–6-μm and thermal infrared 8–14-μm emissivity spectra. *Remote Sensing of Environment, 118*, 95–102.

Ustin, S. L. (2010). Spectral identification of native and non-native plant species. http://www.asdi.com/resource-center/application-notes/spectral-identification-plant-species. Accessed 4 Dec 2013.

Vogelmann, T. C. (1993). Plant tissue optics. *Annual Review of Plant Physiology and Plant Molecular Biology, 44*, 231–251.

Woodman, R. L., & Fernandes, G. W. (1991). Differential mechanical defense: Herbivory, evapotranspiration and leaf hairs. *Oikos, 60*, 11–19.

Yoshimura, H., Zhu, H., Wu, Y., & Ma, R. (2010). Spectral properties of plant leaves pertaining to urban landscape design of broad-spectrum solar ultraviolet radiation reduction. *International Journal of Biometeorology, 54*, 179–191.

# Chapter 7
# Global Boundary Lines of $N_2O$ and $CH_4$ Emission in Peatlands

Jaan Pärn, Anto Aasa, Sergey Egorov, Ilya Filippov, Geofrey Gabiri,
Iuliana Gheorghe, Järvi Järveoja, Kuno Kasak, Fatima Laggoun-Défarge,
Charles Kizza Luswata, Martin Maddison, William J. Mitsch,
Hlynur Óskarsson, Stéphanie Pellerin, Jüri-Ott Salm, Kristina Sohar,
Kaido Soosaar, Alar Teemusk, Moses M. Tenywa, Jorge A. Villa,
Christina Vohla, and Ülo Mander

**Abstract** Predicting $N_2O$ (nitrous oxide) and $CH_4$ (methane) emissions from peatlands is challenging because of the complex coaction of biogeochemical factors. This study uses data from a global soil and gas sampling campaign. The objective is to analyse $N_2O$ and $CH_4$ emissions in terms of peat physical and chemical conditions. Our study areas were evenly distributed across the A, C and D climates of the Köppen classification. Gas measurements using static chambers, groundwater analysis and gas and peat sampling for further laboratory analysis have been conducted in 13 regions evenly distributed across the globe. In each study area at least two study sites were established. Each site featured at least three sampling plots, three replicate chambers and corresponding soil pits and one observation well per plot. Gas emissions were measured during 2–3 days in at least three sessions. A log-log linear function limits $N_2O$ emissions in relation to soil TIN (total inorganic nitrogen). The boundary line of $N_2O$ in terms of soil temperature is semilog linear. The closest representation of the relationship between $N_2O$ and soil moisture is a local regression curve with its optimum at 60–70 %. Semilog linear upper boundaries describe the effects of soil moisture and soil temperature to $CH_4$ best.

---

J. Pärn (✉) • A. Aasa • S. Egorov • J. Järveoja • K. Kasak • M. Maddison • K. Sohar •
K. Soosaar • A. Teemusk • C. Vohla
Institute of Ecology and Earth Sciences, University of Tartu, Vanemuise St. 46, 51014 Tartu, Estonia
e-mail: jaan.parn@ut.ee

I. Filippov
UNESCO Chair of Environmental Dynamics and Climate Change, Yugra State University, Chekhova street 16, Khanty-Mansiysk 628012, Russian

G. Gabiri
Department of Geography, Kenyatta University, P. O. Box 43844, GPO 00100 Nairobi, Kenya

Department of Agricultural Production, College of Agricultural and Environmental Sciences, Makerere University, P. O. Box 7062, Kampala, Uganda

The global $N_2O$ boundary lines revealed a striking similarity with the Southern German $N_2O$ boundary lines, as well as with analogous scattergrams for Europe (Couwenberg et al. 2011) and Southern Queensland (Wang and Dalal 2010). This suggests that local rather than global conditions determine land-use-based greenhouse gas emissions.

Further work will analyse relationships between the environmental factors and the spatial distribution of the main functional genes *nirS*, *nirK* and *nosZ* regulating the denitrification process in the soil samples currently stored in fridge at $-18°$. An additional analysis will study the relationships between the intensity of $CH_4$ emissions and methanogenesis-regulating functional genes *mcrA*, *pmoA* and *dsrAB*.

**Keywords** Bog • Ecosystem • Fen • Histosol • Hydromorphic • Landscape • Methane • Microbiology • Mire • Nitrous oxide • Organic soil

I. Gheorghe
Faculty of Ecology and Environmental Protection, Ecological University of Bucharest, Vasile Milea Bd. no.1G, Section 6, Bucharest, Romania

F. Laggoun-Défarge • J.A. Villa
Institut des Sciences de la Terre d'Orléans, CNRS-Université d'Orléans, Orléans, France

C.K. Luswata • M.M. Tenywa
Department of Agricultural Production, College of Agricultural and Environmental Sciences, Makerere University, P. O. Box 7062, Kampala, Uganda

W.J. Mitsch
Everglades Wetland Research Park, Kapnick Center, Florida Gulf Coast University, 10501 FGCU Blvd South, Fort Myers, FL 33965-6565, USA

H. Óskarsson
Faculty of Environmental Sciences, Agricultural University of Iceland, Hvanneyri, IS - 311, Borgarnes, Iceland

S. Pellerin
Institut de recherche en biologie végétale, Université de Montréal, Jardin botanique de Montréal, 4101 Sherbrooke Est, Montréal, QC H1X 2B2, Canada

J.-O. Salm
Institute of Ecology and Earth Sciences, University of Tartu, Vanemuise St. 46, 51014 Tartu, Estonia

Department of Geography, Estonian Fund for Nature, Lai 29, Tartu 51014, Estonia

Ü. Mander
Institute of Ecology and Earth Sciences, University of Tartu, Vanemuise St. 46, 51014 Tartu, Estonia

Hydrosystems and Bioprocesses Research Unit, National Research Institute of Science and Technology for Environment and Agriculture (Irstea), 1 rue Pierre-Gilles de Gennes CS 10030, F92761 Antony Cedex, France

## 7.1 Introduction

Atmospheric concentrations of the greenhouse gases $N_2O$ (nitrous oxide) and $CH_4$ (methane) have increased due to human activity since the preindustrial era. The concentrations of these greenhouse gases exceeded the year 1750 levels by 20 % and 150 %, respectively, in year 2011. Most $N_2O$ originate from agricultural land use and are attributed to increased fertiliser application. Wetlands are the source of most terrestrial $CH_4$ (Le Mer and Roger 2001) but act as sinks of $N_2O$ (Stocker et al. 2013). Artificial drainage turns these into sources of $N_2O$ (Martikainen et al. 1993). Northern peatlands contain 20–30 % of the globe's nitrogen and carbon pools (Sjörs 1981; Gorham 1991 cit (Martikainen et al. 1993)). Tropical peatlands add a significant amount (Page et al. 2010). This makes the world's peatlands a potentially important $N_2O$ and $CH_4$ source.

$N_2O$ and $CH_4$ emission is largely event based – rainfall (Li et al. 1992), soil moisture, pH (Goodroad and Keeney 1984), freeze-thaw (Koponen and Martikainen 2004), fertilisation, root activity (Christensen 1983) and bubbles (Frenzel and Karofeld 2000) to name a few of the involved phenomena. Therefore, natural variation in the gas emissions is very high regardless of sample size (Sabrekov et al. 2013).

The main biological process that produces $N_2O$ is denitrification, the sequential reduction of $NO_2^-$ (nitrite) and $NO_3^-$ (nitrate) to the NO (nitric oxide), $N_2O$ and $N_2$ (nitrogen) gases through four enzymic complexes. The final of these, the conversion of $N_2O$ to $N_2$, is catalysed by Nos (nitrous oxide reductase). A large share of denitrifying complexes does not contain the complete denitrification pathway (Jones et al. 2008). Several soil conditions including soil pH, moisture, temperature and inorganic nitrogen compounds and vegetation affect the soil microbial community and, hence, the $N_2O$: N ratio (Weier et al. 1993). Marginal conditions for denitrification produce a higher $N_2O$: N ratio (Van Cleemput 1998). The main knowledge gap in methanogenesis regards also the involved microflora (Le Mer and Roger 2001).

The complexity in the factors underlying $N_2O$ and $CH_4$ emissions has led to a number of elaborate process-based models, including DAYCENT (Del Grosso et al. 2000) and DNDC (Li et al. 1992). Even though these are able to regard different management and mitigation options, the models need large amounts of input data and need to be calibrated for different land uses (Topp et al. 2013). Boundary line approaches provide simple and practical alternatives in $N_2O$ emission estimation from soil parameters. TIN (inorganic nitrogen), WFPS (water-filled pore space) and t° (temperature) have been used in the models (Conen et al. 2000; Schmidt et al. 2000; Topp et al. 2013; Wang and Dalal 2010).

The objective of this paper is to analyse site-scale $N_2O$ and $CH_4$ emissions from peatlands in relation to soil physical and chemical conditions.

## 7.2 Methods

### 7.2.1 Field Work and Laboratory Analysis

The data were collected in a global soil and $N_2O$ and $CH_4$ sampling campaign. We sampled 29 study sites evenly distributed across various peatlands under the A, C and D climates of the Köppen classification: Iceland (2011); Romania (2012); Santa Catarina, Brazil (2012); Quebec, Canada (2012); Bashkortostan, Russia (2012); Estonia (2012–2013); Florida, USA (2013); France (2013); West Siberia, Russia (2013); Uganda (2013); and French Guiana (2013) (Fig. 7.1; Table 7.1). Data analysis of the samples from the Tibetan Plateau, China (2012), and Tasmania and New Zealand (2014) is still underway. We will sample bogs and fens in Finland, Sweden, Ireland, the Pyrénées, the Okavango delta, Heilongjiang (China), Kamchatka, Borneo, the Cordillera (Colombia) and Tierra del Fuego in the coming few years.

In each study area at least two transects were established along the groundwater depth gradient, one preferably in an undisturbed and another one in an artificially drained area. Each measured position had three replicate chambers and one observation well. $CO_2$, $N_2O$ and $CH_4$ effluxes were measured with the dark static chamber method. 0.5 m diameter PVC collars were installed into the soil at least 3 h before the sampling. During a 1 h session, air was sampled from three white PVC 0.065 $m^3$ static chambers every 20 min (Teiter and Mander 2005) placed into water-filled rings on top of the collars. The gas samples were collected into 50 mL glass vials during 2–3 days in at least three sessions, brought to the University of Tartu laboratory and analysed by gas chromatography. This totalled to 1063 gas

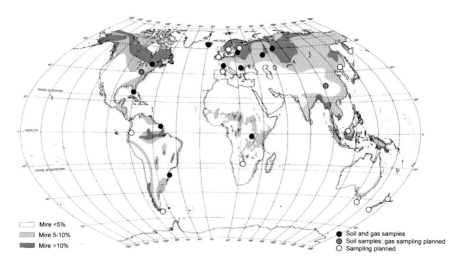

**Fig. 7.1** Study areas at the global mire map (Adapted from the International Peat Society http://www.peatsociety.org/peatlands-and-peat/peat-energy-resource)

fluxes measured from 265 chambers (Table 7.1). Groundwater table, $O_2$ content, oxygen reduction potential and pH were recorded from the wells. Soil temperatures were measured at four depths, 10–40 cm. Peat samples of 100–200 g were taken from below the chambers at two depths: 0–10 cm and 5–15 cm above the groundwater table. The samples were kept cool during the transport to the laboratory. Peat chemical properties (oven-dry mass, pH and C, N, P, C, $NH_4^+$, $NO_3^-$, Ca, K, Mg and S content) were analysed at the laboratories of the Estonian University of Life Sciences.

In addition, organic sediments from the created riverine wetlands in Ohio, USA, in 2009 and relevant gas emission studies have been used in the analyses (Ligi et al. 2014a, b). The peat samples for further pyrosequencing and qPCR analyses are stored in the fridge under −18 °C.

## 7.2.2 Principal Component Analysis

In order to show the interactions between environmental factors and gas fluxes, we performed a PCA (principal component analysis) of the sites using seven variables: soil moisture at 0–10 cm, soil t° at 20 cm, total C and TIN (total inorganic nitrogen) at 0–10 cm, total N, $N_2O$ emission [µg N $m^{-2}$ $h^{-1}$] and $CH_4$ emission [µg N $m^{-2}$ $h^{-1}$]. Soil moisture was calculated as the complementary number of oven-dry mass (100 %–x) but assumed 92 % when the groundwater table was above the ground and 82–92 % when it laid within the range of 0–10 cm. Soil t° at 20 cm depth was chosen as the best correlating depth for gas emissions. TIN was calculated as the sum of $NH_4^+$ and $NO_3^-$.

## 7.2.3 Boundary Lines

Two alternative boundary line applications have been used in $N_2O$ flux modelling, where a boundary may represent either (1) the upper or lower limit of gas emission in relation to one explanatory parameter (Farquharson and Baldock 2008; Schmidt et al. 2000) or (2) a partition of a two-explanatory-variable space into regions according to predicted gas emission (Farquharson and Baldock 2008; Topp et al. 2013; Wang and Dalal 2010).

We explored our data with the first approach. The three-step procedure was as follows: (1) Split the data set into groups, (2) calculate boundary points and (3) fit boundary lines. Step 1 meant two-dimensional scattergrams between the independent variable and the gas flux were plotted. The data set was divided into eight equidistant segments along the independent-variable axis. The number of segments was a compromise between the need for a high group size on one hand and for a sufficient number of boundary points on the other. Data with a log-normal distribution were log-transformed for the procedure. Phase 2 was the calculation of

Table 7.1 Median soil conditions and gas emissions at the study sites

| Site | Coordinates | Climate | Description | No. of chambers | No. of gas measurements | Soil moisture (%) | Soil t° (20 cm depth) | C (%) | TIN (g/kg) | N (%) | CH$_4$ (μg C m$^{-2}$ h$^{-1}$) | N$_2$O (μg N m$^{-2}$ h$^{-1}$) |
|---|---|---|---|---|---|---|---|---|---|---|---|---|
| BAdrain | 55.5476, 55.5980 | Dfb | Drained poor fen | 9 | 27 | 42.7 | 16.8 | 9.70 | 7.97 | 0.82 | 1.65 | 7.09 |
| BAfen | 55.4337, 55.4497 | Dfb | Floodplain fen | 9 | 18 | 39.6 | 13.4 | 15.73 | 18.87 | 1.47 | 0.00 | 117.40 |
| BAhay | 55.4491, 55.4865 | Dfb | Floodplain hay field | 9 | 27 | 30.1 | 17.6 | 13.72 | 23.13 | 1.14 | −16.63 | 7.78 |
| BAslough | 55.4804, 55.9943 | Dfb | Oxbow slough | 9 | 27 | 88.4 | 18.2 | 33.75 | 32.07 | 1.57 | 873.25 | 0.14 |
| EEdrain | 58.4236, 26.5123 | Dfb | Drained floodplain fen | 9 | 27 | 73.3 | 9.2 | 30.40 | 31.95 | 2.78 | −8.66 | 23.46 |
| EEfen | 58.4142, 26.4839 | Dfb | Floodplain fen | 9 | 27 | 79.7 | 8.5 | 56.11 | 1.84 | 2.66 | 45.44 | 1.18 |
| FLswamp | 26.3800, −81.5800 | Dfb | Cypress swamp | 12 | 36 | 75.4 | 20.1 | 46.00 | 77.16 | 2.93 | −18.40 | 13.71 |
| FRbog1 | 47.3200, 02.2300 | Aw | Transitional bog (April) | 14 | 84 | 89.4 | 10.6 | 41.00 | 6.64 | 1.77 | 218.36 | 0.54 |
| FRbog2 | 47.3200, 02.2300 | Cfb | Transitional bog (August) | 14 | 70 | 89.8 | 15.3 | 37.00 | 6.70 | 1.81 | 1,017.52 | 0.24 |
| GFdrain | 5.1688, −52.6595 | Am | Drained tropical peatland | 9 | 54 | 61.4 | 29.1 | 30.00 | 9.66 | 1.14 | 3.90 | 6.94 |
| GFfen | 4.9915, −52.4506 | Am | Tropical fen | 9 | 54 | 78.2 | 26.9 | 33.00 | 46.55 | 1.65 | 3,995.90 | 0.37 |
| ISdrain1 | 64.2872, −15.4645 | Cfc | Drained fen | 12 | 36 | 88.3 | 11.2 | 21.01 | 7.69 | 1.14 | 122.46 | 0.00 |
| ISdrain2 | 64.8332, −22.9160 | Cfc | Drained fen | 12 | 36 | 87.3 | 9.7 | 29.12 | 15.30 | 1.83 | 1,068.50 | 0.00 |
| ISfen1 | 64.3456, −15.4922 | Cfc | Poor fen | 12 | 36 | 92.0 | 11.5 | 4.80 | 1.40 | 0.35 | 1,751.74 | 0.00 |
| ISfen2 | 64.8317, −22.9240 | Cfc | Poor fen | 12 | 36 | 80.1 | 9.2 | 27.45 | 20.93 | 1.64 | 329.93 | 0.00 |
| KHMbog1 | 60.8800, 68.7000 | Dfc | Open bog | 9 | 27 | 86.1 | 16.2 | 44.68 | 0.14 | 0.61 | 87.71 | 0.00 |
| KHMbog2 | 60.8800, 68.7000 | Dfc | Pine and shrub bog | 12 | 18 | 90.0 | 17.8 | 42.27 | 0.39 | 0.57 | 2,190.01 | 0.00 |
| KHMfen | 60.9100, 68.7500 | Dfc | Floodplain fen | 6 | 18 | 85.7 | 12.5 | 18.45 | 7.68 | 1.46 | 2,697.10 | 2.21 |
| QCfen | 45.1273, −74.2191 | Dfb | Floodplain fen | 9 | 54 | 77.7 | 15.9 | 35.00 | 2.94 | 1.00 | 0.00 | 26.65 |
| QChay | 45.1189, −74.2039 | Dfb | Drained hay field | 9 | 54 | 66.7 | 21.3 | 35.00 | 41.39 | 2.00 | 71.39 | 0.47 |
| RObog | 47.2591, 25.3574 | Dfb | Open bog | 9 | 54 | 90.7 | 0.8 | 43.32 | 12.11 | 1.07 | 56.33 | 0.00 |
| ROhay | 47.2539, 25.3501 | Dfb | Hay field; pine bog | 9 | 54 | 86.9 | 1.3 | 41.07 | 71.41 | 2.17 | 0.00 | 0.00 |
| SCdrain | −28.5523, −48.8076 | Cfa | Drained hay field | 9 | 72 | 73.9 | 20.9 | 21.47 | 18.40 | 2.04 | 97.15 | 8.08 |
| SCfen | −28.5527, −48.8079 | Cfa | Floodplain fen | 9 | 45 | 92.0 | 19.0 | 12.82 | 13.42 | 1.17 | 128.63 | 16.47 |
| UGcult1 | −1.1202, 29.8957 | Aw | Freshly hoed potato field | 3 | 9 | 67.9 | 17.9 | 39.64 | 108.47 | 1.79 | 430.05 | 778.21 |
| UGcult2 | −1.3418, 30.0250 | Aw | Freshly hoed potato field | 3 | 9 | 71.0 | 19.4 | 71.82 | 128.05 | 2.49 | 635.06 | 147.78 |

| | | | | | | | | | | |
|---|---|---|---|---|---|---|---|---|---|---|
| UGfen1 | −1.1200, 29.8956 | Aw | Fallow and natural sedge fen | 6 | 18 | 74.2 | 17.0 | 33.72 | 52.75 | 2.55 | 1,434.30 | 119.45 |
| UGfen2 | −1.2354, 29.9689 | Aw | Floodplain fen | 6 | 18 | 67.0 | 16.6 | 12.35 | 17.25 | 1.16 | 1,817.96 | 5.09 |
| UGfen3 | −1.342, 30.0253 | Aw | Fallow and natural sedge fen | 6 | 18 | 76.6 | 17.7 | 37.19 | 86.61 | 2.71 | 1,323.86 | 75.08 |

BA, Birsk, Bashkortostan; EE, Estonia; FL, Corkscrew, Everglades, Florida; FR, La Guette, France; GF, Kourou and Macouria, French Guiana; IS…1, Suðursveit, Iceland; IS…2, Snæfellsnes, Iceland; KHM, Mukhrino, Khanty-Mansiysk, Russia; QC, the Saint Lawrence river, Quebec; RO, Transylvania, Romania; SC, the Tubarão river, Santa Catarina, Brazil; UG, Kigezi highlands, Uganda. Mire categories are presented according to the US EPA Wetland Types (EPA 2013). Climate type based on the Köppen classification: A, tropical; C, temperate; and D, continental boundary points as the upper 0.99 percentile values counted from the top of the segment group. Phase 3 was the calculation of the line best fitting the boundary points (Schmidt et al. 2000).

We used soil TIN (from the 0–10 cm depth), t° (at the 20 cm depth) and soil moisture (at 0–10 cm) (Schmidt et al. 2000). The boundary line approach has not been used for $CH_4$ modelling before, wherefore we made a choice of soil total C content, t°, and soil moisture as commonly used predictors in literature.

## 7.3 Results and Discussion

### 7.3.1 Gas Fluxes in Terms of Environmental Conditions

The highest median $N_2O$ effluxes, 75…778 µg N m$^{-2}$ h$^{-1}$, were shown by the arable peatlands in Uganda as well as the fertile warm fens in Uganda and Bashkortostan, all within 39–74 % soil moisture (Table 7.1; Fig. 7.2). The emissions were an order of magnitude less than some of the highest fluxes recorded in history (South Florida agricultural peatlands, 165 kg N ha$^{-1}$ a$^{-1}$ = 1,884 µg N m$^{-2}$ h$^{-1}$ on average) (Terry et al. 1981). However, the Bashkir and Ugandan gas fluxes exceeded the 95 % upper confidence limit of the IPCC tropical/subtropical grassland and cropland emission factor (IPCC 2013). The

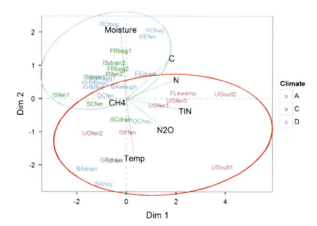

**Fig. 7.2** Principal component plot of sites in relation to environmental variables (vectors magnified 2.6× for visualisation). BA, Birsk, Bashkortostan; EE, Estonia; FL, Corkscrew, Everglades, Florida; FR, La Guette, France; GF, Kourou and Macouria, French Guiana; IS…1, Suðursveit, Iceland; IS…2, Snæfellsnes, Iceland; KHM, Mukhrino, Khanty-Mansiysk, Russia; QC, the Saint Lawrence river, Quebec; RO, Transylvania, Romania; SC, the Tubarão river, Santa Catarina, Brazil; UG, Kigezi highlands, Uganda. Blue ellipse, cool and moist weather (spring, autumn or cool maritime summer); red ellipse, hot and dry weather (tropical or hot continental summer). Mire categories are presented according to the US EPA Wetland Types (EPA 2013). Climate types are according to the Köppen classification groups: A, tropical; C, temperate; and D, continental

highest fluxes from cool (<10°) soil were shown by the drained fen in Estonia (median 23 µg N m$^{-2}$ h$^{-1}$), which were close to the previously recorded emissions in similar Finnish conditions (Alm et al. 2007; Regina et al. 1996) but less than the IPCC lower confidence limit for boreal drained grasslands (IPCC 2013).

The greatest median CH$_4$ fluxes, 1,000...3,995 µg C m$^{-2}$ h$^{-1}$, were recorded in the tropical fens of French Guiana and Uganda and the northern mires of Siberia, France and Iceland. These were generally characterised by a waterlogged condition (with the exception of the Uganda and ISdrain sites). This shows the CH$_4$ potential of tropical fens (Bloom et al. 2010), which has sometimes been considered small (Couwenberg et al. 2010; Hergoualc'h and Verchot 2012). The median emission we measured in the floodplain fens of Siberia, 2,697 µg C m$^{-2}$ h$^{-1}$, practically equals the median of earlier records from the Siberian middle-taiga poor fens (Sabrekov et al. 2013). The median efflux we measured at the open bog types in Siberia (oligotrophic hollows and peat mats) was also within the quartile range of the previous records from analogous ecosystems of the Siberian middle taiga (Sabrekov et al. 2013).

The PCA revealed a positive relationship between soil TIN content and the N$_2$O emissions and soil moisture and the CH$_4$ flux. Soil temperature and moisture are negatively correlated, which is in correspondence with a similar finding from Southern German, Scottish and Southern Queensland agricultural soils (Conen et al. 2000; Schmidt et al. 2000; Wang and Dalal 2010). The tropical sites can be distinguished from the temperate and boreal climates, with the Bashkir samples collected during the extraordinarily hot summer of 2012 as the only exception.

### 7.3.2 Boundary Lines

A log-log linear function describes the upper limit of N$_2$O emissions in relation to soil TIN at 0–10 cm (LogN$_2$O = 1.4075 LogTIN + 0.7431; R$^2$ = 0.88; p = 0.0005; Fig. 7.3). The boundary in terms of soil temperature at 20 cm depth is semilog linear (LogN$_2$O = 1.2387 t° + 0.0821; R$^2$ = 0.71; p = 0.0078; Fig. 7.4). We discontinued the soil t° boundary line at the data gap at 24°. The closest representation of the relationship between N$_2$O and soil moisture is a nonparametric local regression curve with the optimum at 60–70 % (Fig. 7.5), which probably indicates the water level in the microbially active soil horizon. The higher emissions close to 0° show the freeze-thaw effect.

Semilog linear upper boundaries describe the effects of soil moisture (LogCH$_4$ = 0.0852 moisture % − 2.7954; $R^2$ = 0.85; $p$ = 0.001; Fig. 7.6) and soil temperature (Log CH$_4$ = 2.049 t° + 0.0899; $R^2$ = 0.6; $p$ = 0.0231; Fig. 7.7) to CH$_4$ best.

The global N$_2$O boundary lines revealed striking similarities with the Southern German boundary lines for agricultural soils (Schmidt et al. 2000) with noticeable differences in data range only (lower emissions in the peatlands included in this study). In addition, our soil moisture vs. gas flux scattergrams are principally

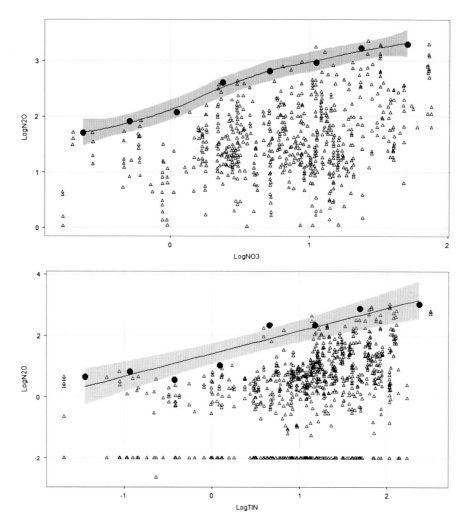

**Fig. 7.3** Local vs. global relationships between inorganic nitrogen and N$_2$O efflux. Above, scattergram and boundary line of soil NO$_3$ plotted against N$_2$O emission in the agricultural soils of Southern Germany (Schmidt et al. 2000). Below, soil TIN [mg kg$^{-1}$] plotted against N$_2$O emissions in our global study [log μg N$_2$O-N m$^{-2}$ h$^{-1}$]. LogN$_2$O = 1.4075 LogTIN + 0.7431; $n = 1{,}063$; $R^2 = 0.88$; $p = 0.0005$. *Black dots* represent 99 % boundary points. The *grey areas* indicate 95 % confidence limits of the boundary lines

similar to the soil mineral nitrogen content vs. N$_2$O efflux plot from Southern Queensland (Wang and Dalal 2010) and pan-European N$_2$O and CH$_4$ emissions plotted against the water table (Couwenberg et al. 2011). However, contrary to Couwenberg & Fritz (Couwenberg and Fritz 2012), our data show no significant difference between the northern and the tropical CH$_4$ fluxes in terms of soil

7 Global Boundary Lines of N₂O and CH₄ Emission in Peatlands

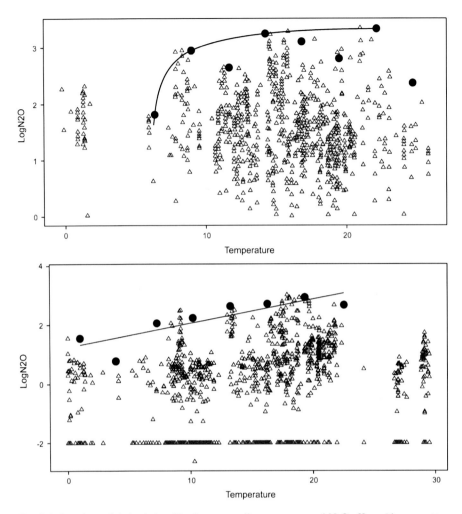

**Fig. 7.4** Local vs. global relationships between soil temperature and N₂O efflux. Above, scattergram and boundary line of soil temperature [C°] plotted against N₂O emission from the agricultural soils in Southern Germany (Schmidt et al. 2000). Below, soil temperature plotted against the N₂O emissions [log µg N₂O-N m$^{-2}$ h$^{-1}$] from our global peatland study. *Black dots* represent 99 % boundary points. LogN₂O = 1.2387 t° + 0.0821; $n = 1{,}063$; $R^2 = 0.71$; $p = 0.0078$

moisture (water table). Hence, local rather than global conditions determine N₂O and CH₄ emissions from peatlands.

This also shows that instead of a default emission factor or a simple correlation for the prediction of N₂O or CH₄ emissions, a local confidence interval should be used for a known set of soil conditions. It must be noted, though, that most of the upper boundary limits are exponential, marking a great efflux range at the more favourable conditions.

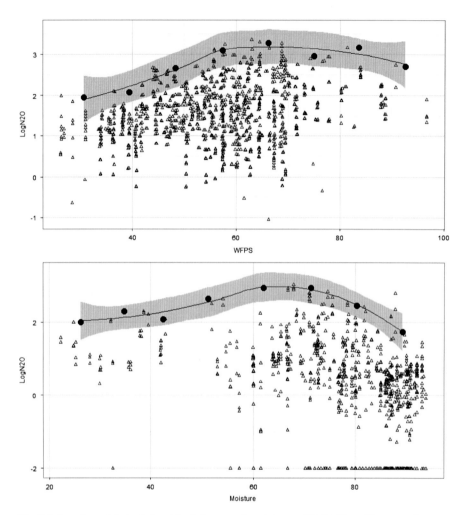

**Fig. 7.5** Local vs. global relationships between soil moisture and N₂O efflux. Above, scattergram and boundary line of soil WFPS (water-filled pore space) % plotted against N₂O emission [log μg N₂O-N m$^{-2}$ h$^{-1}$] from the agricultural soils in Southern Germany (Schmidt et al. 2000). Below, soil moisture % plotted against N₂O emissions from our global peatland study. LogN₂O = 1.4075 LogTIN + 0.7431; $n = 1{,}063$; $R^2 = 0.88$; $p = 0.0005$. *Black dots* represent 99 % boundary points. The *grey areas* indicate 95 % confidence limits of the boundary lines

Neither soil moisture nor water table has explanatory power for CH₄ emission between sites with the water above the ground surface (Couwenberg and Fritz 2012). To address this, we intend to complement the global CH₄ analysis with the percentage cover data of plants with aerenchymatous shunts (Gray et al. 2013).

7 Global Boundary Lines of N$_2$O and CH$_4$ Emission in Peatlands

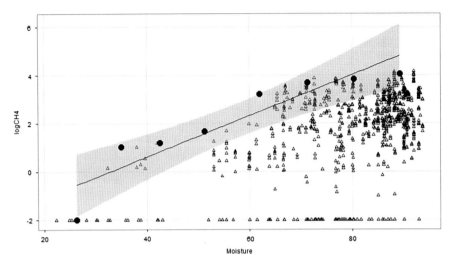

**Fig. 7.6** Global relationship between soil moisture [%] and CH$_4$ efflux [log μg CH$_4$-C m$^{-2}$ h$^{-1}$]. The *grey area* indicates 95 % confidence limits of the regression line. LogCH$_4$ = 0.0852 moisture % + 0.0899; $n = 1{,}063$; $R^2 = 0.6$; $p = 0.0231$

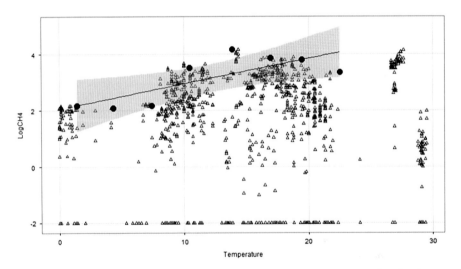

**Fig. 7.7** Global relationship between soil t° [C°] and CH$_4$ efflux [log μg CH$_4$-C m$^{-2}$ h$^{-1}$]. The *grey area* indicates 95 % confidence limits of the regression line. Log CH$_4$ = 2.049 t° + 0.0899; $n = 1{,}063$; $R^2 = 0.6$; $p = 0.0231$

## 7.4 Conclusion

The close match of the global boundary lines of $N_2O$ emission to Southern German and Southern Queensland boundary lines suggests that local rather than global conditions determine the gas emissions. The findings are a development towards capturing the true cumulative $N_2O$ and $CH_4$ emission potential of peatlands in a multivariable space.

## 7.5 Further Perspective

In the future, relationships between the environmental factors and the spatial distribution of the main functional genes, *nirS*, *nirK* and *nosZ*, regulating the denitrification process in the soil samples, currently stored in the refrigerator at $-18°$, will be analysed. Additionally, we will study the relationships between the $CH_4$ emissions and methanogenesis-regulating functional genes *mcrA*, *pmoA* and *dsrAB*.

**Acknowledgements** This study was supported by the IAEA's Coordinated Research Project (CRP) on "Strategic placement and area-wide evaluation of water conservation zones in agricultural catchments for biomass production, water quality and food security", the Ministry of Education and Science of Estonia (grant SF0180127s08), the Estonian Research Council (grant IUT2-16); and the EU through the European Regional Development Fund (Centre of Excellence ENVIRON). We are sincerely grateful to the assistance of Mr. Charles Kizza Luswata in the selection of study sites. Our work benefitted from the contribution of Prof. Jaak Truu, Dr. Marika Truu, Mrs. Teele Ligi and Mr. Kristjan Oopkaup to the perspective microbiological study.

## References

Alm, J., Shurpali, N. J., Minkkinen, K., Aro, L., Hytönen, J., Laurila, T., Lohila, A., Maljanen, M., Martikainen, P. J., Mäkiranta, P., Penttila, T., Saarnio, S., Silvan, N., Tuittila, E. S., & Laine, J. (2007). Emission factors and their uncertainty for the exchange of $CO_2$, $CH_4$ and $N_2O$ in Finnish managed peatlands. *Boreal Environment Research, 12*, 191–209.

Bloom, A. A., Palmer, P. I., Fraser, A., Reay, D. S., & Frankenberg, C. (2010). Large-scale controls of methanogenesis inferred from methane and gravity spaceborne data. *Science, 327*, 322–325.

Christensen, S. (1983). Nitrous-oxide emission from a soil under permanent grass – seasonal and diurnal fluctuations as influenced by manuring and fertilization. *Soil Biology and Biochemistry, 15*, 531–536.

Conen, F., Dobbie, K. E., & Smith, K. A. (2000). Predicting $N_2O$ emissions from agricultural land through related soil parameters. *Global Change Biology, 6*, 417–426.

Couwenberg, J., & Fritz, C. (2012). Towards developing IPCC methane 'emission factors' for peatlands (organic soils). *Mires and Peat, 10*, 1–17.

Couwenberg, J., Dommain, R., & Joosten, H. (2010). Greenhouse gas fluxes from tropical peatlands in south-east Asia. *Global Change Biology, 16*, 1715–1732.

Couwenberg, J., Thiele, A., Tanneberger, F., Augustin, J., Baerisch, S., Dubovik, D., Liashchynskaya, N., Michaelis, D., Minke, M., Skuratovich, A., & Joosten, H. (2011). Assessing greenhouse gas emissions from peatlands using vegetation as a proxy. *Hydrobiologia, 674*, 67–89.

Del Grosso, S. J., Parton, W. J., Mosier, A. R., Ojima, D. S., Kulmala, A. E., & Phongpan, S. (2000). General model for $N_2O$ and $N_2$ gas emissions from soils due to denitrification. *Global Biogeochemical Cycles, 14*, 1045–1060.

EPA. (2013). Wetland types. United States Environmental Protection Agency, http://water.epa.gov/type/wetlands/types_index.cfm

Farquharson, R., & Baldock, J. (2008). Concepts in modelling $N_2O$ emissions from land use. *Plant and Soil, 309*, 147–167.

Frenzel, P., & Karofeld, E. (2000). $CH_4$ emission from a hollow-ridge complex in a raised bog: The role of $CH_4$ production and oxidation. *Biogeochemistry, 51*, 91–112.

Goodroad, L. L., & Keeney, D. R. (1984). Nitrous-oxide production in aerobic soils under varying pH, temperature and water-content. *Soil Biology and Biochemistry, 16*, 39–43.

Gorham, E. (1991). Northern peatlands: Role in the carbon cycle and probable responses to climatic warming. *Ecological Applications, 1*, 182–195.

Gray, A., Levy, P. E., Cooper, M. D. A., Jones, T., Gaiawyn, J., Leeson, S. R., Ward, S. E., Dinsmore, K. J., Drewer, J., Sheppard, L. J., Ostle, N. J., Evans, C. D., Burden, A., & Zieliński, P. (2013). Methane indicator values for peatlands: A comparison of species and functional groups. *Global Change Biology, 19*, 1141–1150.

Hergoualc'h, K. A., & Verchot, L. V. (2012). Changes in soil $CH_4$ fluxes from the conversion of tropical peat swamp forests: A meta-analysis. *Journal of Integrative Environmental Sciences, 9*, 31–39.

IPCC. (2013). 2013 supplement to the 2006 IPCC guidelines for national greenhouse gas inventories: Wetlands. Pre-publication version. http://www.ipcc-nggip.iges.or.jp/home/wetlands.html

Jones, C. M., Stres, B., Rosenquist, M., & Hallin, S. (2008). Phylogenetic analysis of nitrite, nitric oxide, and nitrous oxide respiratory enzymes reveal a complex evolutionary history for denitrification. *Molecular Biology and Evolution, 25*, 1955–1966.

Koponen, H. T., & Martikainen, P. J. (2004). Soil water content and freezing temperature affect freeze-thaw related $N_2O$ production in organic soil. *Nutrient Cycling in Agroecosystems, 69*, 213–219.

Le Mer, J., & Roger, P. (2001). Production, oxidation, emission and consumption of methane by soils: A review. *European Journal of Soil Biology, 37*, 25–50.

Li, C., Frolking, S., & Frolking, T. A. (1992). A model of nitrous oxide evolution from soil driven by rainfall events: 1. Model structure and sensitivity. *Journal of Geophysical Research: Atmospheres, 97*, 9759–9776.

Ligi, T., Oopkaup, K., Truu, M., Preem, J.-K., Nõlvak, H., Mitsch, W. J., Mander, Ü., & Truu, J. (2014a). Characterization of bacterial communities in soil and sediment of a created riverine wetland complex using high-throughput 16S rRNA amplicon sequencing. *Ecological Engineering*.

Ligi, T., Truu, M., Truu, J., Nõlvak, H., Kaasik, A., Mitsch, W. J., & Mander, Ü. (2014b). Effects of soil chemical characteristics and water regime on denitrification genes (*nirS, nirK,* and *nosZ*) abundances in a created riverine wetland complex. *Ecological Engineering*.

Martikainen, P. J., Nykänen, H., Crill, P., & Silvola, J. (1993). Effect of a lowered water table on nitrous oxide fluxes from northern peatlands. *Nature, 366*, 51–53.

Page, S. E., Rieley, J. O., & Banks, C. J. (2010). Global and regional importance of the tropical peatland carbon pool. *Global Change Biology, 17*, 798–818.

Regina, K., Nykänen, H., Silvola, J., & Martikainen, P. J. (1996). Fluxes of nitrous oxide from boreal peatlands as affected by peatland type, water table level and nitrification capacity. *Biogeochemistry, 35*, 401–418.

Sabrekov, A. F., Glagolev, M. V., Kleptsova, I. E., Machida, T., & Maksyutov, S. S. (2013). Methane emission from mires of the West Siberian taiga. *Eurasian Soil Science, 46*, 1182–1193.

Schmidt, U., Thöni, H., & Kaupenjohann, M. (2000). Using a boundary line approach to analyze $N_2O$ flux data from agricultural soils. *Nutrient Cycling in Agroecosystems, 57*, 119–129.

Sjörs, H. (1981). The zonation of northern peatlands and their importance for the carbon balance of the atmosphere. *International Journal of Ecology and Environmental Sciences, 7*, 11–14.

Stocker, T. F., Dahe, Q., & Plattner, G.-K. (2013). *Climate Change 2013: The physical science basis*. Cambridge: IPCC, Cambridge University Press.

Teiter, S., & Mander, Ü. (2005). Emission of $N_2O$, $N_2$, $CH_4$, and $CO_2$ from constructed wetlands for wastewater treatment and from riparian buffer zones. *Ecological Engineering, 25*, 528–541.

Terry, R. E., Tate, R. L., & Duxbury, J. M. (1981). Nitrous-oxide emissions from drained, cultivated organic soils of South Florida. *Journal of the Air Pollution Control Association, 31*, 1173–1176.

Topp, C., Wang, W., Cloy, J., Rees, R., & Hughes, G. (2013). Information properties of boundary line models for $N_2O$ emissions from agricultural soils. *Entropy, 15*, 972–987.

Van Cleemput, O. (1998). Subsoils: Chemo-and biological denitrification, $N_2O$ and $N_2$ emissions. *Nutrient Cycling in Agroecosystems, 52*, 187–194.

Wang, W., & Dalal, R. (2010). Assessment of the boundary line approach for predicting N2O emission ranges from Australian agricultural soils. In *19th World Congress of Soil Science: Soil Solutions for a Changing World* (pp. 1–6). Brisbane, Australia.

Weier, K. L., Doran, J. W., Power, J. F., & Walters, D. T. (1993). Denitrification and the dinitrogen/nitrous oxide ratio as affected by soil water, available carbon, and nitrate. *Soil Science Society of America Journal, 57*, 66–72.

# Chapter 8
# Distribution of Solar Energy in Agriculture Landscape: Comparison Between Wet Meadow and Crops

Hanna Huryna, Petra Hesslerová, Jan Pokorný, Vladimír Jirka, and Richard Lhotský

**Abstract** This study examines the impact of plant cover on water and energy exchange between land and atmosphere in the Třeboň Biosphere Reserve, Czech Republic. Energy fluxes, evapotranspiration and evaporative fraction were determined over typical crops of agriculture landscape and compared with fluxes in an adjacent wet meadow. The results show distinct differences in heat and water exchange between these ecosystems. Diurnal average difference in evapotranspiration rates for days with high irradiance over the wet meadow and arable crops ranged from 1.1 mm day$^{-1}$ to 3.4 mm day$^{-1}$. Furthermore, the evapotranspiration differences between C3 (rapeseed) and C4 (cornfield) was about 2.3 mm day$^{-1}$. Analysis of thermovision pictures showed that temperature variation reached about 9 °C between the ploughed field and meadows at the time of maximum intensity of solar radiation. Heat exchange (sensible heat flux) was greater over arable lands, while water exchange (latent heat flux) was stronger over the wet meadow. The evaporative fraction displayed that more than 100 % of available energy was released by the wet meadow through evapotranspiration due to the advection of dry air from the surroundings. Wetlands show equal or even inverse temperature in vertical profile, whereas corn and wheat show noticeable higher temperature at soil surface in comparison with plant stand surface. Therefore, we suggest that introduction of wetlands to agricultural land is one of the most important instruments for the management of water and heat balance of the landscape.

**Keywords** Wetlands • Arable land • Evapotranspiration • Energy fluxes • Evaporative fraction • Thermo-vision pictures

---

H. Huryna (✉)
Faculty of Science, University of South Bohemia, Branisovska 31a, České Budějovice CZ 370 05, Czech Republic

ENKI, o.p.s, Dukelská 145, Třeboň CZ 379 01, Czech Republic
e-mail: huryna@nh.cas.cz

P. Hesslerová • J. Pokorný • V. Jirka • R. Lhotský
ENKI, o.p.s, Dukelská 145, Třeboň CZ 379 01, Czech Republic

## 8.1 Introduction

Vegetation that is well supplied with water transfers a high amount of solar energy into latent heat of evapotranspiration (Penman 1948; Monteith 1981). The direct role of plants in the distribution of solar energy and in mitigating extremes of local climate has become a topical issue in discussion on climate change (Ripl 2003; Kravčík et al. 2008, Andrich and Imberger 2013). Recommendations of the IPCC (Intergovernmental Panel on Climate Change) to decision-makers focus on correlation between artificially generated average earth temperature, on the one hand, and concentration of specific greenhouse gases, on the other.

Currently, the IPCC (2013) does not engage adequately with the complex interplay of water, sunlight and vegetation. Assessment Report (AR5) by Intergovernmental Panel on Climate Change (IPCC 2013) declared that the surface temperature has increased about 0.85 °C over the period 1880–2012. Greenhouse gases are considered a major source of temperature increase. According to the report, the amount of greenhouse gases in the atmosphere is directly related to the temperature. Increased concentrations of greenhouse gases enhance the temperature of the atmosphere. According to IPCC (2007) the radiative forcing caused by an increase of greenhouse gases in the atmosphere from 1,750 is equal to 1–3 W m$^{-2}$. In the next 10 years, radiative forcing is expected to increase by 0.2 W m$^{-2}$.

Plant stands supplied with water are able to respond to incoming solar energy by intense transpiration to an order of magnitude of hundreds of W m$^{-2}$. Similarly, the amount of water vapour in the atmosphere is one to two orders of magnitude higher than concentration of carbon dioxide. Water vapour is a greenhouse gas and its concentration in the atmosphere changes dramatically in time and space. Furthermore, water exists in three phases and the transitions between these phases are linked with both the uptake and release of high amounts of energy and changes of volume (Pokorný et al. 2010).

Vegetation and the hydrological cycle are closely linked (Kucharik et al. 2000). Plants affect a hydrological cycle through albedo and transpiration. Key characteristics of plant stands are biomass, leaf area, the rooting system, etc. (Milly 1997; Adegoke and Carleton 2000; Eckhardt and Ulbrich 2003). Land-use practices such as deforestation, agriculture, and grazing may impact on a local hydrological cycle and climate (Bryant et al. 1990; Pielke et al. 1991; Vitousek 1994; Pielke et al. 1998; Hesslerova et al. 2013). Recent studies have showed that extensive deforestation has decreased evapotranspiration (Henderson-Sellers et al. 1993; Werth and Avissar 2002; Sanches et al. 2011; Ma et al. 2013). Ge et al. (2013) confirmed that a variation in vegetation exerts a negative effect on regional climate by means of temperature changes. A decrease in vegetation increased the rate of climate warming.

Kedziora and Ryszkowski (1999) stated that "...actively-growing plants (such as cereals in spring or sugar beet in summer) modify the structure of the heat balance to damping atmospheric convection and enhancing the flux of latent heat. Simulated effects of land-use changes show that feedbacks due to plant cover concerning the heat balance are greater under predicted climate conditions than the changes evoked by enhancement of greenhouse effect alone. An increase in landscape structural diversity can compensate for the effect of climate changes in latent and sensible heat flux changes. On the scale of the whole of Poland, conversion of forest into cultivated fields will increase the sensible heat flux and decrease the latent heat in proportion to the percentage of forest in the region".

According to Ryszkowski and Kedziora (2007), the introduction of shelter belts in an agricultural landscape is one of the best tools for managing heat balance and water regime in the landscape. The heat is transferred from cultivated fields to shelter belts by advection-enhanced evapotranspiration rates from trees. This scheme creates conditions in which plants can use solar energy with great efficiency to improve local and regional climate. But they did not focus on the role of wetland ecosystems in arable lands.

In this paper we analyze solar energy flux and associated evapotranspiration in typical crops of agriculture landscape and compare it with fluxes in an adjacent wet meadow. Data from three vegetation seasons (1 May–31 August) were evaluated.

The following hypotheses and questions were formulated:

– What is the temperature difference between different types of land cover? Temperature difference depends on an amount and activity of growing biomass and can reach several $^\circ$C between adjacent areas.
– What is the difference in evapotranspiration and energy fluxes between a wet meadow and crops? During vegetation season, growing vegetation increases evapotranspiration. Wetlands have higher evapotranspiration rates compared with arable lands.
– How can crops and wet meadow impact on a local hydrological cycle? Vertical profiles of vegetation stands will be studied using a thermo-camera; partitioning latent (LE) and sensible (H) fluxes among different ecosystems during the growing seasons and evaporative fraction will be calculated.

## 8.2 Site Description and Methods

The experiments were carried out in the Třeboň Basin Biosphere Reserve (TBBR), in the southern part of the Czech Republic, near the Austrian border (49° 05′ N, 14° 46′ E, 430 m a.s.l). The TBBR was declared in the frame of UNESCO's Man and Biosphere Programme in 1977. The entire Biosphere Reserve (700 km$^2$) was declared as a Protected Landscape Area. The TBBR is a flat or slightly wavy region, dominated mainly by clay and sandy soil. It is characterized by man-made lakes covering a total area of 7,500 ha (about 500 fishponds) which were constructed mostly in the sixteenth century. The reserve belongs to the moderately warm regions of the temperate zone, with an annual mean temperature of 7.8 $^\circ$C and average annual precipitation about 600–650 mm. (Květ et al. 2002).

Continuous monitoring of meteorological parameters was done during three vegetation periods (between 1 May and 31 August) in 2008, 2011 and 2012 when evaporative rates were the highest. A dataset was collected from several localities with different land cover types in the region. The first part of the experiment was conducted in 2008. Two localities were selected:

- A wet meadow (area c. 500 ha) in the Rožemberk fishpond (450 ha) supralittoral. Dominant species were high sedges (*Carex gracilis, Carex vesicaria, Calamagrostis canescens, Phalaris arundinacea, Urtica dioica*). The area surrounding the meteorological station is not managed; tall vegetation is regularly cut only in the close vicinity of the meteorological station. The dryer parts of the wet meadow are mown once a year.
- A winter barley field (22 ha).

Three localities with the following land cover types were chosen for the analysis in 2011:

- A cornfield
- A wheat
- A meadow

Two localities with the following land cover types were chosen in 2012:

- A rapeseed
- A meadow

Data used for the estimation of sensible heat flux, actual evapotranspiration and micrometeorological conditions were obtained from the automatic meteorological stations M4016 placed directly at the study locations (Fig. 8.1).

The automatic meteorological station based on each site measured the air temperature ($T_a$ and $T_c$, °C; T+RH probes, accuracy +/− 0.1 °C), relative air humidity ($RH_a$ and $RH_c$, %; T+RH probes, accuracy +/− 2 %) at the vegetation surface and 2 m above the soil surface, temperature at the soil surface and 0.2 m below soil surface ($T_s$ and $T_{0.2}$, °C; Pt 100, accuracy +/− 0.1 °C), incoming and reflected shortwave (global) radiation ($R_{s\downarrow}$ and $R_{s\uparrow}$, respectively, $W.m^{-2}$; CM3 pyranometers, Kipp & Zonen, the Netherlands, spectral range from 310 to 2800 nm), wind speed and wind direction at 2 m above ground (W12 TM Praha, Czech Republic), atmospheric pressure (PTB 100 A Vaisala sensor, Finland) and volumetric content of liquid water in the soil at 0.05 m underground ($\theta$, %; Wirrib, AMET, Czech Republic, accuracy +/− 0.01 $m^3 m^{-3}$). The data of air temperature, relative air humidity, global radiation, soil temperature and atmospheric pressure were measured every second and stored at 10-min average intervals. Wind speed and wind direction were recorded at 10-min intervals, with an integration time of 30 s. The transmission of measured and archived data from automatic weather stations was done by GPRS transfers in GSM net, and data were stored in the Internet server.

The amount of evapotranspiration was determined: (1) for days with high intense irradiance and (2) for total vegetation period (Table 8.1). Clear days are defined by the daily total income solar radiation over 6 kWh $m^{-2}$.

# 8 Distribution of Solar Energy in Agriculture Landscape: Comparison Between...

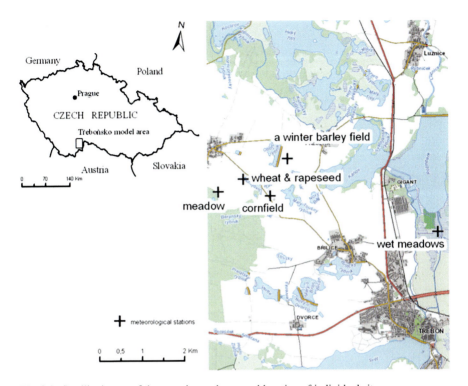

**Fig. 8.1** Satellite image of the experimental area and location of individual sites

**Table 8.1** Daily mean evapotranspiration rates for sunny days and vegetation periods (1 May–31 August)

|                    | Wet meadow | Barley field | Wheat | Meadow          | Corn   | Rapeseed |
|--------------------|------------|--------------|-------|-----------------|--------|----------|
| ET, mm (sunny)     | 5.6[a]     | 4.3[a]       | 3.9[b]| 3.4[b]          | 2.2[b] | 4.5[c]   |
|                    |            |              |       | 3.5[c]          |        |          |
| ET, mm (all days)  | 3.6[a]     | 3.3[a]       | 3.2[b]| 2.5[b]          | 2.1[b] | 3.1[c]   |
|                    |            |              |       | 2.4[c]          |        |          |

[a]Dataset for 2008
[b]Dataset for 2011
[c]Dataset for 2012

The energy balance equation (Eq. 8.1) and Bowen ratio method (Eq. 8.2) were applied for the calculation of dissipation of available energy between latent and sensible heat fluxes using data from the automatic weather stations.

$$R_n - LE - H - G = 0, \tag{8.1}$$

$$\beta = \frac{H}{LE} = \gamma \frac{\Delta T}{\Delta e}. \tag{8.2}$$

In the equations, $R_n$ is the net radiation (W m$^{-2}$), LE is the latent heat flux (W m$^{-2}$), H is the sensible heat flux (W m$^{-2}$), $G$ is the soil heat flux (W m$^{-2}$), $\gamma$ is the psychrometric constant (kPa K$^{-1}$), $\Delta T$ is the temperature differences between the air temperature at 2 m above surface and air temperature of canopy height (0.3 m) (°C) and $\Delta e$ is the difference of water vapour pressure (kPa) in these levels.

Net radiation ($R_n$) was calculated from the incoming ($R_{l\downarrow}$) and outgoing ($R_{s\uparrow}$) shortwave radiation as well as incoming ($R_{l\downarrow}$) and outgoing ($R_{l\uparrow}$) longwave radiation (Eq. 8.3). Shortwave radiation was measured directly by a pyranometer. The longwave radiation received from the atmosphere and emitted by the plant stand surface was calculated from monitored temperature using the Stefan-Boltzmann equation (Eq. 8.4):

$$R_n = R_{s\downarrow} - R_{s\uparrow} + R_{l\downarrow} - R_{l\uparrow}, \tag{8.3}$$

$$R_{l\uparrow} = \varepsilon\sigma(T_c + 273.16)^4, \tag{8.4}$$

where $\varepsilon$ is emissivity, $\sigma$ is the Stefan-Boltzmann constant (W m$^{-2}$ K$^{-4}$) and $T_c$ is air temperature of canopy height (°C). Emissivity was set at 0.98 for all stations. For emissivity values see Gates (1980).

Ground heat flux ($G$) is given by Fourier's law of heat conduction (Eq. 8.5) and was calculated from differences of temperature in a vertical profile of soil (Monteith and Unsworth 1990):

$$G = k\frac{T_s - T_{0.2}}{z_s - z_{0.2}} \tag{8.5}$$

where $k$ is the thermal conductivity of soil (W m$^{-1}$ K$^{-1}$) and $T_s$ and $T_{0.2}$ are the soil temperatures in the depth $z_s$ and $z_{0.2}$.

The Bowen Ratio-Energy Balance (BREB) method was used for the calculation of energy fluxes. Accurate measurement of air temperature and air humidity is a critical factor of reliability of results. The monitored values of relative humidity of air were converted into water vapour pressure according to Eq. 8.6:

$$e = \frac{RH \cdot e_w}{100} \tag{8.6}$$

where $e$ is the actual water vapour pressure (kPa), $RH$ is a relative humidity of air (%) and $e_w$ is the saturation pressure of saturated water (kPa) in the air and at the canopy level, respectively.

The values of saturation pressure ($e_w$) were applied using the modified empirical Magnus-Tetens equation (Eq. 8.7) (Buck 1981):

$$e_w = 0.61121 \exp\left(\left(18.678 - \frac{T}{234.5}\right)\left(\frac{T}{257.14 + T}\right)\right) \quad (8.7)$$

where $T$ is the temperature in 2 m above surface ($T_a$) and at canopy level ($T_c$), respectively. The latent heat flux was calculated by (Bowen 1926) using Eq. 8.8:

$$LE = \frac{R_n - G}{1 + \beta}. \quad (8.8)$$

Then, the sensible heat flux (Eq. 8.9) was computed by difference using the energy balance equation

$$H = \frac{\beta(R_n - G)}{1 + \beta}. \quad (8.9)$$

The latent heat flux was divided by latent heat of vapourization by the following equation (Shuttleworth 2007):

$$L_e = 2.501 - 0.02361 T_a \quad (8.10)$$

where $L_e$ is the latent heat of vapourization (MJkg$^{-1}$) and $T_a$ is the air temperature (°C).

The Bowen ratio ($\beta$) is based on the flux-profile relation of energy and mass exchange and calculated as the ratio of air temperature gradient to water vapour gradient (Eq. 8.2). The values of Bowen ratio ranging between $-0.8$ and $-1.2$ were excluded. Based on the theory of the gradient-flux relationship, the following conditions of BREB calculation were used for the estimation of latent and sensible heat fluxes: $\Delta e < 0$ and $\Delta T < 0$ in the daytime and $\Delta e$ and $\Delta T > 0$ at night. The BREB $> 4$ was also excluded from the further processing (Unland et al. 1996; Foken 2008; Perez et al. 2008).

The sign of convention across this chapter is that the $R_n$ and $G$ are positive downwards during the daytime and negative upwards during the nighttime. The latent and sensible heat fluxes are the positive upwards and negative downwards, opposite to the gradient direction. Thereby the temperature and humidity vertical gradients are always computed as the difference between upper and lower levels.

Evapotranspiration rate (*ET*) was calculated from latent heat flux (*LE*) using Eq. 8.11:

$$ET = \frac{LE}{L_e} \quad (8.11)$$

Daily sums of evapotranspiration rates were expressed in mm.

The ratios of the energy fluxes to net radiation were used for the analysis of the study site's behaviour. Furthermore, evaporative fraction (*EF*, rel.) was used for the analysis of the available energy amount consumed for evaporation of water

(Lhomme and Elguero 1999; Suleiman and Crago 2004; Gentine et al. 2007). *EF* (Eq. 8.12) was computed using the equation

$$EF = \frac{LE}{R_n - G}. \tag{8.12}$$

The mean values reported in Table 8.1 were derived from 24-h data. Linear analysis was used to assess existing correlation between time and individual meteorological parameters and between time and energy fluxes. A confidence interval of 95 % was used and p values <0.05 were considered significant.

All missing data in 10-min intervals owing to the malfunctioning stations were replaced by the average of the two adjacent numbers. Thus, all errors have been excluded.

## 8.3 Results

The total variation of incoming solar radiation ($R_\downarrow$), net shortwave ($R_{ns}$) and net radiation ($R_n$) did not alter much at the localities for three vegetation seasons (Fig. 8.2). The sum of $R_\downarrow$ varied from 640 kWh m$^{-2}$ (wet meadow) to 668 kWh m$^{-2}$ (barley field) in 2008. $R_{ns}$, which is a difference between incoming and outgoing shortwave radiation, did not exceed 540 kWh m$^{-2}$ at both localities. $R_n$ differed by 50 kWh m$^{-2}$ (289 kWh m$^{-2}$ and 339 kWh m$^{-2}$ in the wet meadow and the barley field, respectively). In 2011 (for the 4 months period from 1 May till 31 August), the total fluxes of $R_\downarrow$ at those localities were similar (513 kWh m$^{-2}$, 508 kWh m$^{-2}$ and 504 kWh m$^{-2}$ in the wheat, the meadow and the cornfield, respectively). $R_{ns}$ peaked around 408 kWh m$^{-2}$ in the meadow, 414 kWh m$^{-2}$ in the wheat and 405 kWh m$^{-2}$ in the cornfield. $R_n$ fluctuated between 260 kWh m$^{-2}$ in the meadow and 270 kWh m$^{-2}$ in the wheat. These are about 80 % from $R_{ns}$ and near 50 % of $R_\downarrow$ at both sites.

In 2012 (4 months period, May–August), total $R_\downarrow$ reached a maximum of about 755 kWh m$^{-2}$ and 738 kWh m$^{-2}$ in the rapeseed and the meadow, respectively. Total $R_{ns}$ ranged between 588 kWh m$^{-2}$ (meadow) and 616 kWh m$^{-2}$ (rapeseed). $R_n$ was about 45 % of total $R_\downarrow$ with the values of 320 kWh m$^{-2}$ in the meadow and 358 kWh m$^{-2}$ in the rapeseed.

The total distribution of energy fluxes in three years is presented in Fig. 8.3. The most part of $R_n$ was converted into latent heat flux (LE) during all three monitored periods. In 2008, the values of LE were 263 kWh m$^{-2}$ (barley field) and 300 kWh m$^{-2}$ (wet meadow). Sensible heat flux (H) ranged between −40 kWh m$^{-2}$ (wet meadow) and 54 kWh m$^{-2}$ (barley field). H was negative in the wet meadow, suggesting that some local advection of heat from the surroundings was present. In 2011, LE was up to 145 kWh m$^{-2}$ in the meadow, up to 170 kWh m$^{-2}$ in the wheat and up to 125 kWh m$^{-2}$ in the cornfield. H was a minor part of $R_n$ in the wheat (up to 70 kWh m$^{-2}$). H was virtually similar over the remaining sites and did not

# 8 Distribution of Solar Energy in Agriculture Landscape: Comparison Between... 111

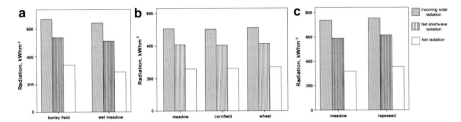

**Fig. 8.2** Total income, net shortwave and net radiation in 2008 (**a**), 2011 (**b**) and 2012 (**c**) for chosen localities measured during the period 1 May–31 August

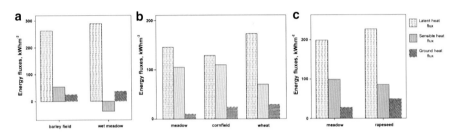

**Fig. 8.3** Distribution of energy fluxes in 2008 (**a**), 2011 (**b**) and 2012 (**c**) measured during the period: 1 May–31 August

exceed 110 kWh m$^{-2}$. In 2012, the values of energy used for evapotranspiration ranged from 199 kWh m$^{-2}$ (meadow) to 227 kWh m$^{-2}$ (rapeseed). For the meadow and the rapeseed, H ranged between 99 kWh m$^{-2}$ and 86 kWh m$^{-2}$, respectively. G varied from 28 kWh m$^{-2}$ (meadow) to 50 kWh m$^{-2}$ (rapeseed).

The seasonal variation of energy flux over the course of the study period is shown in Fig. 8.4. The barley field was characterized by high $R_n$ and $LE$ in June (Fig. 8.4a, b). The sum of $LE$ exceeded $R_n$ in this month. The monthly sum of $H$ was a smaller part of $R_n$ and ranged between $-10$ kWh m$^{-2}$ in June and 40 kWh m$^{-2}$ in August (Fig. 8.4c).

In the wet meadow, sensible heat transcended latent heat in May with the values 35 kWh m$^{-2}$ and 32 kWh m$^{-2}$ for H and LE, respectively (Fig. 8.4b, c). LE was dominant flux over the wet meadow in the following months. The values of monthly sum LE were 79 kWh m$^{-2}$, 97 kWh m$^{-2}$ and 81 kWh m$^{-2}$ in June, July and August, respectively.

The meadow is characterized by the ranging of $R_n$ from 96 kWh m$^{-2}$ in June to 25 kWh m$^{-2}$ in August 2011 (Fig. 8.4a). The maximum values of LE were observed in June (up to 55 kWh m$^{-2}$), while the lowest values were in August (18 kWh m$^{-2}$) (Fig. 8.4b). The maximum H was up to 39 kWh m$^{-2}$ (May) and then H had gradually decreased to 8 kWh m$^{-2}$ (August) (Fig. 8.4c). The G did not exceed 5 kWh m$^{-2}$ and was even negative in August ($-1$ kWh m$^{-2}$) (Fig. 8.4c).

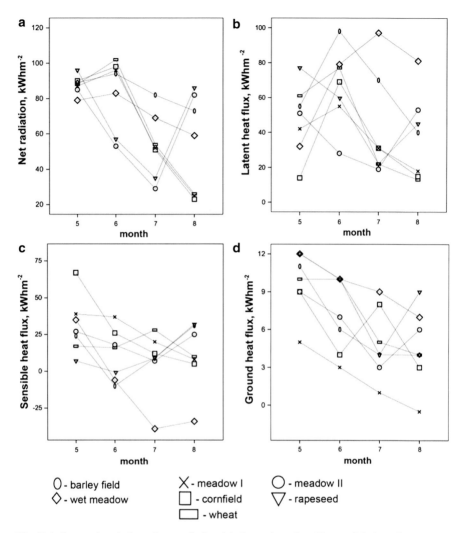

**Fig. 8.4** Seasonal variation of net radiation (**a**), latent heat flux (**b**), sensible heat flux (**c**) and ground heat flux (**d**) for all localities

The cornfield showed a similar pattern in the distribution of $R_n$ as the meadow; $R_n$ ranged from 98 kWh m$^{-2}$ (June) to 23 kWh m$^{-2}$ (August) (Fig. 8.4a). LE was a dominant flux for all months except May and ranged from 70 kWh m$^{-2}$ (June) to 15 kWh m$^{-2}$ (August) (Fig. 8.4b). H fluctuated between 67 kWh m$^{-2}$ in May and 5 kWh m$^{-2}$ in August (Fig. 8.4c). Corn as a C4 plant grows slowly in late spring (May) and accelerates growth rate with temperature increase from June.

In the wheat, the mean monthly sum values of $R_n$ reached a maximum of about 190 kWh m$^{-2}$ in June, and by August the peak of radiation had gradually decreased to about 50 kWh m$^{-2}$ (Fig. 8.4a). LE was a dominant flux for all months except

July. H was 6 kWh m$^{-2}$ higher than LE in July (Fig. 8.4c). G ranged between 10 kWh m$^{-2}$ in May and 4 kWh m$^{-2}$ in August (Fig. 8.4d).

In the meadow, H was a dominant flux in 2012. About 65 % of $R_n$ was converted in H (Fig 8.4b). The monthly sum of LE ranged from 50 kWh m$^{-2}$ in May and August to 20 kWh m$^{-2}$ in July. The monthly sum of H was always lower than LE and fluctuated between 27 kWh m$^{-2}$ and 7 kWh m$^{-2}$ (Fig. 8.4c). G varied between 9 kWh m$^{-2}$ (May) and 3 kWh m$^{-2}$ (July) (Fig. 8.4d). The meadow was mowed twice during the season.

In the rapeseed, $R_n$ ranged from 96 kWh m$^{-2}$ (May) to 37 kWh m$^{-2}$ (July) (Fig. 4a). LE was predominant flux with a maximum of about 77 kWh m$^{-2}$, 60 kWh m$^{-2}$, 22 kWh m$^{-2}$ and 45 kWh m$^{-2}$ in May, June, July and August, respectively (Fig. 8.4b), whereas H was 7 kWh m$^{-2}$, 2 kWh m$^{-2}$, 9 kWh m$^{-2}$ and 32, kWh m$^{-2}$ at the same months (Fig. 8.4c). G exceeded H in May and June (up to 12 kWh m$^{-2}$). In the other cases, G was the smallest part of $R_n$ and did not exceed 10 kWh m$^{-2}$ (Fig. 8.4d). Rapeseed showed high growth rate in spring which is linked with high ET. Rapeseed has the highest ratio of ET among studied crop plants (60 %) (Fig. 8.4).

For each site the energy dissipation was evaluated. More than 90 % of $R_n$ was dissipated as *H* and *LE*, whereas the contribution of *G* was a very small part at all of the sites (Fig. 8.5). *LE* was the primary user of $R_n$. The fraction of $R_n$ consumed by *LE* was very high at the wet meadow (87 %). The magnitude of $LE/R_n$ was quite similar at the meadow for 2 years and exceeded 57 % and 53 % in 2011 and 2012, respectively. The values of $LE/R_n$ for rapeseed were close to the wheat and the cornfield (60–57 %). The highest value of $H/R_n$ was observed at the cornfield (42 %), while the negative value was shown at the wet meadow. Dissipation of ground heat flux did not exceed 15 % at all ecosystems.

The evaporative fraction (EF) (i.e. the ratio of evapotranspiration to available energy at the ground surface (Rn-G)) ranged from 0.75 (barley field) to 1.14 (wet meadow), from 0.58 (meadow) to 0.66 (wheat) and from 0.57 (meadow) to 0.69 (rapeseed) in 2008, 2011 and 2012, respectively (Fig. 8.6).

The daily mean values of evapotranspiration rate for sunny days and for all days during the measurement periods are presented in Table 8.1. The highest daily mean values of actual evapotranspiration (ET) were observed in the wet meadow (5.6 mm day$^{-1}$) in 2008. We estimated that the lowest ET rates of about 2.2 mm day$^{-1}$ were observed during vegetation season in the cornfield in 2011. The daily mean ET rates were 3.4 mm day$^{-1}$ and 3.5 mm day$^{-1}$ for the meadow in 2011 and 2012. The mean ET rate for the barley field, the wheat and the rapeseed were 4.3 mm day$^{-1}$, 3.9 mm day$^{-1}$ and 4.5 mm day$^{-1}$, respectively. The actual ET observed on clear days was always higher than ET for all days and ranged from 2.1 mm day$^{-1}$ (cornfield) to 3.6

Temperature differences between different types of land cover are not too obvious – the highest values of average surface temperatures were recorded for the meadow (25.3 °C), for a field of wheat by 0.7 °C lower. The average surface temperature of the trees is 23.5 °C (Fig. 8.7a; Table 8.2).

A field of wheat has been harvested and ploughed. In the morning differences between average surface temperature of bare ground and vegetated areas are still

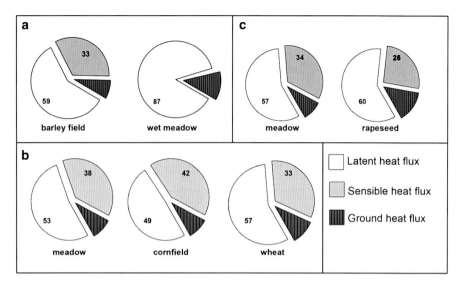

**Fig. 8.5** Variation of the ratio of net radiation dissipation (sensible heat flux, latent heat flux and soil heat flux) in 2008 (**a**), 2011 (**b**) and 2012 (**c**)

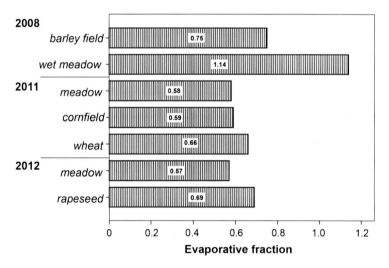

**Fig. 8.6** Variation of the ratio of evaporative fraction in 2008, 2011 and 2012, measured during the period 1 May–31 August

not so obvious. Ploughed field and meadow have nearly the same temperature of 23.6 °C and the trees less than 22 °C (Fig. 8.7b; Table 8.2).

Temperature differences at the time of maximum intensity of solar radiation increase. The surface temperature of ploughed field is almost 40 °C, and in the case

8 Distribution of Solar Energy in Agriculture Landscape: Comparison Between... 115

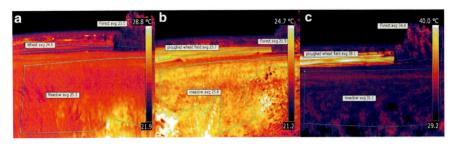

**Fig. 8.7** The thermal images of meadow, forest and wheat field in the background. 7 June 2011 at 11:00 a.m. CET (**a**), 24 August 2011 at 9:00 a.m. CET (**b**) and 24 August 2011 at 1:00 p.m. CET (**c**)

**Table 8.2** Air temperature $T_a$ at 30 cm and 2 m measured in screens in monitoring areas

|  | 7.6.2011 at 11–11:30 a.m. CET | | 24.8.2011 at 9:00–9:30 a.m. CET | | 24.8.2011 at 1–1:30 p.m. CET | |
| --- | --- | --- | --- | --- | --- | --- |
|  | $T_a$ 30 cm (°C) | $T_a$ 2 m (°C) | $T_a$ 30 cm (°C) | $T_a$ 2 m (°C) | $T_a$ 30 cm (°C) | $T_a$ 2 m (°C) |
| Meadow | 29.52 | 22.35 | 26.48 | 24.58 | 35.67 | 31.43 |
| Wheat field | 29.3 | 25.15 | 26.51 | 25.53 | 34.03 | 32.64 |
| Cornfield | 27.95 | 25.13 | 27.58 | 26.95 | 33.28 | 32.28 |

of forest and meadows, temperatures are 34.4 °C and around 31 °C (Fig. 8.7c; Table 8.2).

The average surface temperature of the cornfield was 26.3 °C. Due to intense transpiration, the apex of the maize leaves are the coldest parts of plants (23–26 °C), and the lower part of the stem is around 31 °C. In unploughed crop soil, temperature can reach more than 40 °C. Temperature differences between plants and bare surface reach values of 14 °C (Fig. 8.8).

In a comparison with crops, a vertical profile of forest is characterized by temperature inversions. Temperature of the trunk base and undergrowth is around 23 °C and 28 °C, and the crowns and upper parts of the trunk reach 31 °C and 27 °C (Fig. 8.9). The thermal image also displays an apparent difference in transpiration activity of individual trees. Higher transpiration activity is characterized by deciduous trees, by lower coniferous. In general, this inverse temperature profile is significant for water retention within the growth.

The thermal and visible pictures of floodplain and wetland showed an indication of vertical temperature stratification – warm on the top and cool at the bottom (Fig. 8.10). The temperature differences in floodplain image are only within the range of 2–3 degrees, when compared with cornfield, where the temperature ranges are wider (dozens of degrees) (Fig. 8.8).

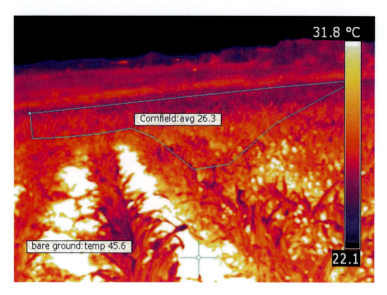

**Fig. 8.8** Cornfield; 7 June 2011 at 11:45 a.m. CET

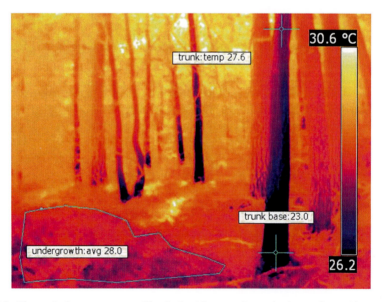

**Fig. 8.9** The vertical temperature profile of mixed forest – pine and oak trees located between the arable land and the meadow; 24 August 2011 at 1:00 p.m. CET

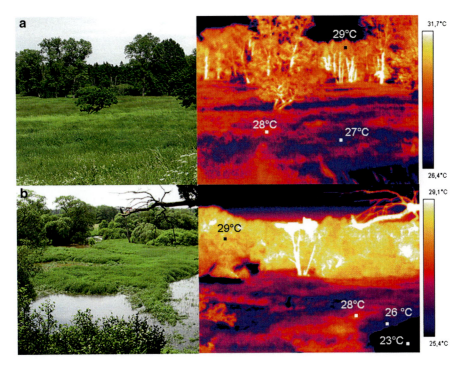

**Fig. 8.10** The thermal and visible pictures of a floodplain (**a**) and a floodplain with open water of a pool showing inversion temperature in a vertical profile (**b**)

## 8.4 Discussion

The focus of this chapter has been placed on the role of vegetation at the energy exchange between the land surface and the atmosphere. Analysis of thermovision pictures showed slight differences in temperature between different types of land cover in June. In August the morning temperature difference was not still so obvious (1.6 °C) while temperature distinction at noon increased. The temperature variation reached about 9 °C between the ploughed field and meadows at the time of maximum intensity of solar radiation in August. Accordingly, vegetation that is well supplied with water can effectively reduce temperature gradient occurring between adjacent areas. The differences between vegetation well supplied with water and drained areas can reach up to 20 °C (Hesslerova et al. 2013).

The diurnal mean evapotranspiration was up to 1.3 mm, 1.7 mm, 3.4 mm and 1.1 mm higher at the wet meadow in comparison with barley field, wheat, cornfield and rapeseed, respectively, for days with high intensity of solar radiation. However, for all vegetation seasons, the difference between total evapotranspiration of wet meadow and arable land ranged between 0.3 mm (barley field) and 1.5 mm (corn). Ryszkowski and Kedziora (1995) presented results of the analysis of evapotranspiration in the growing season for dry and normal years chosen from observations made in the period 1956–1992. Their results showed that the difference between meadow and cultivated fields was about 0.2 mm and 0.3 mm for normal and dry years in the agricultural landscape in Poland, respectively.

The evapotranspiration differences between cornfield and rapeseed were 1 mm day$^{-1}$ and 2.3 mm day$^{-1}$ for the whole vegetation period and for days with high intensity of solar radiation, respectively. Corn as a C4 plant grows slowly in late spring and accelerates growth rate with temperature increase in the midsummer, while the rapeseed has more evapotranspiration releases in the beginning of the growing season.

According to our results, the total distribution of net radiation indicated that the most part of net radiation was converted into latent heat flux over all monitored areas. The wet meadow had dissipation of about 87 % of net radiation into latent heat flux, and sensible heat flux was even negative. On the arable lands the dissipation of $R_n$ accounts for a substantial part of sensible heat flux. A similar result was found at the riparian fen wetland dominated by vascular vegetation and surrounded by agricultural areas and forest in central Denmark during the growing season of 1999 where mean daily evapotranspiration over wetland accounted for 82 % of the available radiant energy (Andersen et al. 2005). During periods with high evapotranspiration demand, warm air is formed over the arable areas and wetland evapotranspiration rates are enhanced by local advection (Rejskova et al. 2010; Huryna et al. 2014).

Cornfield indicates a higher temperature at soil than at the top of the stand. While wetlands and forest have equal inversion temperature distribution in vertical profile, the surface layer of a plant stand has a higher temperature than that of the ground. This temperature inversion may have a significant effect on local climate. Warm air ascends upwards if ground soil is hotter than the boundary layer of atmosphere. It could take away water vapour far from the place where it evaporated. Whereas when temperature at the ground is lower than air temperature, cold air stays at the ground and does not rise up so fast. Hence, the vegetation can keep water; thus, condensation occurs above these plants. Values of evapotranspiration (ET) rates vary widely in the literature. Attarod et al. (2009) reported that average ET rates for the four vegetation seasons in the wheat, the soybean and the cornfield in the experimental farm of Tokyo University of Agriculture and Technology (Japan) were 2.4 mm day$^{-1}$, 3.6 mm day$^{-1}$ and 3.5 mm day$^{-1}$, respectively. Daily ET rates varied from 4.3 mm day$^{-1}$ at the winter wheat to 6.8 mm day$^{-1}$ at the cornfield in the Fars province (Iran) (Shahrokhnia and Sepaskhah 2012). Djaman and Irmak (2013) reported maize ET rates as 2.7 – 3.5 mm day$^{-1}$ at the South Central Agricultural Laboratory, Nebraska, USA. During the growing season, the daily ET of winter wheat was about 1.3 mm day$^{-1}$ in New Delhi, India; 3 mm day$^{-1}$ in Shijiazhuang, China; and 2.5–7.0 mm day$^{-1}$ in North Central Oklahoma, USA (Burba and Verma 2005; Chattaraj et al. 2011; Zhang et al. 2011). In the Czech Republic, daily values of evapotranspiration by rapeseed plant fluctuated from 1.7 mm to 5.2 mm day$^{-1}$, on average 3.2 mm day$^{-1}$; by cornfield 2.2 mm day$^{-1}$ on average; and by spring barley ranged from 1.2 mm day$^{-1}$ to 4.6 mm day$^{-1}$, on average 2.9 mm day$^{-1}$ (Pivec and Brant 2009). It is obvious that the seasonal ET rates of different irrigated crops alter under different climatic conditions, crop type and soil water availability.

Human land use and land cover are associated with the hydrological cycle on local, regional and global scales (Kedziora and Olejnik 2002). However, some scientists focused on the changes of albedo or contribution of carbon emission to greenhouse gas rather than on the direct relationship between land cover and rain precipitation (IPCC 2013). Although land drainage for agriculture is linked with a loss in vegetation cover and forest clearing which results in a reduction of rainfall (van Noordwijk et al. 2014).

Andrich and Imberger (2013) focused on the effect of land-use change on rainfall and the consequences of reductions of rainfall on surface water in Southwest Western Australia. Based on the available data, the authors declared that warming effects were contributing to rainfall decline, while the results of the inland to coastal analyses suggested that land-use change has been the dominant factor affecting rainfall decline in this region. The results indicated that deforestation causing native vegetation to be reduced from 60 % to 30 % of the coastal strip correlates with a 15–18 % decline in annual winter rainfall relative to rainfall at the coast. It was also found that land-use change has reduced stream flow by around 300 GL/year.

Agriculture is a major consumer of water, and thereby improvement of agricultural management is essential. The introduction of wetlands to agricultural land is one of the most important instruments for the management of water and heat balance of the landscape. Excess heat and lack of water considerably reduce the crop yield. Wetlands or ecosystems well supplied with water can provide nearby ecosystems with sufficient amounts of water in vegetation periods.

Importance of wetlands in the energy balance of an agricultural landscape was studied by Huryna et al. (2014). The results document a striking difference in the seasonal energy balance. Under hot weather conditions, the wet meadow showed a high capacity of energy conversion into latent heat of water vapourization. Wetlands can compensate for temperature rises due to warm dry air by increasing their evapotranspiration rate, which can be even higher than the value of net radiation at the wetland site. Different ET rates result in differences in land cover surface temperatures at vegetation (canopy) height. The excess of heat and the lack of water over arable fields dramatically reduce crop yields, while wetlands play a very important role in moderating the water cycle and retaining water within the catchment. The structure of agricultural landscapes bears important linkages with the local climate. Restoring wetlands next to agricultural plots could be an effective measure for managing the heat balance of the landscape.

Makarieva and Gorshkov (2007) formulated a theory of a biotic pump which explains the role of forest transport of wet air from oceans into continents. They showed that large drained overheated areas of agriculture land block transport of wet air from the ocean, and they function as water donors. In terms of a biotic pump, large-scale deforestation and drainage change acceptor areas of native functioning vegetation into donor drying landscape. The task for science is to explain the role of forests, wetlands and native vegetation in local climate and hydrological functions and propose principles and strategies for designing a cultured landscape which would function like native vegetation in water cycling. The question is if and under what conditions a dense vegetation (e.g. forest, littoral vegetation) would be able to function as a biotic pump attracting humid air from the surroundings.

## 8.5 Conclusion

Our results show that (1) the surface temperature of wetland and forest is lower and more vertically balanced than adjacent arable lands; (2) more than 87 % of net radiation was consumed in the evapotranspiration process at the wet meadow; (3) the evaporative fraction demonstrated that more than 100 % of available energy was released by the wet meadow through evapotranspiration due to advection of dry air from the surroundings; (4) diurnal rates of ET in wetlands are more than 2–3 mm higher in comparison with agricultural lands; and (5) possessing high cooling capacity, wetlands play an important role in dampening temperature and improving the local hydrological cycle.

**Acknowledgements** The study was supported by the MSMT RVO ENKI and South Bohemia University grant GAJU 152/2010/Z.

## References

Adegoke, J. O., & Carleton, A. M. (2000). Warm season land surface - Climate interactions in the United States Midwest from mesoscale observations. *Linking Climate Change to Land Surface Change, 6*, 83–97.

Andersen, H. E., Hansen, S., & Jensen, H. E. (2005). Evapotranspiration from a riparian fen wetland. *Nordic Hydrology, 36*, 121–135.

Andrich, M. A., & Imberger, J. (2013). The effect of land clearing on rainfall and fresh water resources in Western Australia: A multi-functional sustainability analysis. *International Journal of Sustainable Development and World Ecology*. doi:10.1080/13504509.2013.850752.

Attarod, P., Aoki, M., & Bayramzadech, V. (2009). Measurement of the actual evapotranspiration and crop coefficients of summer and winter seasons crop in Japan. *Plant and soil environment, 55*(3), 121–127.

Bowen, I. S. (1926). The ratio of heat losses by conduction and by evaporation from any water surface. *Physical Review, 27*(6), 779–787. doi:10.1103/Physrev.27.779.

Bryant, N. A., Johnson, L. F., Brazel, A. J., Balling, R. C., Hutchinson, C. F., & Beck, L. R. (1990). Measuring the effect of overgrazing in the Sonoran Desert. *Climatic Change, 17*(2–3), 243–264. doi:10.1007/Bf00138370.

Buck, A. L. (1981). New equations for computing vapor pressure and enhancement factor. *Journal of Applied Meteorology, 20*, 1527–1532. doi:10.1175/1520-0450(1981)-020.

Burba, G. G., & Verma, S. B. (2005). Seasonal and interannual variability in evapotranspiration of native tallgrass prairie and cultivated wheat ecosystems. *Agricultural and Forest Meteorology, 135*(1-4), 190–201. doi:10.1016/j.agrformet.2005.11.017.

Chattaraj, S., Chakraborty, D., Garg, R. N., Singh, R., Singh, G. P., Sehgal, V. K., Sahoo, R. N., Singh, S., Gupta, V. K., & Chand, D. (2011). Evaluating the effect of irrigation on crop evapotranspiration in wheat (*Triticum aestivum* L.) by combining conventional and remote sensing methods. *Journal of Agricultural Physics, 11*, 35–52.

Djaman, K., & Irmak, S. (2013). Actual crop evapotranspiration and alfalfa- and grass-reference crop coefficients of maize under full and limited irrigation and rainfed conditions. *Journal of Irrigation and Drainage Engineering-Asce, 139*(6), 433–446. doi:10.1061/(Asce)Ir.1943-4774.0000559.

Eckhardt, K., & Ulbrich, U. (2003). Potential impacts of climate change on groundwater recharge and streamflow in a central European low mountain range. *Journal of Hydrology, 284*(1–4), 244–252. doi:10.1016/j.jhydrol.2003.08.005.

Foken, T. (2008). *Micrometeorology*. Berlin: Springer.

Gates, D. M. (1980). *Biophysical ecology*. New York: Springer.

Ge, Q., Zhang, X., & Zheng, J. (2013). Simulated effects of vegetation increase/decrease on temperature changes from 1982 to 2000 across the Eastern China. *International Journal of Climatology*. doi:10.1002/joc.3677.

Gentine, P., Entekhabi, D., Chehbouni, A., Boulet, G., & Duchemin, B. (2007). Analysis of evaporative fraction diurnal behaviour. *Agricultural and Forest Meteorology, 143*(1–2), 13–29. doi:10.1016/j.agrformet.2006.11.002.

Henderson-Sellers, A., Dickinson, R. E., Durbidge, T. B., Kennedy, P. J., Mcguffie, K., & Pitman, A. J. (1993). Tropical deforestation – modeling local-scale to regional-scale climate change. *Journal of Geophysical Research-Atmospheres, 98*(D4), 7289–7315. doi:10.1029/92jd02830.

Hesslerova, P., Pokorny, J., Brom, J., & Rejskova-Prochazkova, A. (2013). Daily dynamics of radiation surface temperature of different land cover types in a temperate cultural landscape: Consequences for the local climate. *Ecological Engineering, 54*, 145–154. doi:10.1016/j.ecoleng.2013.01.036.

Huryna, H., Brom, J., & Pokorny, J. (2014). The importance of wetlands in the energy balance of an agriculture landscape. *Wetlands Ecology and Management*. doi:10.1007/s11273-013-9334-2.

IPCC. (2007). Climate changes – synthesis report. In R. K. Pachauri & A. Reisinger (Eds.), http://www.ipcc.ch. Accessed 16 Feb 2010.

IPCC. (2013). Climate Change 2013: The physical science basis. http://www.ipcc.ch. The final draft report. Dated 7 June 2013.

Kedziora, A., & Olejnik, J. (2002). Water balance in agricultural landscape and options for its management by change in plant cover structure of landscape. In L. Ryszkowski (Ed.), *Landscape ecology in agroecosystems management* (pp. 57–11). Boca Raton: CRC Press.

Kedziora, A., & Ryszkowski, L. (1999). Does plant cover structure in rural areas modify climate change effects? *Geographia Polonica, 72*(2), 65–88.

Kravčík, M., Pokorný, J., Kohutiar, J., Kovac, M., & Toth, E. (2008). *Water for the recovery of the climate. A new paradigm*. Kosice: Municipali and TORY Consulting.

Kucharik, C. J., Foley, J. A., Delire, C., Fisher, V. A., Coe, M. T., Lenters, J. D., & Gower, S. T. (2000). Testing the performance of a Dynamic Global Ecosystem Model: Water balance, carbon balance, and vegetation structure. *Global Biogeochemical Cycles, 14*(3), 795–825. doi:10.1029/1999gb001138.

Květ, J., Lukavská, J., & Tetter, M. (2002). Biomass and net primary production in graminoid vegetation. In *Freshwater wetlands and their sustainable future: A case study of Trebon Basin Biosphere Reserve, Czech Republic* (Man and the biosphere series). Paris: UNESCO.

Lhomme, J. P., & Elguero, E. (1999). Examination of evaporative fraction diurnal behaviour using a soil-vegetation model coupled with a mixed-layer model. *Hydrology and Earth System Sciences, 3*(2), 259–270.

Ma, E. J., Liu, A. P., Li, X., Wu, F., & Zhan, J. Y. (2013). Impacts of vegetation change on the regional surface climate: A scenario-based analysis of afforestation in Jiangxi Province, China. *Advances in Meteorology*. doi:10.1155/2013/796163.

Makarieva, A. M., & Gorshkov, V. G. (2007). Biotic pump of atmospheric moisture as driver of the hydrological cycle on land. *Hydrology and Earth System Sciences, 11*, 1013–1033.

Milly, P. C. D. (1997). Sensitivity of greenhouse summer dryness to changes in plant rooting characteristics. *Geophysical Research Letters, 24*(3), 269–271. doi:10.1029/96gl03968.

Monteith, J. L. (1981). Evaporation and surface-temperature. *Quarterly Journal of the Royal Meteorological Society, 107*(451), 1–27. doi:10.1256/Smsqj.45101.

Monteith, J. L., & Unsworth, M. H. (1990). *Principles of environmental physics*. London: Edward Arnold Press.

Penman, W. R. (1948). Intrathecal ephedrine sulfate anesthesia in obstetrics. *American Journal of Medicine, 5*(4), 621–621. doi:10.1016/0002-9343(48)90123-5.

Perez, P. J., Castellvi, F., & Martínez-Cob, A. (2008). A simple model for estimating the Bowen ratio from climatic factors for determining latent and sensible heat flux. *Agricultural and Forest Meteorology, 148*, 25–37.

Pielke, R. A., Dalu, G. A., Snook, J. S., Lee, T. J., & Kittel, T. G. F. (1991). Nonlinear influence of mesoscale land-use on weather and climate. *Journal of Climate, 4*(11), 1053–1069. doi:10.1175/1520-0442(1991)004<1053:Niomlu>2.0.Co;2.

Pielke, R. A., Avissar, R., Raupach, M., Dolman, A. J., Zeng, X. B., & Denning, A. S. (1998). Interactions between the atmosphere and terrestrial ecosystems: Influence on weather and climate. *Global Change Biology, 4*(5), 461–475. doi:10.1046/j.1365-2486.1998.t01-1-00176.x.

Pivec, J., & Brant, V. (2009). The actual consumption of water by selected cultivated and weed species of plants and the actual values of evapotranspiration of the stands as determined under field conditions. *Soil and Water Research, 4*(2), 39–48.

Pokorný, J., Brom, J., Čermák, J., Hesslerová, P., Huryna, H., Nadyezhdina, N., & Rejšková, A. (2010). Solar energy dissipation and temperature control by water and plants. *International Journal of Water, 5*, 311–336. doi:10.1504/IJW.2010.038726.

Rejšková, A., Čížkova, H., Brom, J., & Pokorný, J. (2010). Transpiration, evapotranspiration and energy fluxes in a temperate wetland dominated by Phalaris arundinacea under hot summer conditions. *Ecohydrology, 5*, 19–27. doi:10.1002/eco.184.

Ripl, W. (2003). Water: The bloodstream of the biosphere. *Philosophical Transactions of the Royal Society of London Series B-Biological Sciences, 358*(1440), 1921–1934. doi:10.1098/rstb.2003.1378.

Ryszkowski, L., & Kedziora, A. (1995). Modification of the effects of global climate change by plant cover structure in an agricultural landscape. *Geographia Polonica, 65*(2), 5–35.

Ryszkowski, L., & Kedziora, A. (2007). Modification of water flows and nitrogen fluxes by shelterbelts. *Ecological Engineering, 29*(4), 388–400. doi:10.1016/j.ecoleng.2006.09.023.

Sanches, L., Vourlitis, G. L., Alves, M. D., Pinto, O. B., & Nogueira, J. D. (2011). Seasonal patterns of evapotranspiration for a Vochysia divergens forest in the Brazilian Pantanal. *Wetlands, 31*(6), 1215–1225. doi:10.1007/s13157-011-0233-0.

Shahrokhnia, M. H., & Sepaskhah, A. R. (2012). Evaluation of wheat and maize evapotranspiration determination by direct use of the Penman-Monteith equation in a semi-arid region. *Archives of Agronomy and Soil Science, 58*(11), 1283–1302. doi: 10.1080/03650340.2011.584216.

Shuttleworth, W. J. (2007). Putting the 'vap' into evaporation. *Hydrology and Earth System Sciences, 11*(1), 210–244.

Suleiman, A., & Crago, R. (2004). Hourly and daytime evapotranspiration from grassland using radiometric surface temperatures. *Agronomy Journal, 96*, 384–390.

Unland, H. E., Houser, P. R., Shuttleworth, W. J., & Yang, Z. L. (1996). Surface flux measurement and modelling at a semi-arid Sonoran Desert site. *Agricultural and Forest Meteorology, 82*, 119–153.

van Noordwijk, M., Namirembe, S., Catacutan, D., Williamson, D., & Gebrekirstos, A. (2014). Pricing rainbow, green, blue and grey water: Tree cover and geopolitics of climatic teleconnections. *Current Opinion in Environmental Sustainability, 6*(0), 41–47. doi:http://dx.doi.org/10.1016/j.cosust.2013.10.008

Vitousek, P. M. (1994). Beyond global warming – ecology and global change. *Ecology, 75*(7), 1861–1876. doi:10.2307/1941591.

Werth, D., & Avissar, R. (2002). The local and global effects of Amazon deforestation. *Journal of Geophysical Research-Atmospheres, 107*(D20),55-1–55-8. doi:10.1029/2001jd000717.

Zhang, Y. C., Shen, Y. J., Sun, H. Y., & Gates, J. B. (2011). Evapotranspiration and its partitioning in an irrigated winter wheat field: A combined isotopic and micrometeorologic approach. *Journal of Hydrology, 408*(3–4), 203–211. doi:10.1016/j.jhydrol.2011.07.036.

# Chapter 9
# Surface Temperature, Wetness, and Vegetation Dynamic in Agriculture Landscape: Comparison of Cadastres with Different Types of Wetlands

Petra Hesslerová and Jan Pokorný

**Abstract** The aim of this study is to demonstrate and emphasize the importance of wetlands and permanent vegetation in an agriculture landscape. We compared functional aspects (surface temperature, wetness, and vegetation) of seven different cadastres (administrative units) in two distinctive regions in the Czech Republic. In terms of land cover, one is represented by heterogeneous mosaic, the other by intensive agriculture. Functional aspects were calculated for each cadastre and then within the region compared among themselves. The comparison was realized during the growing season, i.e., in the months of May to September.

The study, based on Landsat satellite data assessment, confirmed that areas with a higher proportion of forest and wetlands can provide a more balanced temperature-moisture regime of the landscape throughout the growing season, low temperatures, and high humidity. In these areas, the solar radiation is transformed into latent heat which leads to landscape cooling, minimizing erosion, and loss of water, matter, and nutrients. In the landscape where crops dominate, temperature-moisture regime is characterized by high temperature and its amplitude and low humidity. The exception is the beginning of the growing season when the landscape is cooled by intensive crop growth. In periods of high stage of maturity or after harvest, the agricultural landscape is characterized by high surface temperature. At the end of this chapter, we present suggestions for sustainable management of agricultural landscape, which is represented by construction of wetlands.

**Keywords** Land cover • Solar energy dissipation • Sustainable landscape management • Satellite data

P. Hesslerová (✉) • J. Pokorný
ENKI, o.p.s, Dukelská 145, Třeboň 379 01, Czech Republic
e-mail: hesslerova@enki.cz

## 9.1 Introduction

The long-term goal with respect to landscape should be ensuring dissipation of solar energy through the water cycle, i.e., through evapotranspiration. This process is accompanied by decreasing of matter losses transported by water, compensation for temperature differences, and balance of rainfall and water runoff. The concept of sustainable landscape management is based on the dissipation of solar energy in the water cycle and closure cycle of nutrients (Ripl 1995, 2003). Vegetation cover in the landscape plays an important role in transformation of solar energy at the surface of the Earth into individual energy fluxes, i.e., sensible heat, latent heat of evaporation, and ground heat flux.

Surface temperature is determined by the action of solar energy dissipation. A low surface temperature and small temperature differences within a landscape are manifestations of effective distribution of solar energy through the water cycle; they result in a greater ability to retain water and nutrients; in the case of a high surface temperature and its fluctuations, it is the opposite. In areas with higher values of surface temperature (the absence of vegetation well supplied with water is typical), solar energy is converted into sensible heat, which causes overheating of the surface, rise of temperature with consequences such as increased erosion, soil drying, increased runoff, and nutrient losses. If the surface temperature is lower, it usually indicates that the surface is covered by vegetation, well stocked with water. In such areas, solar radiation is used for water vaporization. Energy is transformed into latent heat, which cools down the landscape. These relations have been described and are supported by research and data measurements in the following studies (Monteith 1981; Ryszkowski and Kedziora 1987, 2007; Ripl 2003, 2010; Hesslerová et al. 2012, 2013; Pokorný et al. 2010; Eiseltová et al. 2012; Brom et al. 2012; Procházka et al. 2009; Huryna and Pokorný 2010; Makarieva et al. 2013).

Forests sufficiently stocked with water, wetlands, peatlands, floodplains, etc., show the best cooling capacity, the lowest nutrient and matter losses, and closeness of both cycles. Conversely, cropland devoid of water and permanent vegetation has high losses and low cooling capacity. In this regard, drained and sealed urban and industrial areas fare the worst. Drainage of large areas leads to large temperature differences and changes in air flow and local climate. Agriculture in the temperate zone is based on growing crops that do not tolerate flooding. Agricultural soils are drained, often sealed, with deteriorated infiltration ability; rapid alternation of water logging and subsequent drying leads to rapid mineralization of soil and nutrient losses. Therefore, there is a need for revitalization and encouragement of nonproductive functions of agriculture, focused on improving the dissipation of solar energy, i.e., mitigation of surface temperatures and their differences, reduction of nutrient losses, and closing water cycle.

The aim of the study is to evaluate functioning of two distinctive landscape types in the Czech Republic – intensively farmed (represented by the Žatecko region in Central Bohemia) and heterogeneous landscape with a significant representation of forests and wetlands (represented by Třeboňsko, South Bohemia). In the area of interest, three cadastres in Žatecko and four in Třeboňsko were selected. The cadastres differ among themselves in various representations of land cover categories. The evaluation of landscape functioning is based on the assessment surface temperature, humidity, and the amount of green biomass, with respect to the type of land cover. The assessment is targeted on the growing season, i.e., months from May to September. A key factor is represented by the surface temperature, which is affected by the type of land cover and water content. We assume that the area in which the arable land predominates will have higher surface temperatures and lower humidity. These factors may vary seasonally depending on the amount of biomass, i.e., the degree of maturity of crops. In contrast, in areas with heterogeneous landscape cover, forested, with a higher proportion of wet meadows and wetlands, we expect lower surface temperatures, higher humidity, and lower biomass seasonality. The assessment is based on Landsat satellite data processing and interpretation. The increasing problem of drought and declining yields in former fertile areas has made manifest the need for this study to be worked out for the Ministry of Agriculture. One of the ways in which the drought and loss of nutrients can be faced is construction of wetlands. However, such landscape intervention should be supported by comparative studies.

## 9.2 Model Area, Data, and Methods

### 9.2.1 Třeboňsko

Třeboňsko is a rather flat basin located in South Bohemia (Fig. 9.1), proclaimed as biosphere reserve and landscape protected area. Originally a peaty marsh, the reserve is nowadays dominated by man-made lakes. Its extraordinary high diversity of habitats and species has led to its designation as a Ramsar Site of International Importance and a UNESCO Man and Biosphere Programme. The model area covers 518.5 km$^2$ and represents a rich mosaic of different types of land cover. Forests dominate by almost 50 %; arable land, 21 %; and meadows and pastures with natural vegetation, 17.5 %. Water bodies occupy 6 % and wetland communities 2.5 %. Built-up areas cover less than 3 %. Within the model area, the following types of cadastres were selected:

- Agriculture (cadastre Dunajovice: 8.42 km$^2$) – is mainly used as farmland (arable land 64 %). The southern part is forested (28 %). There is a negligible proportion of meadows and pastures (2 %).
- Agriculture/water (cadastre Domanín: 12.5 km$^2$) – arable land takes 41 %, but a significant proportion of the area is forested (28 %). Water bodies occupy 13.2 %, and agricultural areas with natural vegetation cover 11 %.

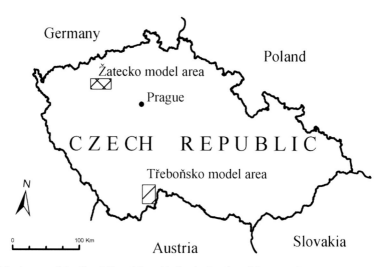

**Fig. 9.1** A map of the Czech Republic with the depicted model areas – heterogeneous landscape of Třeboňsko and agriculture landscape of Žatecko

- Heterogeneous (cadastre Třeboň: 24.01 km$^2$) – forests cover 40.5 %, less than 16 % is grassland, and a significant proportion of the water surface represents almost 11 %. Arable land covers 13 % of the territory.
- Forested/wetlands (cadastre Halámky: 7.1 km$^2$) – the dominant land cover type is forest (71.4 %), and meadows and pastures (17.7 %). More than 10 % is occupied by wetlands and natural floodplain.

## 9.2.2 Žatecko

A significant part of land cover in the Žatecko model area (Fig. 9.1) (454 km$^2$) covers arable land (63 %), orchards and hop gardens occupy less than 4 %, and agricultural areas with natural vegetation 9.3 %; 16.4 % of the territory is forested. There are no any large water bodies. For landscape functioning assessment, three cadastres were selected:

- Agriculture_1 (cadastre Letov: 4.95 km$^2$) and agriculture_2 (cadastre Milčeves: 3.5 km$^2$) are cadastres characterized by a high proportion of arable land (97 %), and 3 % fall on discontinuous urban areas.
- Forested (F, cadastre Lhota u Nečemic: 6.3 km$^2$) cadastre is heterogeneous – almost 77 % of the territory is forested, and 6.4 % occupies agricultural areas with natural vegetation. Arable land comprises only 15 % and urban area 1.3 %.

To characterize functioning of the landscape on a regional scale, following parameters that can be obtained from satellite data, we selected the following:

- The relative amount of green biomass, which is determined on the basis of spectral index NDVI (Normalized Difference Vegetation Index), which correlates with the amount of green biomass (Tucker 1979)
- Relative moisture of land cover – evaluated using the spectral index NDWI (Normalized Difference Water Index) which correlates with the water content (liquid water) in vegetation canopies and upper soil layer (Gao 1996)
- Surface temperature – calculated from the thermal channel TM6

For each model area – Třeboňsko and Žatecko, values of the spectral indices NDVI and NDWI were calculated, based on the following equations (Eq. 9.1):

$$\text{NDVI} = (\text{NIR} - \text{RED})/(\text{NIR} + \text{RED}) \quad (9.1)$$
$$\text{NDWI} = (\text{NIR} - \text{SWIR})/(\text{NIR} + \text{SWIR})$$

where RED, NIR, and SWIR are values of spectral reflectance measurements acquired in the red visible (600–700 nm), near-infrared (700–900 nm), and shortwave infrared (1,550–1,750 nm) regions of electromagnetic spectra.

Both indices take values in the interval $(-1, 1)$. The NDVI values less than 0 indicate bare surfaces, 0–0.4 surfaces with sparse vegetation, above 0.5 dense vegetation. The index correlates with the content of green biomass. Values of NDWI $\leq 0$ indicate dry surfaces, and the increasing value of the index correlates with increasing liquid water content. Surface temperature was calculated from the thermal channel TM6 which records wavelengths from 10.4 to 12.5 μm. The calculation was based on a mono-window algorithm (Sobrino et al. 2005).

As the model areas are not located on the same satellite scene (which implies different dates and years of data acquisition), it is not relevant to compare them among themselves. Comparison of individual cadastres was done only within the appropriate model area. Due to the fact that the data for the month were acquired in different years, under different imaging conditions (atmospheric, effects of seasonality, etc.), the values were standardized using a method of standard scores (z-scores) to make them comparable.

A formula (Eq. 9.2) used for data standardization:

$$As = \frac{Ai - \overline{A}}{SD} \quad (9.2)$$

where $A_s$ is standardized value, $A_i$ is value of the parameter (NDVI, NDWI, or surface temperature) of pixel $i$, $\overline{A}$ is a mean value of the parameter in model the area, and $SD$ is the standard deviation.

The model area has a mean of standardized data equal to 0, and the standard deviation is 1. The data range is usually in the range of $-3$ to 3, and others can be considered as outliers. The method uses an approach introduced by Brom et al. (2012).

Not only was the surface temperature standardized but also the values of both spectral indices NDVI and NDWI.

Comparison of cadastral was performed as follows:

- For each model area (Třeboňsko and Žatecko), the values of both indices NDVI, NDWI, and surface temperature were calculated.
- Standardization of all three parameters was performed. Individual standardized parameters for the model area are equal to zero; however, the standard of each model area is given by the distribution of values of individual parameters, which are determined mainly by the land cover in the model area. Therefore, the standard of Třeboňsko does not represent the same values as the standard of Žatecko.
- The values of parameters for individual cadastres were retrieved by masking.
- For a comparison and interpretation, differences between cadastres within the model area, time series of box plots representing seasonal changes of the parameters, were created.
- Standardized values of NDVI in cadastre:

    >0 indicate a higher amount of biomass than is the mean of the model area (and vice versa).

- Standardized NDWI index values in cadastre:

    >0 indicate less moisture than is the mean of the model (and vice versa).

- Standardized values of surface temperature in cadastre:

    >0 indicate a higher surface temperature than is the mean of the model (and vice versa).

The assessment is based on data from satellites Landsat 5 and Landsat 7. Třeboňsko is situated on satellite scene number 191–026, and Žatecko on 192–025. To evaluate the parameters of biomass, humidity, and surface temperature in terms of their seasonality, it was necessary to obtain data from the months of May, June, July, August, and September. Only scenes without clouds, or a maximum cloudiness of 20 %, were searched in the data archives. This demand made it impossible to obtain data from one year. Therefore, time series representing growing seasons consists of different years. Daytime of imaging is constant for all dates 9:38 UTC +1. An emphasis was put on obtaining the newest data from the requested month. Data were downloaded from the data archive of the United States Geological Survey (USGS) (http://glovis.usgs.gov/) and the University of Maryland – Earth Science Data Interface (ESDI) at the Global Land Cover Facility http://glcfapp.glcf.umd.edu:8080/esdi/). Information about land cover was taken from a European database CORINE Land Cover (2006).

# 9 Surface Temperature, Wetness, and Vegetation Dynamic in Agriculture...

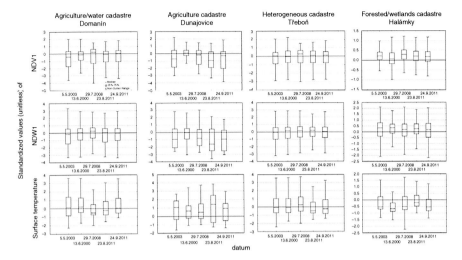

**Fig. 9.2** Seasonal course of standardized values of indices NDVI (correlation with green biomass), NDWI (correlation with liquid water content in vegetation and soil), and surface temperature in *Třeboňsko* model area. The model area has a mean of standardized data equal to 0. Positive standardized values of: NDVI indicate higher amount of green biomass in the cadastre than is the standard of model area; NDWI indicate higher moisture in the cadastre than is the standard of model area; surface temperature indicate higher surface temperature in the cadastre than is the standard of model area. For negative values, the interpretation is reversed

## 9.3 Results

The assessment and data interpretation are based on the following:

- During the period from May to September, we assessed the average standardized values of the parameters NDVI, NDWI, and surface temperature of individual cadastres in the appropriate model area. As a statistical basic value for comparison of the individual cadastres, median was used.
- Assessment of differences between individual cadastral within the model area.
- The differences between cadastres and seasonal variations of the parameters show box plots (Figs. 9.2 and 9.3).

Standardized values of the individual parameters for the whole model area reach zero value. The standard for both model areas represents different values of the parameters. The values in the individual cadastres range within the interval (−3, 3).

Positive standardized values of:

- NDVI indicate higher amount of green biomass in the cadastre than is the standard of model area
- NDWI indicate higher moisture in the cadastre than is the standard of model area
- Surface temperature indicate higher surface temperature in the cadastre than is the standard of model area

For negative standardized values, the interpretation is reversed.

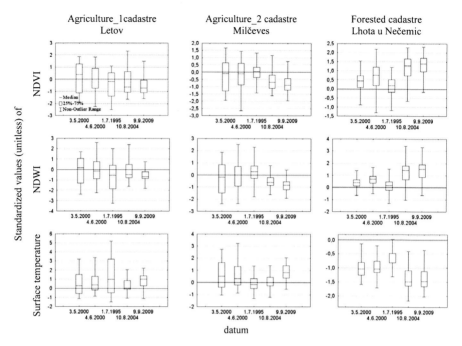

**Fig. 9.3** Seasonal course of standardized values of indices NDVI (correlation with green biomass), NDWI (correlation with liquid water content in vegetation and soil), and surface temperature in *Žatecko* model area. The model area has a mean of standardized data equal to 0. Positive standardized values of: NDVI indicate higher amount of green biomass in the cadastre than is the standard of model area; NDWI indicate higher moisture in the cadastre than is the standard of model area; surface temperature indicate higher surface temperature in the cadastre than is the standard of model area. For negative values, the interpretation is reversed

## 9.3.1 Třeboňsko

Agriculture/water cadastre Domanín is characterized by relatively balanced values of NDVI and NDWI – there are no significant differences in the model area. NDVI values reach its minimum in May (−0.47) and June (−0.17), peaks in July (0.21); in August and September, the values are almost identical to the model area. Moisture characteristics almost correspond with the mean of the model area, except May (−0.18), and are relatively balanced throughout the season. The higher moisture content (0.25) corresponds with the period of maximum biomass in July. The standardized surface temperatures are slightly higher at the beginning of the season in May (0.11) and June (0.18), and significant decrease fall in July (−0.51) and August; in September, the temperature is approaching an average of the model area.

The parameters in the heterogeneous cadastre of Třeboň follow a similar trend as in the previous case. The values of NDVI and NDWI are slightly above average for the model area; maximums are reached in July (0.25) or in August (0.21). The lowest values of standardized surface temperatures were recorded in August

(−0.47), alternatively in September; just below average are in May and June; maximum is achieved in July (0.26).

Both cadastral areas are characterized by heterogeneity of land cover, relatively equal representation of forests, grassland, water, and agricultural land. They are typical representatives of the Třeboňsko model area.

Forest/wetlands cadastre Halámky represents a considerably different area. All three parameters are significantly different not only from other cadastres but also from the whole model area. NDVI values are higher in all five terms. Maximum is reached in July (0.29); in May, August, and September values are relatively balanced (0.17 and 0.21); minimum was recorded in June (0.07). Similarly, behavior of a wetness index values in all five terms is higher than the average of the model area, almost balanced (between 0.2 and 0.33), with a minimum in September (0.17). The temperature values are also the lowest from all studied cadastres in the model area. They have a slightly fluctuating trend, with a peak in August (−0.21), and minimum in June (−0.66). Cadastre is mostly forested, with a significant proportion of meadows, pastures, wetland areas, and natural floodplain. A high proportion of permanent biomass, well supplied with water through the whole season, is positively reflected in temperature-moisture parameters. Of the four cadastres here is recorded the lowest surface temperature, highest humidity, and amount of permanent biomass.

The opposite is agriculture cadastre Dunajovice, with a high proportion of arable land (over 60 %). In all three parameters, the cadastre is significantly different not only from the others but also from the model area. The values of NDVI (green biomass) and NDWI (humidity) are significantly lower in all months, and the minimums are reached in August (−0.96 and −1.69, respectively). Standardized temperature follows this course, the maximum temperature is also reached in August (1.43). The course of individual parameters fluctuates. The cadastre is characterized by the highest values of surface temperatures and lowest humidity of land cover. The results are displayed as Fig. 9.2.

## 9.3.2 Žatecko

Three cadastres in the Žatecko model area were selected. The two (Letov and Milčeves) are intensively agriculturally used areas, with nearly 100 % share of arable land. Therefore, standardized values NDVI, NDWI and surface temperatures are nearly identical, as well as the seasonal course. NDVI values are in months of May, June, July in both cadastres almost identical to the standard of model area; however, in August and September they are significantly lower (−0.7 and −0.87). A similar course is typical of wetness values (NDWI). In the period of May-July both cadastres move around the standard of the model area, it falls significantly below, as the biomass values, in August and September (−0.45 to −0.85). Forested cadastre Lhota u Nečemic is characterized by a heterogeneous landscape – forests dominate, arable land is represented by only 15 %. This is reflected in the values of both indices – through the whole season they are higher than the standard of the model area; their course is relatively fluctuating. Minimum values of NDVI and NDWI are reached in July (0.2 and 0.17); significant peaks appear in August and September, which is in contrast to the other cadastres. NDVI values reached 1.28 and 1.39; NDWI 1.43 and 1.54. So there is a noticeable effect on permanent vegetation.

Standardized temperature of both agricultural cadastres is slightly higher than the standard. The values range between 0 and 0.8. Higher temperatures were recorded in the agriculture_1 cadastre Letov in July and September (1.01 and 1). In the forested cadastre Lhota, the temperature values are under the model area standard throughout the whole growing season; in May and June they are around $-1$; in July ($-0.82$) and the lowest values are reached in August and September ($-1.49$ and $-1.47$). The results are displayed in Fig. 9.3.

## 9.4 Discussion Aimed at Suggestions for Sustainable Farmland Management and Draught Precautions

The study confirmed that the most effective tool for the mitigation of temperature and its differences in the landscape is a well-structured land cover with a significant proportion of permanent vegetation. Studies and realized restoration projects showed that wetlands retain and accumulate biomass, i.e., carbon and nutrients. Less attention is paid to direct effect of wetlands on surface temperature and local climate (Ryszkowski and Kedziora 1995, 2007). Biomass growths, nutrient retention in wetlands are linked with evapotranspiration which regulates temperature and local climate – these are inseparable processes. We propose to adopt these principles of sustainable landscape management as formulated by Ripl (1995) and reward farmers in certain areas according to water quality (i.e., by nutrient losses), which flows from the catchment that the farmer manages. The current criteria of support for nonproductive functions of agriculture are based on the soil quality classification, altitude, and slope. It does not stimulate sustainable uses of the landscape.

We propose that the evaluation of restoration and management interventions should take into account the functional criteria of the landscape, i.e., the relationship between rainfall and runoff, the amount of matter and nutrient losses, together with a reduction in surface temperature and temperature differences in the landscape. Specific recommendations on how to achieve this goal rests in the creation and restoration of four types of wetlands and ponds. These are the types:

(a) Short-retention-time wetlands established to capture the N, P, and suspended solids

They are natural wetlands type in the excavated areas with gravel and other coarse-grained material or a type of constructed wetlands with foil backing and regulated outflow or ponds and small lakes for phosphorus removal. This type of wetland is widespread in Sweden (Tonderski et al. 2005).

This type of wetlands we propose to establish under the larger farm blocks, in the place of concentrated outflows, or in the outflows of drainage water. The aim is to retain runoff water and capture some of the nutrients in the wetland. Wetlands should have a surface area of 0.5–1 % of the total area of agricultural catchment. It is not recommended to build such wetlands for catchments smaller than 50 ha, and the retention time of water should be over 2 days (average per year), which may be insufficient in case of torrential rains (experience from Sweden and Finland). Another way is to create pools, wetlands, and small water reservoirs (polders) at the end of amelioration systems (before the outlet into the stream) in order to retain nutrient outflow from farmland.

(b) Small wetlands in soil blocks spontaneously and repeatedly emerging in former springs in terrain depressions or in the places of disturbed drainage systems

The aim is to retain water in the agricultural landscape, increase eco-stabilization surfaces in contact with arable land (greening), and retain nutrient outflowing of agricultural crops on arable land.

(c) Construction of water holes for cattle in pastures

The purpose of the measure is to retain water on agricultural land, which will have both technological sense (source of water for grazing cattle) and eco-stabilizing sense – creating the aquatic (close to natural state) habitat within agricultural areas that can provide living conditions for animals (e.g., amphibians, birds), which would otherwise not occur here. For a given group of animals, this measure will increase the permeability of the agricultural landscape.

(d) Flooded wet meadows and floodplains of small streams

These habitats flooded at high water, moderate flood wave, retaining nutrients. In these biotopes, flood-tolerant plants dominate, so flooding does not lead to necrosis of vegetation. It is possible to effectively utilize nutrients from biomass of wetland plants (especially grasses) and process biomass pellets.

(e) Restoration and construction of extinct ponds and use of some pond systems to water retention and nutrient recycling

The objective is again concentrated on water and nutrient retention in farmland, increasing eco-stability surfaces in contact with arable land, increasing biodiversity, and promoting of short water cycle.

- In the food chain with a balanced fish stock, it is possible to recycle nutrients and maintain biodiversity.
- Eutrophic fishponds with anaerobic bottom dispose nitrate (denitrification to $N_2$).
- Fishponds with aerobic bottom bind phosphorus.

The aim is to restore ponds which in the area had been in the past and for some reason have disappeared (e.g., were converted to cropland).

## 9.5 Conclusions

The study has shown how to evaluate the functioning of landscape by the use of satellite images. It confirmed that the heterogeneous landscape with cover, with a prevalence or significant proportion of forests and wetlands, can provide a more balanced temperature-moisture regime of the landscape throughout the whole growing season, low temperature, and high humidity. If the surface temperature is lower, it is mostly related to the surface covered by permanent vegetation, well stocked with water; solar radiation is transformed mainly into latent heat, which is used for water vaporization, which cools the landscape. This is accompanied by minimizing erosion and loss of nutrients and matter.

When crops in the landscape predominate, temperature-moisture regime varies considerably; it is characterized by high temperatures and amplitude and low humidity. Exception is partially represented by the beginning of the season, and temperature-moisture characteristics are more balanced. However, this period covers only 1–2 months of the year. In the rest of the growing season, in the high stage of maturity of crops or after harvest (harvest field = bare soil), the farmland is characterized by high values of surface temperature. On these surfaces with the absence of vegetation and lack of water, solar radiation is converted especially into sensible heat, which causes of the surface, with such consequences as increased erosion, soil drying, increased runoff, and nutrient losses.

Permanent vegetation (forests, wetlands) and water retention in the landscape are important tools for sustainable development and management of the landscape (Ryszkowski and Kedziora 1987, 1995, 2007; D'Souza and Lobo 2004; Pijnappels and Dietl 2013; Makarieva and Gorshkov 2007). Intensification of agriculture and urbanization contribute significantly to the disruption of energy flows in the landscape. Drainage, removal of permanent vegetation (and not only deforestation but also decrease of sparse tree and shrub vegetation, drainage of wetlands, wet meadows, and floodplains), and preference of thermophilous crops lead to overheating of the landscape and its degradation, resulting in increased erosion of matter and nutrients. The surface temperature of farmland in a crops maturity period, and especially after harvest, shows similar values to those of urban, industrialized, and mining areas (Hesslerová and Pokorný 2010). People who manage and farm the landscape, including those who create the concept of land-use practices and participate in decision-making processes and landscape management (landowners, fishermen, foresters, farmers), should be responsible for the distribution of solar energy in the landscape, closed water and nutrient cycles, and the creation of local climate, with all the consequences. An effective tool is the retention and accumulation of water in the landscape and well-structured land cover, with the representation of permanent vegetation.

**Acknowledgements** This paper was supported by projects of The Ministry of Education, Youth and Sports of the Czech Republic. "The latest remote sensing technologies in the service of research, education and application for regional development" CZ.1.07/2.4.00/31.0213.

# References

Brom, J., Nedbal, V., Procházka, J., & Pecharová, E. (2012). Changes in vegetation cover, moisture properties and surface temperature of a brown coal dump from 1984 to 2009 using satellite data analysis. *Ecological Engineering, 43*, 45–52.

D'Souza, M., & Lobo, C. (2004). Watershed development, water management and the millennium development goals. http://pdf.wri.org/ref/dsouza_04_watershed_develop.pdf. Accessed 15 Oct 2013.

Eiseltová, M., Pokorný, J., Hesslerová, P., & Ripl, W. (2012). Evapotranspiration – A driving force in landscape sustainability. In A. Irmak (Ed.), *Evapotranspiration – Remote sensing and modeling* (pp. 305–328). InTech Publishing.

Gao, B. (1996). A normalized difference water index for remote sensing of vegetation liquid water from space. *Remote Sensing of Environment, 58*(3), 257–266.

Hesslerová, P., & Pokorný, J. (2010). The synergy of solar radiation, plant biomass, and humidity as an indicator of ecological functions of the landscape: A case study from Central Europe. *Integrated Environmental Assessment and Management, 6*(2), 249–259.

Hesslerová, P., Chmelová, I., Pokorný, J., Šulcová, J., Kröpfelová, L., & Pechar, L. (2012). Surface temperature and hydrochemistry as indicators of land cover functions. *Ecological Engineering, 49*, 146–152.

Hesslerová, P., Pokorný, J., Brom, J., & Procházková-Rejšková, A. (2013). Daily dynamics of radiation surface temperature of different land cover types in a temperate cultural landscape: Consequences for the local climate. *Ecological Engineering, 54*, 145–154.

Huryna, H., & Pokorný, J. (2010). Comparison of reflected solar radiation, air temperature and relative air humidity in different ecosystems. In J. Vymazal (Ed.), *Water and nutrient management in natural and constructed wetlands* (pp. 309–326). Dordrecht: Springer.

Makarieva, A. M., & Gorshkov, V. G. (2007). Biotic pump of atmospheric moisture as driver of the hydrological cycle on land. *Hydrology and Earth System Science, 11*(2), 1013–1033.

Makarieva, A. M., Gorshkov, V. G., & Li, B.-L. (2013). Revisiting forest impact on atmospheric water vapor transport and precipitation. *Theoretical and Applied Climatology, 11*, 79–96.

Monteith, J. L. (1981). Evaporation and surface temperature. *Quarterly Journal of the Royal Meteorological Society, 107*, 1–27.

Pijnappels, M., & Dietl, P. (Eds). (2013). The Adaptation Inspiration Book, 22 implemented cases of local climate change adaptation to inspire European citizens. University of Lisbon, Wageningen UR. http://www.circle-era.eu/np4/%7B$clientServletPath%7D/?newsId=432&fileName=BOOK_150_dpi.pdf. Accessed 15 Oct 2013.

Pokorný, J., Brom, J., Čermák, J., Hesslerová, P., Huryna, H., Nadyezhdina, N., & Rejšková, A. (2010). How water and vegetation control solar energy fluxes and landscape heat. *International Journal of Water, 5*(4), 311–336.

Procházka J., Brom, J., Pechar, L. (2009). The comparison of water and matter flows in three small catchments in the Šumava Mountains. *Soil and Water Research* 4 (Special Issue 2), S75-S82.

Ripl, W. (1995). Management of water cycle and energy flow for ecosystem control: the energy-transport-reaction (ETR) model. *Ecological Modelling, 78*, 61–76.

Ripl, W. (2003). Water: the bloodstream of the biosphere. *Philosophical Transactions of the Royal Society London B, 358*, 1921–1934.

Ripl, W. (2010). Loosing fertile matter to the sea: How landscape entropy affects climate. *International Journal of Water, 5*(4), 353–364.

Ryszkowski, L., & Kedziora, A. (1987). Impact of agricultural landscape structure on energy flow and water cycling. *Landscape Ecology, 1*, 85–94.

Ryszkowski, L., & Kedziora, A. (1995). Modification of the effects of global climate change by plant cover structure in an agricultural landscape. *Geographia Polonica, 65*, 5–34.

Ryszkowski, L., & Kedziora, A. (2007). Modification of water flows and nitrogen fluxes by shelterbelts. *Ecological Engineering, 29*, 388–400.

Sobrino, J. A., Jiménez-Muñoza, J. C., & Paolini, L. (2005). Land surface temperature retrieval from LANDSAT TM 5. *Remote Sensing of Environment, 90*, 434–440.

Tonderski, K., Arheimer, B., & Pers, B. C. (2005). Measured and modeled effect of constructed wetlands on phosphorus transport in South Sweden. *Ambio, 34*(7), 544–551.

Tucker, C. J. (1979). Red and photographic infrared linear combinations for monitoring vegetation. *Remote Sensing of Environment, 8*, 127–150.

# Chapter 10
# Agricultural Runoff in Norway: The Problem, the Regulations, and the Role of Wetlands

**Anne-Grete Buseth Blankenberg, Ketil Haarstad, and Adam M. Paruch**

**Abstract** Agricultural runoff contains amounts of sediments, nutrients, pesticides, and microbes causing water quality problems. Best Management Practices (BMP) are necessary but often insufficient to achieve the requirements set by the environmental goals in the Water Framework Directive (WFD) and to achieve good water status for all water bodies within 2020.

Here we discuss the problems associated with agricultural runoff in Norway, the regulations, and the role of the wetlands. Despite the fact that constructed wetlands (CWs) in Norway are small, studies have shown that in Nordic climate CWs and vegetated buffer zones in first- and second-order streams are good supplements to BMP when it comes to removing and retaining sediments, nutrients, and pesticides. Having a system of agricultural subsidies allows the authorities some leverage in pursuing environmental and other objectives that would otherwise be difficult.

CWs are an important measure for agriculture runoff, and during the last 20 years, more than 900 CWs were built in Norway. The water quality in agricultural areas is regulated mainly by the Norwegian Pollution Act and the Water Framework Directive. The regulation of pesticides is based on action plans and incentives. Studies have shown that for Nordic climatic conditions wetlands can be as effective as advanced treatment systems in removing pesticides from water. The CWs needs adjustment to local conditions and to the specific pollution to be treated.

**Keywords** Agricultural runoff • Constructed wetlands • Diffuse pollution • Sediments • Nutrients • Phosphorus • Nitrogen • Pesticides • Pathogens

A.-G.B. Blankenberg • K. Haarstad (✉) • A.M. Paruch
Bioforsk, Norwegian Institute for Agricultural and Environmental Research, Frederik A. Dahls vei 20, N-1430 Aas, Norway
e-mail: ketil.haarstad@bioforsk.no

## 10.1 What Is the Problem?

Increased intensity in modern agriculture combined with removal of natural buffer systems, such as wetlands, small streams, and vegetative buffer zones, has led to increased and intensified runoff, erosion, and loss of nutrients and other pollutants from agricultural areas to watercourses and downstream rivers, where flooding might occur. These problems are additionally intensified by climatic fluctuations, in particular short-lasting but extreme events of precipitation, or abrupt thawing and snowmelting affecting both the quantity and quality of storm water and natural surface water runoffs.

Runoff and diffuse pollution from agricultural areas are now one of the major anthropogenic sources of both nitrogen (N), phosphorus (P), and sediment inputs to Norwegian surface waters (Borgvang and Tjomsland 2001; Solheim et al. 2001; Selvik et al. 2006). Nitrogen is quantitatively the most important nutrient for plant growth, but it is also a common pollutant (Korseth 2001), and potentially responsible for algal blooms. The loss of soil, particles, nutrients, and pesticides in agricultural watersheds largely depends on source factors such as soil type and crop management; climatic factors such as temperature and precipitation, for example, warm and dry vs. cold and wet periods; and transport factors such as runoff, erosion, drainage, and channeling. N leaching increases significantly with the amount of N applied (Simmelsgaard 1998), but the average N loss also varies greatly among different catchments (Vagstad et al. 2004). Quality problems of freshwater in Norway and the regulation of wastewater treatment are largely caused by or motivated from high P inputs from the catchment area. With increased focus on reducing eutrophication and algae blooms in lakes and rivers, large efforts have been made to implement measures aimed at reducing the runoff of P and sediments. In Norway and other Nordic countries, the problem with runoff and erosion is the greatest in winter, especially during periods of snowmelt (e.g., Øygarden 2000; Grønsten et al. 2007; Søvik and Syversen 2008).

Pesticide losses to the environment should also be avoided due to possible environmental hazards. In Norway, the annual consumption of pesticides is about 900 t, most of being used in conventional agriculture in the southern part of the country. Another big consumer is the vegetable and greenhouse industry. In a study in Germany (von der Ohe et al. 2011) ranging over 500 priority hazardous substances (PHS), 75 compounds were classified as category 1, with observed predicted environmental concentration (PEC) being higher than the predicted no-effect concentration (PNEC), indicating potential hazard, of which two thirds were pesticides. In a review of pollutants in CWs (Haarstad et al. 2012), where more than 500 compounds had been detected, the number of detected pesticides was also high, with more than 30 referenced papers reporting findings of pesticides, of which CWs in agricultural areas made up a considerable portion. Table 10.1 shows, not unexpectedly, that nearly all detected compounds have rather high water solubility, the three highest with a low $pK_a$ showing that they occur ionized and

**Table 10.1** Pesticides found in high concentrations (Conc.) in wetlands (Haarstad et al. 2012), and compounds used in potato productions, their characteristics*, and monitoring ranking (von der Ohe et al. 2011)

| Pesticide | Type | Conc. µg/l | Kow Log | T$_{1/2}$ Days | Solubility µg/l | pKa – | EQS – N µg/l | EQS – Ger µg/l | Rank |
|---|---|---|---|---|---|---|---|---|---|
| *In wetlands* | | | | | | | | | |
| Dichloroaniline | M | 340 | 2.78 | 96 | 620,000 | 2.00 | | | – |
| Atrazin | H | 259 | 2.50 | 45 | 33,000 | 1.74 | 0.40 | 0.40 | 50 |
| Mecoprop | H | 230 | 0.10 | 21 | 860,000 | 3.11 | 44 | | – |
| Diazinon | I | 49 | 3.30 | 27 | 60,000 | | 0.0034 | 0.0011 | 1 |
| Prometryn | H | 13 | | | 40,000 | | | 0.016 | 45 |
| Endosulfan | I | 10 | 3.13 | 150 | 330 | | 0.05 | 0.005 | 5 |
| *In potato productions* | | | | | | | | | |
| Azoxystrobin | F | | 2.76 | 110 | 6,000 | | 0.95 | 0.11 | 2 |
| Cyazofamid | F | | 3.20 | 16 | 107 | | | | – |
| Dimethomorph | F | | | | 50,000 | | | | – |
| Fenamidon | F | | | 1,120 | 8000 | | | | – |
| Imazalil | F | | 3.82 | 150 | 180,000 | | 3 | | – |
| Mandipropamid | F | | 3.3 | 13 | 4,200 | | | | – |
| Pencycuron | F | | | 72 | 300 | | | | – |
| Diquat | H | | 5** | | 700,000 | | | | – |

*K$_{ow}$ = octanol water partition, T$_{1/2}$ = half-life, water solubility, pK$_a$ = acid dissociation constant, Environmental Quality Standards, EQS-N = Norwegian, EQS-Ger = German. Ranking = 1 means most risky pesticide
**Koc = organic carbon partition

mobile in the environment, but these compounds are not necessarily ranked high for monitoring due to their low toxicity.

Some agricultural productions are temporarily intensive in their use of pesticides, such as vegetables, strawberries, and flowers, both in the open and in greenhouses. For example, the production of potatoes in Norway includes the use of the fungicides: azoxystrobin, cyazofamid, dimethomorph, fenamidon, imazalil, mandipropamid, pencycuron, tolklofos-methyl, and zoxamide, and the herbicide diquat, used as frequently as once a week toward the end of the growing season, increasing the possibility of being applied during heavy rain and runoff.

Zoonotic pathogens are another water pollution problem, e.g., from livestock and horses, and also to plant pathogens from the production of grain crops, fruits, and vegetables. A recent Norwegian study (Blankenberg et al. 2012) shows how rainfall affects the quantity of fecal indicator bacteria (*Escherichia coli*, intestinal enterococci, and *Clostridium perfringens*) and parasitic protozoa (*Cryptosporidium* and *Giardia*) in runoff from grazing areas for horses, cattle, sheep, and goats. There was an interrelation between increased precipitation and increased occurrence of fecal microorganisms and potential pathogens in runoff from grazing areas for livestock reaching surface water bodies.

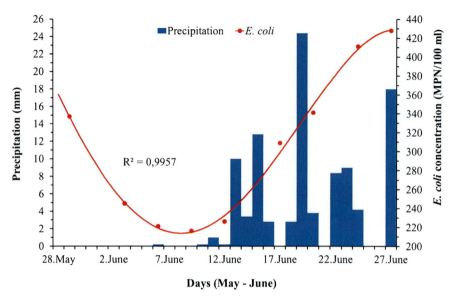

**Fig. 10.1** Changes in *E. coli* concentrations in irrigation water (right axis) throughout the dry and wet periods (precipitation, left axis)

Another Norwegian study demonstrated how irrigation water can be affected by microbial pollution coming from agricultural lands after high rainfall episodes. The study was conducted in a creek used for irrigation of vegetables, such as potato, carrot, parsley, cabbage, onion, and leek. Fecal indicator bacteria in terms of *E. coli* concentration was measured over a period of one month, during which two weeks were relatively dry and warm climatic condition and another two weeks were wet and cold. These two climatic fluctuations revealed an impact on *E. coli* concentrations in the creek (Fig. 10.1). It was found that concentration of the bacteria doubled during the cold and wet climatic conditions, while a dry and warm season caused a decreasing trend in *E. coli* concentrations. The creek is an open water body without any buffered/vegetated zone along it; thus, runoff from agricultural areas cannot be treated in a natural way, and hence, the elimination of microbial pollution was limited. The outcomes of the study revealed that surface water used for agricultural irrigation is very vulnerable to weather variations; hence, even periodical weather fluctuations may influence the hygienic quality of the water. Thus, focus should be given to the sources of irrigation water and its microbial quality. With regard to human and environmental health, it can be suggested that irrigation water should first pass through some natural purification barriers, such as buffer zones or CWs, before its application on crops and soils.

In Norway, surface water bodies are used extensively for agricultural irrigation and drinking water supply. In fact, Norway is one of the countries in Europe, which to a large extent depends on surface water bodies as sources for drinking water, where approximately 85% is obtained from surface water. Therefore, appropriate measures in the catchment area, good protection of vulnerable water sources, and effective drinking water management are important.

## 10.2 What Are the Regulations?

Water quality in agricultural areas in Norway is regulated mainly by the Norwegian Pollution Act (Forurensningsloven 1981) and the Water Framework Directive (WDF, Vannforskriften 2006). In addition to supporting regulations, the agricultural subsidies are directed to obtain several environmental objectives, laid down in, e.g., regional plans for the environment, action plan for the handling of pesticides, and others. The WDF also sets minimum requirements (EQS) for irrigation water. Norway performed a voluntary implementation of the WFD in selected subdistricts across the country from 2007 until 2009, thus gaining the experience of River Basin Management Planning (RBMP) covering the entire country, and which is to be implemented by 2015, synchronized with the time schedule of the second cycle of implementation in the EU. The first edition of guidelines for measures in agricultural landscapes is now available online at www.bioforsk.no/tiltak. This guide provides information on measures to limit water pollution in rural areas and are prepared to contribute to the local authorities' work with water regulations. Also the environmental plans and approaches will be focused at priority locations and achieved through contracts with each farmer rather than through general rules and regulations. The implementation of measures is also stimulated through subsidies, which also applies to CWs, buffer zones, and other natural-based water treatment systems.

The WFD aims at achieving a good status of European surface and groundwaters by 2015, assessed according to a limited set of 33 PHS. If other chemicals are discharged in significant amounts, they shall also be considered under the ecological status assessment. This includes pesticides despite their control by a rather strict premarket approval process. The concept of "significant discharge" is defined based on the predicted environmental concentration concept compared to the predicted no-effect concentration (PEC/PNEC) being greater than one. There is, however, insufficient data on many substances. Where there are missing data the PEC/PNEC ratio can be modeled. Significant discharges of pesticides will have an effect on the ecological status of the receiving water body, also in groundwater.

The use of irrigation water in agriculture is regulated by the Norwegian System for Classification of Environmental Quality in Freshwater (SFT 1997). There are four classes of water quality (Table 10.2) and three categories of usage of irrigation water, depending on the specific group of crops: category A, fruits and vegetables that can be eaten raw without peeling; category B, food that is peeled and/or warmed up before eaten, e.g., potatoes, cabbage, etc.; and category C, grain crops and others that are dried and/or undergo ensilage.

There are no restrictions for irrigations with class 1 water. Class 2 water can be used for crops category A within two weeks before harvesting or until harvesting in

**Table 10.2** Quality parameters (µg/l or no./100 ml) and classes of irrigation water in Norway (SFT 1997)

| Quality parameters | Classes of irrigation water | | | |
|---|---|---|---|---|
| | 1. Very good | 2. Good | 3. Less good | 4. Unacceptable |
| Phosphorus | <11 | 11–20 | 20–50 | >50 |
| Chlorophyll a | <4 | 4–8 | 8–20 | >20 |
| CB | <20 | 20–200 | 200–1,000 | >1,000 |
| TCB | <2 | 2–20 | 20–100 | >100 |

*CB* coliform bacteria, *TCB* thermotolerant coliform bacteria

case of drip irrigation, otherwise no restrictions. Class 3 water cannot be used for irrigation of crops category A and until two weeks before harvesting of crops category B, otherwise no restrictions. Class 4 water should not be used for irrigation in Norway.

Nowadays, there are other quality parameters of significant importance to human and environmental health; thus, these guidelines require certain updating. For instance, the hygienic quality should be defined by more precise agents, such as *E. coli*, which has been proven to be the most accurate indicator of fecal contamination in the environment (Paruch and Mæhlum 2012).

## 10.3 What Is the Role of Wetlands?

The introduction of Best Management Practices (BMP) in agriculture, such as restrictions on manure spreading, reduced use of fertilizer, and reduced tillage during nongrowing seasons, is necessary but often insufficient to achieve water quality requirements set by environmental goals and quality standards in the WFD (Direktoratsgruppa 2009). In addition, there is a widespread reintroduction of buffer systems in the landscape to reduce agricultural sediment, nutrient, pesticide, and microbe losses, both at source areas and along different pathways. Measures such as vegetated buffer zones and small CWs in first- and second-order streams are established to reduce downstream loads of sediments, nutrients, pesticides, and microbes through several mechanical, chemical, and biological processes, including sedimentation, sorption, uptake by vegetation, photodegradation, and microbial activities.

During the last 20 years more than 900 small CWs have been established along streams in Norway (Kollerud, 2013, personal communication). The CWs are located as close to the source as possible in order to optimize their functions. The CWs have different shapes and various components designed according to the specific problem discovered at each location.

Since Norwegian freshwater quality problems largely are caused by high P inputs in the catchment area, the CWs are designed mainly to remove P and sediments. They generally include a deeper sedimentation pond at the inlet with depth about 1.5–2 m, followed by one or more shallower vegetated zones. The focus on P removal is also motivated by controlling the growth of toxic algae and eutrophication of water bodies. The sedimentation pond decreases the water velocity to allow particles to settle, while the vegetated shallower parts filter the particles escaping from the sedimentation pond and prevent trapped sediments from resuspension by stabilizing them with roots. The plants also utilize the nutrients for growing. It is well documented that vegetated ponds are able to remove much finer particles compared to non-vegetated ponds (Braskerud et al. 2005).

Due to the typical small-scale Norwegian agriculture and the rough topography, CWs are often less than 0.1 % of the catchment area. It is therefore important to improve and optimize the retention processes in the CW. Norwegian studies show that total P retention, both particulate and dissolved P, increases with increasing area of the CW (Braskerud et al. 2005). The retention of sediments, nutrients, and pesticides in CWs varies due to factors like design, soil type, hydraulic load, and location (Braskerud and Blankenberg 2005; Blankenberg et al. 2007, 2008; Elsaesser et al. 2011). Through the years of operation, the sedimentation pond and also the vegetated filters will be filled up with sediments. The CWs need to be emptied periodically to maintain optimal treatment capacity (Blankenberg et al. 2013).

An experimental CW was constructed in 2002 in Lier 40 km south of Oslo in the southeastern part of Norway for optimization studies. The size of the catchment area is 0.8 km$^2$ of which 19% is Christmas tree production, 25% is vegetable production, 44% is cereal production, and 13% is urban area. The CW is about 1,200 m$^2$ and covers 0.15% of the catchment area. The experimental CW consists of eight different types of CW filters (Fig. 10.2). Due to problems with the sampling point at the inlet of the sedimentation pond, this treatment part was unfortunately not included in the calculated removal rates. The results presented here are from the samples collected at the end of the sedimentation pond, the outlet of the eight CW

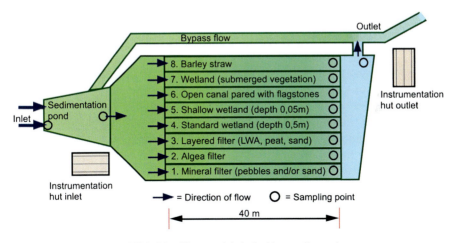

**Fig. 10.2** Experimental CW in Lier, Norway, labeled with sampling points

filters, and the outlet of the entire experimental CW. Filter 4 represents a standard Norwegian wetland filter, and it serves as a reference.

The coarse sand filter (Filter 1) showed the largest removal rate of both total P and dissolved reactive P (DRP) (Braskerud and Blankenberg 2005). In addition, DRP was probably sorbed to presumably Fe-rich filter surfaces. The algae filter (Filter 2) did not improve the P retention compared to the reference (Filter 4), nor did the mixture of the organic and mineral filters and the organic filter (Filter 2 and Filter 8). It is possible that the filters clogged due to sediment transport. The reference (Filter 4) performed well, and the average total P retention was 45% and 30% for the May to September period in 2003 and 2004, respectively. The reduced efficiency in 2004 was probably caused by a higher hydraulic load compared to 2003. For P retention, the ranking of the filters was sand filter (Filter 1) > vegetation filter (Filters 4, 5, and 7) > algae, mixed, and straw filter (Filters 2, 3, and 8) > flagstones, i.e., shallow filter without vegetation (Filter 6).

The total N concentration in the inlet varies with runoff and time of the year. The average TN retention in the experimental wetland was 17% for the period of May to September in 2003, but only 2% in the same period in 2004 (Blankenberg et al. 2007). Again this is probably caused by a higher hydraulic load in 2004. In 2004, there were even periods with leakage of N after large runoff episodes.

The organic filters performed better than the mineral filters with respect to N retention (Filters 8 and 5). The organic filters did also perform better than the reference (Filter 4). The barley straw filter (Filter 8) had the highest TS retention, but seems to be less efficient with time. This filter makes the water anaerobic, and as a result, oxygen from the $NO_3$ was used for mineralization of organic matter. A mixture of the three vegetated CWs (Filters 4, 5, and 7) and the barley straw filter (Filter 8) is needed to improve the N retention in small CWs. The retention time in the CW plays an important role regarding N retention, and if space is limiting, the depth of the CW filter must be adjusted according to the runoff.

All pesticides used in the watershed were found in the experimental wetland, with peak concentrations shortly after spraying (Blankenberg et al. 2007). In 2003, the pesticide retention varied from 11% to 42%, and in 2004, the retention varied from 19% to 56%. Filter 6 and Filter 8, with flagstones and straw, respectively, had higher total pesticide retention than the standard Norwegian wetland (filter 4). When the compounds were treated separately, however, the picture was different. Statistical analyses showed that the treatments were significantly different from zero in six of the wetlands for removal of propachlor; for removal of metalaxyl, none were significantly different; and for removal of chlorfenvinphos, four treatments were significantly different. For the three compounds, none of the relative treatments were significantly different from filter 4. Chemical properties of the pesticides could explain some of the behavior in the watershed and in the wetland. In another study, it was found that systems of natural treatment methods with aeration ponds and wetland can have nearly as good treatment efficiency as advanced technical systems such as reverse osmosis (Haarstad and Mæhlum 2008), and although for some compounds, especially the most water-soluble ones, the concentrations in the output can be high.

Surface flow CWs have an ability to reduce peak concentrations and adverse effects of pesticides in agricultural runoff. The reduction of peak concentrations, total loss, and potential environmental effects varies due to hydraulic conditions, concentration levels, pesticide properties, and vegetation. Elsaesser et al. (2011) showed that hydraulic modification of two wetland cells with dense vegetation improved the reduction of peak concentrations significantly to approximately 90%, although the study also showed a 72% reduction of peak concentration in a cell with no vegetation.

Vegetation also plays a significant role in the retention of micropollutants, in particular metals accumulated in the rhizosphere (Doyle and Otte 1997; Ye et al. 2001). It has also been demonstrated that the bacterial distribution on roots of macrophytes growing in CWs has a steep decrease in numbers within the first few meters along the horizontal flow of the vegetated filters (Vymazal et al. 2001). The root system that expanded throughout the vegetated filter plays a substantial role in the transport of contaminants, serving pathways for gases and moving particles (Scholz et al. 2002). Furthermore, rhizomes of reeds, which are the most common vegetation in CWs, create a natural barrier for the outflow of parasite eggs and thus allowing the antagonistic organisms to destruct the eggs (Paruch 2010). Therefore, CWs have been referred to as very effective in removing parasite eggs, demonstrating a reduction rate of 99.9% (WHO 2006). A similar rate can be achieved in the reduction of fecal indicator bacteria: however, the high efficiency was proven during treatment of wastewater form point sources. There is still a research gap regarding the assessment of CWs in reduction of microbes from agricultural runoff in Norway.

## 10.4 Conclusions

Runoff from agriculture contains significant amounts of particles, nutrients, pesticides, and pathogens capable of doing harm to the environment and human health. The emissions are regulated by several laws, regulations, and guidance, but in practice most of the efforts to reduce the emissions are managed through subsidies according to more or less broad environmental targets rather than trying to meet specific objectives controlled by monitoring. The results presented here show that CWs can play a significant role in the reduction of agricultural pollutants. However, it is difficult to comprehensively assess the reduction of pathogenic organisms as studies performed so far focused mostly on either CWs for point source pollution or on monitoring addressing the problem with pathogens runoff to open and unprotected water bodies without buffer zones. Therefore, there is a shortage of relevant data demonstrating a practical effect of natural systems in reduction of microbial pollution from agricultural runoff.

In order to achieve an efficient treatment, the wetlands need to be adapted to the specific emissions defined at each and every location, determined by the runoff, the local climate conditions, and the range of particular pollutants. It is also important to properly maintain the CW to achieve optimal retention in the CWs.

# References

Blankenberg, A.-G. B., Haarstad, K., & Braskerud, B. C. (2007). Pesticide retention in an experimental wetland treating non point source pollution from agriculture run-off. *Water Science and Technology, 55*(3), 37–44.

Blankenberg, A.-G. B., Haarstad, K., & Søvik, A.-K. (2008). Nitrogen retention in constructed wetland filters treating diffuse agriculture pollution. *Desalination, 226,* 114–120.

Blankenberg, A.-G. B., Tryland, I., Paruch, A., & Robertson, L. (2012). Virkningen av økt nedbør, en følge av klimaendring, på avrenning av tarmbakterier og parasitter fra beiteområder (The effect of heavy rainfall, a consequence of climate change, on runoff of bacteria and parasites from pastures). *Vann, 47*(1), 28–38. (In Norwegian).

Blankenberg, A.-G. B., Deelstra, J., Øgaard, A. F., & Pedersen, R. (2013). Phosphorus and sediment retention in a constructed wetland. In M. Bechmann & J. Deelstra (Eds.), *Agriculture and environment – Long term monitoring in Norway* (pp. 299–314). Trondheim: Akademika Publishing.

Borgvang, S.-A., & Tjomsland, T. (2001). Tilførsler av næringssalter til Norges kystområder, beregnet med tilførselsmodellen TEOTIL (Addition of nutrients to the coast of Norway estimated with the model TEOTIL). NIVA report 815/01, TA-1783. (In Norwegian).

Braskerud, B. C., & Blankenberg, A.-G. B. (2005). Phosphorus retention in the Lier wetland. Is living water possible in agricultural areas? Jordforsk book nr. 48/05. 145:126–128. ISSN/ISBN:82-7467-537-1.

Braskerud, B., Tonderski, K., Wedding, B., Bakke, R., Blankenberg, A.-G. B., Ulen, B., & Koskiaho, J. (2005). Can constructed wetlands reduce the diffuse phosphorus loads to eutrophic water in cold temperate regions? *Journal of Environmental Quality, 34*(6), 2145–2155.

Direktoratsgruppa. (2009). Vanndirektivet: Veileder 01:2009. Klassifisering av miljøtilstand i vann. Økologisk og kjemisk klassifiseringssystem for kystvann, innsjøer og elver i henhold til vannforskriften. 188 pp. (in Norwegian).

Doyle, M. O., & Otte, M. L. (1997). Organism-induced accumulation of iron, zinc and arsenic in wetland soils. *Environmental Pollution, 96*(1), 1–11.

Elsaesser, D., Blankenberg, A.-G. B., Geist, A., Mæhlum, T., & Schulz, R. (2011). Assessing the influence of vegetation on reduction of pesticide concentration in experimental surface flow constructed wetlands: Application of the toxic units approach. *Ecological Engineering, 37*(6), 955–962.

Forurensningsloven. (1981). LOV 1981-03-13 nr 06: Lov om vern mot forurensninger og om avfall (in Norwegian). http://www.lovdata.no/all/hl-19810313-006.html. Accessed 16 Oct 2013.

Grønsten, H. A., Øygarden, L., & Skjevdal, R. (2007). Jordarbeiding til høstkorn- effekter på erosjon og avrenning av næringsstoffer. Bioforsk rapport Vol. 2 Nr. 60/2007. 71 pp.

Haarstad, K., & Mæhlum, T. (2008). Pesticides in Norwegian landfill leachate. *The Open Environmental and Biological Monitoring Journal, 1,* 8–15.

Haarstad, K., Bavor, J., & Mæhlum, T. (2012). Organic and metallic pollutants in water treatment wetlands: A review. *Water Science and Technology, 65*(1), 76–99.

Korseth, A. (2001). Nitrogen dynamics of agro-systems: combinations of modeling and experiments at different spatial and temporal scales. PhD theses 2001:16. ISSN:0802–3220. ISBN 82-575-0464-5, Agriculture University of Norway, Ås.

Øygarden, L. (2000). Seasonal variations in soil erosion in small agricultural catchments in southeastern Norway. In L. Øygarden (Ed.), *Soil erosion in small agricultural catchments, southeastern Norway*. PhD thesis 200:8, Agricultural University of Norway.

Paruch, A. M. (2010). Possible scenarios of environmental transport, occurrence and fate of helminth eggs in light weight aggregate wastewater treatment systems. *Reviews in Environmental Science and Biotechnology, 9*(1), 51–58.

Paruch, A. M., & Mæhlum, T. (2012). Specific features of *Escherichia coli* that distinguish it from coliform and thermotolerant coliform bacteria and define it as the most accurate indicator of faecal contamination in the environment. *Ecological Indicators, 23*, 140–142.

Scholz, M., Hohn, P., & Minall, R. (2002). Mature experimental constructed wetlands treating urban water receiving high metal loads. *Biotechnology Progress, 18*(6), 1257–1264.

Selvik, J. R., Tjomsland, T., Borgvang, S. A., & Eggestad, H. O. (2006). Tilførsler av næringsstoffer til Norges kystområder i 2005, beregnet med tilførselsmodellen TEOTIL2. NIVA-Report 5330 (In Norwegian).

SFT. (1997). Norwegian system for classification of environmental quality in freshwater. (In Norwegian). Veiledning 97:04, Statens forurensningstilsyn, Oslo, Norway.

Simmelsgaard, S. E. (1998). The effect of crop, N-level, soil type and drainage on nitrate leaching from Danish soil. *Soil Use and Management, 14*, 30–36.

Solheim, A. L., Vagstad, N., Kraft, P., Løvstad, Ø., Skoglund, S., Turtumøygard, S., & Selvik, J. R. (2001). Tiltaksanalyse for Morsa (Vansjø-Hobølvassdraget) – Sluttrapport. (Remediation strategies for Morsa (the Vansjø-Hobøl watercourse) – the final report). NIVA-Report 4377. (In Norwegian).

Søvik, A. K., & Syversen, N. (2008). Videreutvikling av vegetasjonssoner som rensefilter for overflateavrenning – Effekt av ulik vegetasjon og variasjon i renseeffekt over tid. Bioforsk rapport, Vol 3, Nr 2, 2008.

Vagstad, N., Stålnacke, P., Andersen, H.-E., Deelstra, J., Jansons, V., Kyllmar, K., Loigu, E., Rekolainen, S., & Tumas, P. (2004). Regional variations in diffuse nitrogen losses from agriculture in the Nordic and Baltic regions. *Hydrology and Earth System Sciences, 8*(4), 651–662.

Vannforskriften. (2006). FOR 2006-12-15 nr 1446: Forskrift om rammer for vannforvaltningen (in Norwegian). http://www.lovdata.no/cgi-wift/ldles?doc=/sf/sf/sf-20061215-1446.html. Accessed 16 Oct 2013.

Von der Ohe, P. C., Dulio, V., Slobodnik, J., De Deckere, E., Kuhne, R., Ebert, R.-U., Ginebreda, A., De Cooman, W., Schuurmann, G., & Brack, W. (2011). A new risk assessment approach for the prioritization of 500 classical and emerging organic microcontaminants as potential river basin specific pollutants under the European Water Framework Directive. *Science of the Total Environment, 409*, 2064–2077.

Vymazal, J., Balcarova, J., & Dousova, H. (2001). Bacterial dynamics in the sub-surface constructed wetland. *Water Science and Technology, 44*(11–12), 207–209.

WHO. (2006). *Guidelines for the safe use of wastewater, excreta and greywater. Volume 3 wastewater and excreta use in aquaculture*. Geneva: World Health Organization.

Ye, Z. H., Whiting, S. N., Lin, Z.-Q., Lytle, C. M., Qian, J. H., & Terry, N. (2001). Removal and distribution of Fe, Mn, Co, and Ni within a constructed wetland treating coal combustion by-product leachate. *Journal of Environmental Quality, 30*, 1464–1473.

# Chapter 11
# Subsurface Flow Constructed Wetland Models: Review and Prospects

Roger Samsó, Daniel Meyer, and Joan García

**Abstract** Numerical models are recognized nowadays as a powerful tool to increase the understanding of the internals of constructed wetlands and to help improve their design. Over the last decade many models have been developed, and many simulation studies have been published. Despite diversity is generally a positive thing, having so many different models can be confusing for potential users and may also hinder further development of the existing ones. The aim of this paper is to summarize the state of the art of this discipline, focussing the attention on the most feature-rich process-based models for constructed wetlands for urban wastewater treatment. Their description is combined with a feature comparison in a tabular format to facilitate the selection of one or another based on the specific needs of the potential user. Moreover, a discussion is made regarding the advantages of each reviewed model regarding features, licencing and expected evolution of each of them. Later in the document, we describe the essential phenomena, parameters and processes that we believe that future generation of constructed wetlands models should incorporate, to guide further research on this discipline. Although this paper is focused on models used in academic circles, a model developed to optimize the design of combined sewer overflow wetlands is presented as an example of the potential of design-focused wetlands. At the end of the paper we provide an overview of the past, present and future of constructed wetlands models and analyse were we stand and which is the way to go and the main goals in the near future.

**Keywords** Simulation • Mathematical model • Mechanistic • Biokinetik • Wastewater treatment

R. Samsó • J. García (✉)
GEMMA – Group of Environmental Engineering and Microbiology, Department of Hydraulic, Maritime and Environmental Engineering, Universitat Politècnica de Catalunya-BarcelonaTech, c/Jordi Girona, 1-3, Building, D1 E-08034 Barcelona, Spain
e-mail: joan.garcia@upc.edu

D. Meyer
IRSTEA, Freshwater Systems, Ecology and Pollution Research Unit, 5 rue de la Doua, CS70077, Villeurbanne Cedex 69626, France

## 11.1 Introduction

For decades constructed wetland (CW) models have been considered a promising tool to increase understanding of the simultaneous physico-chemical and biological processes involved in the treatment of wastewater with this technology. This belief has translated into an increase of the number of publications on the development and utilization of these models over time. Moreover, the great variety of wetland typologies and applications and the infinity of processes taking place within them have created a comparably wide diversity of models in terms of the number and the type of processes they intend to describe.

At least six review papers have been published in recent times to summarize the state of the art of CWs models (Rousseau et al. 2004; Langergraber 2008; Langergraber et al. 2009b; Langergraber 2010; García et al. 2010; Kumar and Zhao 2011). However, based on available reviews, it is still difficult to select the best suited model for every specific requirement. That results from the fact that these reviews mostly consist of descriptions of the models features and no critical in-depth comparison is made between them. Moreover, over the last three years, there have been several relevant publications on CWs models which were not included in the previous reviews.

In this context, the first objective of the current document is to update the state of the art of subsurface flow (SSF) CWs models and to make an in-depth review of six of these models, previously selected among the rest using predefined criteria. Furthermore, a tabular feature comparison of the selected models and an extensive discussion of the main advantages and disadvantages of all of them are presented.

Secondly, we describe the essential phenomena, parameters and processes that we believe future generations of CWs models should incorporate in order to provide a fairly comprehensive description of wetlands functioning.

Finally, and based on that layout/theoretical background, we give an overview of the past, present and future of CWs models, analyse where we stand and determine which way to go and what our main goals should be in the near future. Moreover, we provide an example of the current development of a "simplified" design support model dedicated to CWs for combined sewer overflow (CSO) treatment.

We expect that the present document will serve the purpose of facilitating the choice of the most suitable model for every specific need and will also help to identify areas in which further research is required and where the science of constructed wetland models is going.

## 11.2 Model Selection

There exist arguably as many models for CWs as there are types of wetlands, water pollutants and processes that take place within these systems. Indeed, a general search for the words "constructed wetland model" on common databases of scientific papers brings a limitless number of publications on this topic. Among them, a very general distinction can be made: those focusing on simulation of the hydraulics (Dittmer et al. 2005; Fan et al. 2008; Galvão et al. 2010; Korkusuz et al. 2007; Kotti et al. 2013), the hydrodynamics and clogging (or any of them individually) (Brovelli et al. 2009b; Giraldi et al. 2009, 2010; Hua et al. 2013; Knowles and Davies 2011; Suliman et al. 2006) and those focusing on the removal of a specific pollutant or a set of pollutants (which generally also include hydraulic and hydrodynamic models of diverse complexity). Among the latter, the most commonly targeted pollutants are organic compounds (Akratos et al. 2008; Henrichs et al. 2007; Liolios et al. 2012; Toscano et al. 2009), nitrogen (Akratos et al. 2009; Henrichs et al. 2009; Mayo and Bigambo 2005; McBride and Tanner 2000; Meyer et al. 2006, 2011; Morvannou et al. 2013; Moutsopoulos et al. 2011; Toscano et al. 2009), sulphur (Lloyd et al. 2004), phosphorous (Hafner and Jewell 2006), heavy metals and mine drainage (Goulet 2001; Lee and Scholz 2006; Mitsch and Wise 1998), arsenic (Llorens et al. 2013), pesticides (Krone-Davis et al. 2013) and emerging pollutants (Hijosa-Valsero et al. 2011).

For the sake of brevity and clarity, in this chapter, only the most recent mechanistic models, applied to simulate the treatment of urban wastewater and those able to provide new insight into the functioning of CWs, will be reviewed. In addition, the RSF_Sim model is described as an example of future development of engineering tools to support the design and operation of CWs.

According to these criteria, six models are selected for review: PHWAT (Brovelli et al. 2007, 2009a, b, c), FITOVERT (Giraldi et al. 2009, 2010), HYDRUS-2D-CW2D (Langergraber and Šimůnek 2005), HYDRUS-2D-CWM1 (Langergraber and Šimůnek 2012), CWM1-RETRASO (Llorens et al. 2011a, b), BIO_PORE (Samsó and García 2013a, b) and AQUASIM-CWM1 (Mburu et al. 2012).

## 11.3 Model Description

From the six selected models, the only one that does not use either the biokinetics models CW2D or CWM1 is the FITOVERT model. Provided the importance of the biokinetic models within CWs models, the description of the selected models will be preceded by a brief description of CW2D and CWM1. Tables 11.1, 11.2, 11.3 and 11.4 present a feature comparison of the selected models.

### 11.3.1 CW2D

Constructed wetlands 2D (CW2D) (Langergraber 2001) is a biokinetic model, based on a mathematical formulation of the ASM models (Henze et al. 2000). This model was specifically conceived to simulate the most common biokinetic

**Table 11.1** General model description

| | FITOVERT | PHWAT | CWM1-RETRASO | AQUASIM | HYDRUS-2D-CW2D | HYDRUS-2D-CWM1 | BIO_PORE | RSF_Sim |
|---|---|---|---|---|---|---|---|---|
| Main publications | Giraldi et al. (2009, 2010) | Brovelli et al. (2007, 2009a, b, c) | Llorens et al. (2011a, b, 2013), Mburu et al. (2013) | Mburu et al. (2012) | Korkusuz et al. (2007), Langergraber and Šimůnek (2005, 2012), Langergraber (2007), Toscano et al. (2009) | Langergraber (Langergraber and Šimůnek 2012), Pálfy and Langergraber (2013) | Samsó and García (2013a, b) | Meyer (2011), Meyer et al. (in press) |
| Simulation Platform | MATLAB® | PHWAT | RetrasoCodeBright (RCB) | AQUASIM | HYDRUS-2D | HYDRUS-2D | COMSOL Multiphysics™ | Open (MS Excel®) |
| Biokinetic model | Their own | CW2D | CWM1 | CWM1 | CW2D | CWM1 | CWM1 | Their own |
| Commercially available | No | No | No | No | Yes | Yes | No | In preparation |
| Free of charge | No | For noncommercial uses | Yes | Yes | No | No | No | For noncommercial uses |
| Dedicated graphical user interface for CW modelling | Yes | No | No | No | Yes | Yes | No | No |
| Model Dimensions | 1D | 1D, 2D and 3D | 2D | 0D | 2D | 2D | 2D | 1D |
| Calibrated | Hydraulics and hydrodynamics | Hydrodynamics and clogging | Yes | Yes | Yes | No | Yes | Yes |

(–) Data not available/not specified by the authors/not applicable

# 11 Subsurface Flow Constructed Wetland Models: Review and Prospects

**Table 11.2** Hydraulic and hydrodynamic description

| | FITOVERT | PHWAT | CWM1-RETRASO | AQUASIM | HYDRUS-2D-CW2D | HYDRUS-2D-CWM1 | BIO_PORE | RSF_Sim |
|---|---|---|---|---|---|---|---|---|
| Hydraulics/hydrodynamics | Richards/transport eq. | Darcy/transport eq. | Darcy/transport eq. | CSTR | Richards + transport eq. | Richards + transport eq. | Darcy + Transport eq. | series of CSTRs |
| Saturation conditions | Variably saturated | Saturated | Saturated | Saturated | Variably saturated | Variably saturated | Saturated with variable water table | Variably saturated |
| VF/HF CWs | VF | HF/VF | HF | CSTR | VF | HF/VF | HF | VF |
| Feeding strategy | Feeding–emptying cycles | – | Continuous | Batch (20 days incubation) | Batch | Continuous | Continuous | Event driven (CSO) |
| Evapotranspiration | Evapotranspiration | – | No | Greenhouse conditions | Yes | Transpiration | No | In progress |
| Surface flow/ponding | Yes | No | No | No | Yes | No | No | Yes |
| Clogging | Yes | Yes | No | No | No | No | No | No |

(–) Data not available/not specified by the authors/not applicable

Table 11.3 Biokinetic model

| | FITOVERT | PHWAT | CWM1-RETRASO | AQUASIM | HYDRUS-2D-CW2D | HYDRUS-2D-CWM1 | BIO_PORE | RSF_Sim |
|---|---|---|---|---|---|---|---|---|
| C | Yes | Yes | Yes | Yes | Yes | Yes | Yes | Yes |
| N | Yes | Yes | Yes | Yes | Yes | Yes | Yes | Yes |
| P | No | Yes | No | Yes | Yes | No | No | In progress |
| O | Yes | Yes | Yes | Yes | Yes | Yes | Yes | No |
| S | – | No | Yes | Yes | No | Yes | Yes | No |
| Monod-type | Yes | Yes (CW2D) | Yes (CWM1) | Yes (CWM1) | Yes (CW2D) | Yes (CWM1) | Yes (CWM1) | – |
| Functional bacterial groups | – | 3 | 6 | 6 | 3 | 6 | 6 | – |
| Bacterial growth | Yes | Yes | No | Yes | Yes | Yes | Yes | – |
| Biomass description | Attached | Attached/suspended | Suspended | Suspended | Attached | Attached | Attached | Sediment accumulation |
| Growth limitations | Substrates | Temperature, substrates and logistic expression | – | Temperature and substrates | Temperature and substrates | Temperature and substrates | Substrates, temperature, logistic growth and accumulated solids | – |

(–) Data not available/not specified by the authors/not applicable

# 11 Subsurface Flow Constructed Wetland Models: Review and Prospects

Table 11.4 Physico-chemical processes

| | FITOVERT | PHWAT | CWM1-RETRASO | AQUASIM | HYDRUS-2D-CW2D | HYDRUS-2D-CWM1 | BIO_PORE | RSF_Sim |
|---|---|---|---|---|---|---|---|---|
| Atmospheric oxygen transfer | Yes | Yes | Yes | Yes | Yes | Yes | Yes | No |
| Gas transport | Yes | Yes | Yes | Yes | Yes | Yes | No | No |
| pH | No | Yes | No | No | No | No | No | No |
| Redox | No | Yes | No | No | No | No | No | No |
| Chemical equilibrium | No | Yes | Yes | No | No | No | No | Yes |
| Transport of particulate components | Yes | Yes | No | – | No | No | Yes | Yes |
| Filtration/sedimentation | Filtration | Attachment/detachment of biomass | No | – | No | No | Filtration | Yes |
| Sorption | No | – | No | Adsorption and desorption | Yes | Yes | No | Yes |

(–) Data not available/not specified by the authors/not applicable

processes taking place in vertical flow (VF) CWs. The biochemical components defined in CW2D include dissolved oxygen ($O_2$), three fractions of organic matter (CR, CS and CI), four nitrogen compounds ($NH_4$, $NO_2$, $NO_3$ and N2N), inorganic phosphorus (IP) and heterotrophic and autotrophic microorganisms. Organic nitrogen and organic phosphorous are modelled as part of the COD. Heterotrophic bacteria (XH) are assumed to be responsible for hydrolysis, mineralization of organic matter (aerobic growth) and denitrification (anoxic growth). On the other hand, autotrophic bacteria (XANs and XANb) are assumed to be responsible for nitrification, which is modelled as a two-step process. Microorganisms are assumed to be immobile. Lysis is considered to be the sum of all decay and loss processes. The temperature dependence of all process rates and diffusion coefficients is described using the Arrhenius equation.

### 11.3.2 CWM1

The biokinetic model Constructed Wetland Model Number 1 (CWM1) (Langergraber et al. 2009a) is another general conceptual model based on the ASM series for activated sludge (Henze et al. 2000) and the ADM for anaerobic processes (Batstone et al. 2002), to describe biochemical transformation and degradation processes for organic matter, nitrogen and sulphur in subsurface flow constructed wetlands. The main objective of CWM1 is to predict effluent concentrations from either VF or HF CWs without predicting gaseous emissions (Langergraber et al. 2009a).

This model considers 17 processes and 16 components (eight soluble and eight particulate). In terms of notation and structure, CWM1 is described in a way similar to the presentation of the ASMs. As in the ASMs, concentrations of dissolved components are referred to as *Si* and particulate components as *Xi*. Among the dissolved components there are dissolved oxygen (SO), ammonia and nitrate nitrogen (SNH and SNO), sulphate and dihydrogensulphide sulphur (SSO4 and SH2S), soluble fermentable COD (SF), fermentation products as acetate (SA) and soluble inert COD (SI). Organic nitrogen is considered as a fraction of organic matter (COD). Among the particulate components, there are six functional bacteria groups, including heterotrophic, nitrifying, fermenting, methanogenic, sulphate reducing and sulphide-oxidising bacteria (XH, XA, XFB, XAMB, XASRB and XSOB, respectively) and two particulate fractions of COD (XI and XS). Such as in the IWA ASMs, the kinetic expressions of CWM1 are based on switching functions (hyperbolic of saturation terms and Monod equations).

## 11.3.3 FITOVERT

FITOVERT (Giraldi et al. 2009, 2010) is a 1D code developed in MATLAB® and expressly designed to simulate subsurface VF CWs. This software is able to describe the water flow through unsaturated porous media as well as evapotranspiration and surface ponding. It is also able to simulate transport of dissolved and particulate components and clogging produced by bacterial growth and solids filtration.

The vertical water flow through porous media in unsaturated conditions is described using the volumetric water content form of the Richards equation. The constitutive relationships between pressure head, hydraulic conductivity and water content are handled using van Genuchten–Mualem functions (van Genuchten 1980), the parameters of which were obtained from a previous experimental study (Giraldi et al. 2009).

To describe the root water uptake, a sink term was added to the Richards equation. The model is also able to automatically handle the ponding on the surface of the vertical bed by changing the hydraulic boundary conditions.

The biochemical module, based on the ASM1 (Henze et al. 2000), describes the degradation of both organic matter and transformation of nitrogen. Thirteen components are taken into account, seven of which are dissolved and six are particulate. Neither the features of the biochemical module nor the components and processes considered are described in the original paper.

The advection–dispersion transport in the liquid phase for dissolved components is described according to Bresler's equation (Bresler 1973). Neither the uptake of nutrients and metals by the plants nor adsorption is considered in the existing version of FITOVERT. On the other hand, the transport and filtration of particulate components is described with a scheme based on the work of Iwasaki (1937) for the numerical analysis of the sand filtration process in saturated conditions.

FITOVERT is also able to handle the porosity reduction due to bacteria growth and filtration of particulate components. The effect of pore size reduction on the hydraulic conductivity is considered using a modified version of the Carman–Kozeny's equation (Boller and Kavanaugh 1995).

The oxygen transport is modelled using the same equations as for the rest of the dissolved components, and the diffusive exchange of oxygen with the gas phase is included in the reaction term and described using Fick's law. Oxygen transfer by plants from the atmosphere to their roots is not implemented.

The hydraulic model was calibrated by Giraldi et al. (2009), by comparing model outputs for inert components with experimental breakthrough curves of a pilot plant. Unfortunately, the biochemical and transport modules were not calibrated.

## 11.3.4 HYDRUS-2D-CW2D

The multicomponent reactive transport module CW2D (constructed wetlands 2D), (Langergraber 2001; Langergraber and Šimůnek 2005; Langergraber 2008; Langergraber and Šimůnek 2012), was developed as an extension of the HYDRUS-2D variably saturated flow and solute transport programme (Šimůnek et al. 1999). CW2D was developed to model the biochemical degradation and transformation processes for organic matter, nitrogen and phosphorus in vertical subsurface flow constructed wetlands.

The variable saturated flow is described using the Richards equation. The constitutive relationships between pressure head, hydraulic conductivity and water content are handled using van Genuchten–Mualem functions. Two additional boundary conditions are implemented to represent surface ponding in the vertical bed during wastewater loadings that exceed the infiltration capacity.

The transport of solutes is described using the advection–dispersion–diffusion equation which includes several sources and sinks to simulate adsorption/desorption and nutrients uptake by plant roots. The exchange of $O_2$ from the gas phase into the aqueous phase is described using the equation of Gujer and Boller (1990).

The effect of plants uptake on the removal of organic matter and nutrients in subsurface flow constructed wetlands was tested in Langergraber (2005). The model for plant uptake implemented describes nutrient uptake coupled to water uptake, which is an intrinsic capability of HYDRUS-2D. Literature values were used to calculate potential water and nutrient uptake rates.

Nowadays HYDRUS-2D-CW2D only considers dissolved wastewater compounds and therefore is currently unsuitable for investigating clogging phenomena (García et al. 2010).

## 11.3.5 PHWAT

PHWAT (Brovelli et al. 2007, 2009a, b, c; Mao et al. 2006) is a 3D macroscale code that uses MODFLOW (McDonald and Harbaugh 1988) to solve saturated/variably saturated water flow, MT3DMS (Zheng and Wang 1999) to simulate transport processes and PHREEQC-2 (Parkhurst and Appelo 1999) to describe biochemical reactions. Aerobic processes are based on CW2D, while anaerobic processes are based on the model formulation by Maurer and Rittmann (2004). Aside from the biokinetic reactions, full water chemistry and sediment–water interactions can be modelled using PHREEQC. PHWAT is also able to simulate bioclogging, bacteria attachment and flow-induced biofilm detachment. This model includes a growth-limiting expression to account for the reduction of porosity caused by bacterial growth.

## 11.3.6 HYDRUS-2D-CWM1

In Langergraber and Šimůnek (2012), a new version of the wetland module for HYDRUS-2D was presented. This new version adds the possibility to choose between the already implemented CW2D and the newly implemented CWM1 biokinetic models. The only change of HYDRUS-2D-CWM1 with respect to HYDRUS-2D-CW2D is thus the biokinetic model.

In Langergraber and Šimůnek (2012), simulations were run to make a numerical verification of the implementation of the two biokinetic models in HYDRUS. The authors compared results with simplified versions of the two biokinetic models in a 20 by 20 cm vertical domain. The results demonstrated that the two biokinetic models were implemented correctly in HYDRUS-2D. HYDRUS-2D-CWM1 was also used to recreate the simulations performed by Llorens et al. (2011a, b), obtaining different results.

## 11.3.7 AQUASIM-CWM1

Mburu et al. (2012) implemented the biokinetic model CWM1 into AQUASIM software (Reichert 1998) to simulate the fate of organic matter, nitrogen and sulphur within 16 batch-operated subsurface flow wetland mesocosms planted with three different plant species (for a detailed description of the mesocosms, the reader is referred to Allen et al. (2002) and Stein et al. (2006)). The mixed reactor compartment configuration in AQUASIM was used and the mesocosms were described as constantly stirred tank reactors (CSTR). In addition to the biokinetic reactions of CWM1, this model considers physical re-aeration, adsorption and desorption of COD and ammonium as well as a complex description of plant-related processes. Indeed, the plant model includes five processes, namely, growth, decay/senescence, physical degradation, oxygen leaching and nutrients uptake. The growth rate of microorganisms and plants is made temperature dependent by means of Arrhenius relationships. In this model, the growth of bacterial communities is only limited by substrates.

The model was calibrated and validated with different sets of experimental data from the mesocosms and sensitivity analysis, parameter estimation and uncertainty analysis were carried out.

## 11.3.8 CWM1-RETRASO

The CWM1-RETRASO model (Llorens et al. 2011a, b) is a 2D simulation model obtained from the implementation of CWM1 (Langergraber et al. 2009a) in RetrasoCodeBright (RCB) code (Saaltink et al. 2004) to simulate the hydraulics and hydrodynamics as well as the main biodegradation and transformation processes in horizontal SSF CWs.

RCB enables the simulation of the reactive transport of inorganic dissolved and gaseous species in non-isothermal saturated and unsaturated problems by finite elements. The transport of solutes in water is modelled by means of advection, dispersion and diffusion, together with chemical reactions. This model considers the wetland as a saturated porous media, and thus, the advective flux and dispersive and diffusive fluxes are computed by means of Darcy's and Fick's laws, respectively.

The implementation of the biochemical processes within RCB code consisted on adding the rates relevant to CWM1 to the reaction term of the RCB transport equations. The reactive transport model of the present study basically consists of 19 reactions or processes instead of the 17 described by CWM1.

Physical oxygen transfer from the atmosphere to the water was included in the model. Oxygen leaking from macrophytes, plant uptake, biofilm development and processes linked to clogging were not considered either. A multiplicative exponential function similar to that used by Ojeda et al. (2008) for the hydrolysis process was added to avoid total COD overestimations. In the paper by Llorens et al. (2011b), some changes to CWM1 formulation and to its parameters were proposed, and both the hydraulic and biochemical models were calibrated and validated, comparing it experimental data.

In CWM1-RETRASO bacterial concentrations are defined as inflow concentrations and travel through the wetland just as dissolved components do. Therefore, with this model, only stationary simulations can be performed, as the growth of bacterial populations cannot be simulated.

### 11.3.9 BIO_PORE

BIO_PORE (Samsó and García 2013a, b) is a 2D model built using COMSOL Multiphysics$^{TM}$ which implements fluid flow and transport equations coupled with the biokinetic expressions of CWM1. As a distinctive feature from the original CWM1's formulation, in this model slowly biodegradable and inert particulate COD (XS and XI, respectively) are divided into aqueous ($XS_m$ and $XI_m$) and solid ($XS_f$ and $XI_f$) phases (Samsó and García 2013a). Thus, it considers 18, instead of the 16 components described in the original version of CWM1.

The BIO_PORE model includes attachment and detachment processes for influent particulate components, which permits us to simulate solids accumulation in the gravel media. It also uses a complex hydraulic sub-model which calculates the exact location of the water table level at each time, preventing the growth of bacteria in the dry areas of the bed (above the water level). The unrealistic growth of biomass in areas with high substrate concentrations is avoided using a macroscopic biofilm sub-model, which also takes into account the effects of solids accumulation on bacterial growth. This sub-model includes two empirical parameters $X_{bio\_max}$ and $X_{cap}$ which correspond to the maximum concentration of bacteria that can be sustained within the porosity due to substrate availability (carrying capacity) and the maximum amount of inert solids that can fit a cubic metre of granular media, respectively. These latter additions also facilitate the convergence of numerical solutions.

The calibration of the model was presented in Samsó and García (2013a) using datasets of measured influent and effluent pollutant concentrations, flow rates and water temperatures from the first year of operation of a pilot wetland named C2 in García et al. (2004a, b). During calibration, some biokinetic parameters were modified from those given in CWM1, and others that were missing in the original source were included.

The BIO_PORE model was also used to determine the time taken for bacterial communities to stabilize as well as their distribution and their total biomass Samsó and García (2013a, b).

## 11.4 Discussion

### 11.4.1 Advantages and Disadvantages of the Reviewed Models

#### 11.4.1.1 Feature Comparison

The most relevant aspect of FITOVERT is that it is able to simulate the effects of bacterial growth and solids accumulation on the hydraulic properties of the granular media. It is also able to handle bed surface ponding by adapting the hydraulic boundary conditions. This model was specifically designed to simulate VF CWs only, and it includes a dedicated graphical user interface (GUI). It is the only of the reviewed models using neither CW2D nor CWM1, although its biokinetic model is also based on the ASM formulation. This model was only calibrated for the hydraulics and hydrodynamics, while the biokinetic model lacks calibration. To our knowledge, only two publications (Giraldi et al. 2009, 2010) are available on the application of FITOVERT.

As for FITOVERT, the main advantage of HYDRUS-2D-CW2D/CWM1 models is that they include a dedicated GUI to simulate CWs. This fact has ended in a good adoption of the models among the scientific community which has given place to several scientific publications (Table 11.1). Another advantage is that HYDRUS-2D can use different biokinetic models depending on whether a VF or a HF CW is to be simulated. However, the model has only been applied and calibrated for short simulation time frames, since bacterial growth is not limited and hence the high bacterial concentrations reached after a short simulation time prevents model convergence (Samsó and García 2013a). Another unresolved issue of this model is its inability to simulate the transport and retention of particulate components and thus clogging (Langergraber and Šimůnek 2012).

On the other hand, the main advantages of PHWAT are that it includes a large number of physico-chemical and biological processes. Its modularity is also a positive point, since it facilitates the task of adding or removing features. It is also one of the few available models for CWs able to simulate bioclogging together with biomass attachment–detachment processes and also considers growth limitations for biomass. Although bacteria distribution obtained with PHWAT was qualitatively compared with results obtained with other models for the same experimental setup, this model could only match measured permeability decrease for the initial 28 days out of a 283-day long experiment (Brovelli et al. 2009a). Another weak point of PHWAT for CWs is that little information about model equations is given in the few publications available and so all work is difficult to reproduce.

The most relevant feature of AQUASIM-CWM1 is the inclusion of a sophisticated model for plant-related processes. On the other hand, the main drawback of this model resides in the fact that it uses the mixed reactor compartment hydraulics description, and thus the wetlands are described as CSTR and this model cannot be applied to simulate full-scale systems. Also, this model does not consider clogging nor bacterial growth limitations and thus it cannot be used to simulate continuously fed CWs.

The most relevant features of the BIO_PORE are that it considers growth limitations for bacterial communities and a simple filtration model for particulate solids, which allows it to perform long-term simulations of continuously fed CWs. Moreover, this model has been satisfactorily calibrated with experimentally measured effluent concentrations of COD and ammonium and ammonia nitrogen for a period of 1 year. It has also been used to study the evolution of bacterial communities over a 3-year period. Another advantage of this model is that it is in current development and the modelling platform is very versatile and does not limit progress. However, so far it uses a simplified flow description (Darcy equation with variable water table) and it does not consider the changes in the hydrodynamic properties of the granular media as a result of clogging.

Regarding the model by Llorens et al. (2011a, b), it is a powerful model in terms of geochemical reactions, but its main disadvantage resides in the fact that bacterial communities are not static, but travel with the flow.

### 11.4.1.2 Licensing

Among the platforms used to build the reviewed models, HYDRUS is the only one offering a compiled package/module specific for CWs simulations which can be acquired at a price. The rest of the models reviewed are implementations of different sets of mathematical expressions describing biokinetic reactions and other physical–chemical processes into multipurpose simulation platforms. However, HYDRUS-CW2D/CWM1 is a closed-source piece of software, and thus any modifications to the code can only be applied by their owners.

One of the advantages of the implementation of AQUASIM-CWM1 is that the AQUASIM platform itself can be downloaded free of charge (although it is not open source), and so the distribution of the implementation of the CWM1 made by Mburu et al. (2012) only depends on the author's willingness to share it. Likewise, the equations describing the different physico-chemical and biological processes are easily customizable.

On the other hand, BIO_PORE model is built in COMSOL Multiphysics™ (CM) which is a very versatile piece of mathematical software which is widely used around the world and in many disciplines, but it is expensive and closed source. However, the implementation of CWM1 on this platform can be shared and modified in a way similar as for AQUASIM-CWM1, as long as a CM licence is available. The same applies for FITOVERT, but in this case, the user requires a licence of MATLAB® software.

RetrasoCodebright (RCB) (Saaltink et al. 2003) can be downloaded free of charge, although its source code is not available. So the model of Llorens et al. (2011a, b) was built on top using the available functionalities of the software, much in the way of BIO_PORE, AQUASIM-CWM1 and FITOVERT. However, RCB lacks a graphical user interface, which makes it significantly less versatile and intuitive to the user.

PHWAT is a modular piece of software developed in FORTRAN 90 and C programming languages, and the author makes it available for free for noncommercial use. Brovelli et al. (2007, 2009a, b, c) developed specific modules to simulate CWs, although no reference is made about their availability.

### 11.4.1.3 Code Evolution

HYDRUS-CWM1/CW2D is by far the platform with the most adopters, and the one that has produced more publications (Table 11.1), and thus it should be the one with the most chances of evolving the fastest. However, the fact that the biokinetic models are packaged as closed-source software and that only the authors can modify the code could delay its evolution.

To our knowledge, the last publication using FITOVERT is that of Giraldi et al. (2010), which seems to indicate that its development has been discontinued. The same happens for publications on CWs modelling using PHWAT platform.

Provided AQUASIM can be acquired free of charge, everything indicates that AQUASIM-CWM1 could increase adoption if a more sophisticated description of wetland hydraulics and hydrodynamics was used.

A recent paper by Mburu et al. (2013) has continued the use of CWM1-RETRASO, although the code has not evolved since Llorens et al. (2011a, b) and so fixed biomass cannot yet be simulated. However, Llorens et al. (2013) introduced new reactions related to Arsenic retention in wetlands (precipitation, adsorption, uptake and accumulation in plants) in RCB's code.

On the other hand, BIO_PORE model is receiving increasing attention, since it is the only one that can simulate long-term wetland functioning and the only one that has been used to study bacterial dynamics in CWs. Also, the ease of use and potential of COMSOL Multiphysics$^{TM}$ will certainly result in a fast evolution of the code.

### 11.4.2 What Should CWs Models for Academic Use Aim At?

In our opinion, for a complete and sound description of CWs functioning, at least the following phenomena and processes need to be included in the future generations of the scientific mechanistic models:

- A complete hydraulic description (variably saturated subsurface flow and surface flow and ponding). Transport of dissolved, particulate and gaseous components
- Filtration, sedimentation, precipitation, volatilization, dissolution, adsorption and desorption, hydrolysis of both slowly biodegradable and inert organic fractions and physical re-aeration
- Bacterial processes:
  - Growth and decay (suspended and in the form of biofilms).
  - Attachment and detachment
  - Influence of substrates concentration, available space, diffusion limitations, temperature, pH and redox potential on bacterial growth
- Plant processes (above and belowground growth, decay/senescence, nutrients uptake, evapotranspiration, oxygen release through the rhizosphere and carbon exudates)
- Clogging by accumulate solids, bacterial growth, precipitates and plant roots and its effects on the hydraulics and hydrodynamics of the granular media

The ability to perform long-term simulations, though neither being a process nor a phenomenon, is already reckoned as an essential feature of the next generations of CWs models.

### 11.4.3 Evolution of Models for Constructed Wetlands

At first (in the 1990s) the idea was to develop simple models that could be used to improve design, based on first-order decay expressions and neglecting the real processes leading to pollutants removal. However, wetlands are very complex systems that interact with the environment, and the experience using these models demonstrated that they were not sufficient to reproduce the enormous variability observed in their performance in different environments and for different wastewater compositions. Therefore, these initial ideas and models were later substituted with the use of more sophisticated models, the so-called scientific models (SMs), which intend to describe the widest possible range of the internal processes taking place within wetlands with deterministic mathematical expressions. However, the expectations around these models progressively decreased due to their complexity and their inability to generate consistent results for long-term scenarios, which hindered their use by designers, operators and the scientific community. In fact, so far these models have only been used to match effluent pollutant concentrations for short periods of time, while paying little or no attention to the internal processes. However, a new tendency has been growing in recent times, which accepts the embrionary stage of the SMs to be used for design, and rather it recognizes their immense potential in terms of improving the understanding of wetlands functioning. The current trend is thus to increase even more the complexity and the number of processes considered by the SMs and to try to produce long-term simulations and to focus not only on matching effluent pollutant concentrations but also on helping to understand the internal processes (scientific purposes).

In parallel to the development of SMs, a separate branch has started developing less complex models, the so-called design support models (DSMs), which have a low complexity and can be used by designers and operators. In this way the motivation of modelling comes back to the roots, but now with a massive background of SMs application and corresponding fundamental research. Figure 11.1 shows the evolution of the complexity of numerical models for CWs since the early

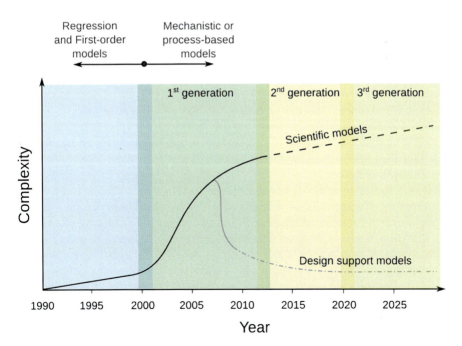

**Fig. 11.1** Evolution of the complexity of scientific (SMs) and design support (DSMs) CWs models over the years. A possible prediction is described with dotted lines

nineties until today. A prediction of the future of CWs models is also presented. In the following lines, a brief description of the different stages shown in Fig. 11.1 is made.

### 11.4.3.1 First-Generation Models

After around 20 years of CWs model development, the current (first-generation) SMs are able to confidently simulate saturated or variably saturated subsurface flow (horizontal or vertical), transport of dissolved pollutants and the temperature and substrates-dependent growth and decay of different bacterial groups at the macroscale. Most models are also able to describe other relevant processes such as sorption, evapotranspiration and plant nutrients uptake and oxygen release. At the current stage, these models can be used to direct research, to explain experimental results and to increase the understanding of their internal functioning.

On the other hand, the first generation of DSMs for CWs are still under heavy development. The first stage was to drastically reduce the number of processes so as to only consider those with the biggest impact on the functioning and application of CWs. A good example of that is the model RSF_Sim which was developed for combined sewer overflows (Meyer 2011). The development of RSF_Sim is based on a fundamental research study (Dittmer and Schmitt 2011) but also on experiences in the application of HYDRUS-2D/CW2D. Since 2002, the relatively new software was found to be generally applicable to simulate laboratory retention soil filters (RSF) column operation. Single feeding events as well as short event series could be simulated with good fittings of measured and simulated pollution outflow curves, but extended dry periods exceeded the ability of reasonable nutrient/biomass balances due to the missing description of particulate organic matter degradation (Dittmer et al. 2005; Henrichs et al. 2007, 2009).

In 2007 the focus was moved to a simplified description in order to support design and dimensioning. The main aim of the engineering tool RSF_Sim can be found in long-term estimations of RSF performances. This seems to be useful due to the fact that RSF design is always requiring sewer system simulation (Meyer et al. 2013). Whereas nowadays hydraulic loads are considered as main design criteria, future approaches should take the pollutant loads and treatment capabilities into account. The according new engineering tool required simple usage, even with a sophisticated background originated from the application of mechanistic models.

RSF_Sim works with three complete stirred tanks in vertical series: [1] the retention layer provides the water storage on top of the process layer (ponding), [2] the process layer describes the sand/gravel layer (saturated during feeding, drained afterwards) in which treatment occurs, and [3] the drainage layer improves the volume balances, but can also be used as a water reservoir under permanent saturation (to avoid water stress, Meyer et al. 2013). Up to date, descriptions of treatment performances for COD and $NH_4$-N are kept basically simple: the total COD needs to be separated into two fractions: particulate COD becomes reduced by filtration; dissolved COD is reduced to be a treatment efficiency factor, which can vary due to influencing factors like temperature, inflow concentrations or the duration of previous dry periods (most important). The retention of $NH_4$-N during the filter passage is calculated with a steady-state two-stage linear isotherm; nitrification is considered only after re-aeration by first-order kinetics.

Successful applications of RSF_Sim are based on the existing high number of calibrations and validations. The number of required input parameters was kept very small, and initial values can be found easily in literature sources. The model code consists out of simple mass and volume balances with "if", "and" and "or" functions and appears thereby to be transferable in any computational language. RSF_Sim can be used with measured data as well as with input originated from pollution load sewer system simulations. In cooperation with the developers, it is available for free for scientific purposes and also generally applicable for commercial needs.

The feedback of the presentation of RSF_Sim is generally positive. Researches in applied sciences, as well as engineers dealing with constructed wetland design, appreciate the trend of practically accessible simulation tools to support their daily work. Due to this responds, ideas of "simplified" design tools for other specific wetland types are growing.

### 11.4.3.2 Second-Generation Models

Possibly the most easily achievable and most significant evolution of the current CWs SMs will come from the development of sophisticated clogging models, which will include the pore space and subsequent hydraulic conductivity reduction caused by the accumulation of solids (organic and inorganic) and by bacterial growth. To that end, the first step will be to improve the description of the transport of particulate components (attachment and detachment). Furthermore, to make a comprehensive description of the whole clogging processes, the subsurface flow equations (Darcy/Richards equations) will need to be coupled to surface-flow equations (e.g., shallow water equations or Saint-Venant equations) to represent surface-flow and ponding. These clogging models will need to be calibrated and validated first with experimental data coming from lab-scale experiments (due to the lack of experimental data in real cases) and later for full-scale systems.

It is already well accepted that bacterial communities play a major role in the functioning of CWs (Samsó and García 2013a, b), and thus better SMs for CWs will necessarily come from more advanced biokinetic models. CWM1 and CW2D are both based on the initial versions of ASM models, which have received continuous upgrades over the years. On the contrary, neither CWM1 nor CW2D has received any modifications since their inception. Moreover, having two different biokinetic models for CWs only makes sense in terms of saving computational power (fewer dependent variables), since the biokinetic processes are mostly the same (in different proportions) in all subsurface flow CWs. However, with the ever-increasing computational power, having fewer dependent variables is ever less of an advantage. In fact, a unified biokinetic model would reduce model divergence, and by joining all efforts on a unique and common platform, faster development could be achieved.

We also envisage that future models will incorporate chemical equilibrium equations for several key components as well as the inclusion of pH and redox potential (Rolle et al. 2008) as controlling factors of the growth of the different functional bacterial groups.

Regarding the second generation of DSMs, it will result from a progressive reduction of the number of considered processes and from extensive validation with experimental data. The main idea can be found in a shift away from a process description towards a treatment effect description. The biggest challenge will be the definition of limits for the extrapolation of performance assumptions. This will only be achieved by creating a freely available database of experimental data from a wide range of systems, especially from those showing malfunctions. In this way, a strong link between researchers and industrial partners is essential, because the stakeholders of interest in design support models are the stakeholders of the required database in the same time.

#### 11.4.3.3 Third-Generation Models

The third generation of mechanistic models will come from a better description of biofilm structure and processes through microscale models and possibly with the inclusion of more sophisticated above and belowground plant models. However, the stochastic nature of plant roots distribution will delay the inclusion of this phenomenon on the clogging description.

The evolution of mathematical models has to go side by side with progresses through experimental studies and fundamental research (empirical evidence). The scientific community can keep increasing the complexity of the models in a parallel to the increased knowledge. There might be a limit of complexity that could be reached, which resides in our ability to understand the model output.

On the other hand, a third generation of design support models will have already been validated in many different scenarios. These models could incorporate optimization modules that should give the optimum wetland configuration and dimensioning for specific cases based only on small sets of monitoring data from the corresponding locations. An easy-to-use GUI should be available for engineers in daily practice. Risks can be seen in inadequate applications, since a model can only support but not replace engineering.

## 11.5 Conclusions

In this chapter, we have made an in-depth review of six of the most feature-rich models for subsurface flow CWs. A feature comparison of the selected models has also been presented in a tabular format, and the advantages and disadvantages of each of them have been discussed. Moreover, we have defined the essential processes and phenomena that future mechanistic models used for scientific purposes should aim at describing. A timeline of past, present and future goals and the evolution of the complexity of CWs models has also been presented. An explanation of the main stages of this timeline has also been made.

CWM1 is the most widely implemented biokinetic model (HYDRUS-2D, AQUASIM, RCB, COMSOL Multiphysics$^{TM}$), followed by CW2D, which has been implemented in HYDRUS-2D and in PHWAT. This is possible due to the fact that in contrast to CW2D, which can only be applied for VF CWs, CWM1 can be used to describe biokinetic processes of both VF and HF CWs. Nevertheless, the implementation of CW2D into HYDRUS-2D has been source of the largest number of publications, since this model has had good acceptance in academia circles based on its early inception and similarities to activated sludge simulations.

In terms of modelling platform, HYDRUS-2D is the most utilized CFD model in the field of CWs, although its inability to simulate particulate transport and limitations of bacterial growth hinders further advances. On the other hand, we have seen that the platforms better positioned to facilitate further CWs model development are AQUASIM and COMSOL Multiphysics$^{TM}$; the main advantage of AQUASIM being that it can be obtained free of charge and can therefore facilitate access to a wider developing community, while COMSOL Multiphysics$^{TM}$ is seen as the most powerful and versatile simulation platform, although it is not free of charge. The closed-source nature of the biokinetic models implemented in HYDRUS-2D hinders a fast development of the code.

Two different trends have evolved after the switch from the original first-order or regression models to the most recent mechanistic models: those designed for scientific purposes, which aim at better understanding the functioning of wetlands (scientific models), and those developed for engineering purposes (design support models), which aim at becoming supporting tools for the design and operation of CWs. An example of the second type of models is RSF_Sim. This model was developed as a consequence of the limitations of HYDRUS-2D-CW2D to produce long-term simulations and with the aim of producing an engineering tool to support the design of CWs for CSO treatment. This model includes only the essential processes that need to be considered for treatment performance estimations of CSO CWs in a long-term view.

We have also discussed that subsequent advances in the mechanistic models for scientific purposes will initially come from a better description of both clogging and chemical process and also from the development of a unified biokinetic model, which will help set common objectives and reduce model divergence and fasten progress. The third generation of mechanistic models will result from a better description of biofilm and plant processes.

Regarding the design support models, the future advances can be found in a wide and accessible application range. Whereas scientific tools with increasing complexity require increased user skills, the "simplified" approaches are addressed to support practical engineering. In this way, the background of scientific modelling gives benefits in design and operation of constructed wetlands via easily applicable tools.

**Acknowledgements** The authors are also grateful to the European Commission for the financial support of the SWINGS project (Grant Agreement No.: 308502).

# References

Akratos, C., Papaspyros, J., & Tsihrintzis, V. (2008). An artificial neural network model and design equations for BOD and COD removal prediction in horizontal subsurface flow constructed wetlands. *Chemical Engineering Journal, 143*(1–3), 96–110.

Akratos, C. S., Papaspyros, J. N. E., & Tsihrintzis, V. (2009). Total nitrogen and ammonia removal prediction in horizontal subsurface flow constructed wetlands: Use of artificial neural networks and development of a design equation. *Bioresource Technology, 100*(2), 586–596.

Allen, W. C., Hook, P. B., Biederman, J. A., & Stein, O. R. (2002). Temperature and wet- land plant species effects on wastewater treatment and root zone oxidation. *Journal of Environmental Quality, 31*(3), 1010–1016.

Batstone, D., Keller, J., Angelidaki, R. I., Kalyuzhnyi, S. V., Pavlostathis, S. G., Rozzi, A., Sanders, W. T. M., Siegrist, H., & Vavilin, V. A. (2002). *Anaerobic digestion model no. 1 (ADM1)*. London: IWA Publishing.

Boller, M. A., & Kavanaugh, M. C. (1995). Particle characteristics and headloss increase in granular media filtration. *Water Research, 29*(4), 1139–1149.

Bresler, E. (1973). Simultaneous transport of solutes and water under transient unsaturated flow conditions. *Water Resources Research, 9*(4), 975–986.

Brovelli, A., Baechler, S., Rossi, L., Langergraber, G., & Barry, D. A. (2007). Coupled flow and hydro-geochemical modelling for design and optimization of horizontal flow constructed wetlands. In Ü. Mander, M. Kóiv, C. Vohla (Eds.), *Proceedings international symposium on "Wetland pollutant dynamics and control WETPOL 2007"* (pp. 393–395). Tartu, Estonia: Tartu University.

Brovelli, A., Baechler, S., Rossi, L., & Barry, D. A. (2009a). Comprehensive process-based modelling of sand filters and subsurface flow constructed wetlands. In *Proceedings of the 3rd international symposium on wetland pollutant dynamics and control* (WETPOL 2009), Barcelona, Spain, 20–24 Sept 2009. Abstract n. P-018

Brovelli, A., Malaguerra, F., & Barry, D. A. (2009b). Bioclogging in porous media: Model development and sensitivity to initial conditions. *Environmental Modelling and Software, 24*(5), 611–626.

Brovelli, A., Rossi, L., & Barry, D. A. (2009c). Mechanistic understanding and prediction of bioclogging in sand filters and subsurface flow constructed wetlands. In *Proceedings of the 3rd international symposium on wetland pollutant dynamics and control* (WETPOL 2009), Barcelona, Spain, 20–24 Sept 2009. Abstract n. O-051.

Dittmer, U., & Schmitt, T. G. (2011). Purification Processes in Biofilter Systems for CSO Treatment. In: *Proceedings 12th international conference on urban drainage*, Porto Alegre, Brazil.

Dittmer, U., Meyer, D., & Langergraber, G. (2005). Simulation of a subsurface vertical flow constructed wetland for CSO treatment. *Water Science and Technology, 51*(9), 225–232.

Fan, L., Reti, H., Wang, W., Lu, Z., & Yang, Z. (2008). Application of computational fluid dynamic to model the hydraulic performance of subsurface flow wetlands. *Journal of Environmental Sciences, 20*(12), 1415–1422.

Galvão, A. F., Matos, J. S., Ferreira, F. S., & Correia, F. N. (2010). Simulating flows in horizontal subsurface flow constructed wetlands operating in Portugal. *Ecological Engineering, 36*(4), 596–600.

García, J., Chiva, J., Aguirre, P., Álvarez, E., Sierra, J., & Mujeriego, R. (2004a). Hydraulic behaviour of horizontal subsurface flow constructed wetlands with different aspect ratio and granular medium size. *Ecological Engineering, 23*(3), 177–187.

García, J., Aguirre, P., Mujeriego, R., Huang, Y., Ortiz, L., & Bayona, J. M. (2004b). Initial contaminant removal performance factors in horizontal flow reed beds used for treating urban wastewater. *Water Research, 38*, 1669–1678.

García, J., Rousseau, D. P. L., Morató, J., Lesage, E., Matamoros, V., & Bayona, J. M. (2010). Contaminant removal processes in subsurface-flow constructed wetlands: A review. *Critical Reviews in Environmental Science and Technology, 40*(7), 561–661.

Giraldi, D., de' Michieli Vitturi, M., Zaramella, M., Marion, A., & Iannelli, R. (2009). Hydrodynamics of vertical subsurface flow constructed wetlands: Tracer tests with rhodamine WT and numerical modelling. *Ecological Engineering, 35*(2), 265–273.

Giraldi, D., de Michieli Vitturi, M., & Iannelli, R. (2010). FITOVERT: A dynamic numerical model of subsurface vertical flow constructed wetlands. *Environmental Modelling and Software, 25*(5), 633–640.

Goulet, R. (2001). Test of the first-order removal model for metal retention in a young constructed wetland. *Ecological Engineering, 17*(4), 357–371.

Gujer, W., & Boller, M. (1990). A mathematical model for rotating biological contactors. *Water Science and Technology, 22*, 53–73.

Hafner, S., & Jewell, W. (2006). Predicting nitrogen and phosphorus removal in wetlands due to detritus accumulation: A simple mechanistic model. *Ecological Engineering, 27*(1), 13–21.

Henrichs, M., Langergraber, G., & Uhl, M. (2007). Modelling of organic matter degradation in constructed wetlands for treatment of combined sewer overflow. *Science of the Total Environment, 380*, 196–209.

Henrichs, M., Welker, A., & Uhl, M. (2009). Modelling of biofilters for ammonium reduction in combined sewer overflow. *Water Science and Technology, 60*(3), 825–831.

Henze, M., Gujer, W., Mino, T., & van Loosdrecht, M. C. M. (2000) *Activated sludge models ASM1, ASM2, ASM2D and ASM3*. IWA scientific and technical report 9. London: WA Publishing.

Hijosa-Valsero, M., Sidrach-Cardona, R., Martín-Villacorta, J., Cruz Valsero-Blanco, M., Bayona, J. M., & Bécares, E. (2011). Statistical modelling of organic matter and emerging pollutants removal in constructed wetlands. *Bioresource Technology, 102*(8), 4981–4988.

Hua, G. F., Li, L., Zhao, Y. Q., Zhu, W., & Shen, J. Q. (2013). An integrated model of substrate clogging in vertical flow constructed wetlands. *Journal of Environmental Management, 119*, 67–75.

Iwasaki, I. (1937). Some notes on sand filtration. *Journal of American Water Works Association, 29*, 1591–1602.

Knowles, P. R., & Davies, P. A. (2011). A finite element approach to modelling the hydrological regime in horizontal subsurface flow constructed wetlands for wastewater treatment. In J. Vymazal (Ed.), *Water and nutrient management in natural and constructed wetlands* (pp. 85–101). Dordrecht: Springer.

Korkusuz, E. A., Meyer, D., & Langergraber, G. (2007). CW2D simulation results of lab-scale vertical flowfilters filled with special media and loaded with municipal wastewater. In *Proceedings international symposium on "Wetland pollutant dynamics and control WETPOL 2007"* (pp. 448–450).Tartu, Estonia:Tartu University.

Kotti, I. P., Sylaios, G. K., & Tsihrintzis, V. A. (2013). Fuzzy logic models for BOD removal prediction in free-water surface constructed wetlands. *Ecological Engineering, 51*, 66–74.

Krone-Davis, P., Watson, F., Los Huertos, M., & Starner, K. (2013). Assessing pesticide reduction in constructed wetlands using a tanks-in-series model within a Bayesian framework. *Ecological Engineering, 57*, 342–352.

Kumar, J. L. G., & Zhao, Y. Q. (2011). A review on numerous modeling approaches for effective, economical and ecological treatment wetlands. *Journal of Environmental Management, 92*(3), 400–406.

Langergraber, G. (2001). *Development of a simulation tool for subsurface flow constructed wetlands* (Wiener Mitteilungen 169, 207p.). Vienna. ISBN 3-85234-060-8.

Langergraber, G. (2005). The role of plant uptake on the removal of organic matter and nutrients in subsurface flow constructed wetlands: A simulation study. *Water Science and Technology, 51* (9), 213–223.

Langergraber, G. (2007). Simulation of the treatment performance of outdoor subsurface flow constructed wetlands in temperate climates. *The Science of the Total Environment, 380*(1–3), 210–219.

Langergraber, G. (2008). Modeling of processes in subsurface flow constructed wetlands: A review. *Vadose Zone Journal, 7*(2), 830–842.

Langergraber, G. (2010). Water and nutrient management in natural and constructed wetlands. In J. Vymazal (Ed.), *Process based models for subsurface flow constructed wetlands* (pp. 21–36). Dordrecht: Springer.

Langergraber, G., & Šimůnek, J. (2005). Modeling variably saturated water flow and multicomponent reactive transport in constructed wetlands. *Vadose Zone Journal, 4*(4), 924.

Langergraber, G., & Šimůnek, J. (2012) Reactive transport modeling of subsurface flow constructed wetlands using HYDRUS Wetland Module. *Vadose Zone Journal* 11(2). Special Issue Reactive Transport Modeling.

Langergraber, G., Rousseau, D. P. L., García, J., & Mena, J. (2009a). CWM1: A general model to describe biokinetic processes in subsurface flow constructed wetlands. *Water Science and Technology, 59*(9), 1687–1697.

Langergraber, G., Giraldi, D., Mena, J., Meyer, D., Peña, M., Toscano, A., Brovelli, A., & Korkusuz, E. A. (2009b). Recent developments in numerical modelling of subsurface flow constructed wetlands. *The Science of the Total Environment, 407*(13), 3931–3943.

Lee, B.-H., & Scholz, M. (2006). Application of self-organizing map (SOM) to assess the heavy metal removal performance in experimental constructed wetlands. *Water Research, 40*, 3367–3374.

Liolios, K. A., Moutsopoulos, K. N., & Tsihrintzis, V. A. (2012). Modeling of flow and BOD fate in horizontal subsurface flow constructed wetlands. *Chemical Engineering Journal, 200–202*, 681–693.

Llorens, E., Saaltink, M. W., Poch, M., & García, J. (2011a). Bacterial transformation and biodegradation processes simulation in horizontal subsurface flow constructed wetlands using CWM1-RETRASO. *Bioresource Technology, 102*, 928–936.

Llorens, E., Saaltink, M. W., & García, J. (2011b). CWM1 implementation in RetrasoCodeBright: First results using horizontal subsurface flow constructed wetland data. *Chemical Engineering Journal, 166*(1), 224–232.

Llorens, E., Obradors, J., Alarcón-Herrera, M. T., & Poch, M. (2013). Modelling of arsenic retention in constructed wetlands. *Bioresource Technology, 147C*, 221–227.

Lloyd, J. R., Klessa, D. A., Parry, D. L., Buck, P., & Brown, N. L. (2004). Stimulation of microbial sulfate reduction in a constructed wetland: Microbiological and geochemical analysis. *Water Research, 38*, 1822–1830.

Mao, X., Prommer, H., Barry, D., Langevin, C., Panteleit, B., & Li, L. (2006). Three-dimensional model for multi-component reactive transport with variable density groundwater flow. *Environmental Modelling and Software, 21*(5), 615–628.

Maurer, M., & Rittmann, B. E. (2004). Modeling intrinsic bioremediation to interpret observable biogeochemical footprints of BTEX biodegradation: mathematical modeling and examples. *Biodegradation, 15*, 419–434.

Mayo, A. W., & Bigambo, T. (2005). Nitrogen transformation in horizontal subsurface flow constructed wetlands I: Model development. *Physics and Chemistry of the Earth, Parts A/B/C, 30*(11–16), 658–667.

Mburu, N., Sanchez-Ramos, D., Rousseau, D. P. L., van Bruggen, J. J. A., Thumbi, G., Stein, O. R., Hook, P. B., & Lens, P. N. L. (2012). Simulation of carbon, nitrogen and sulphur conversion in batch-operated experimental wetland mesocosms. *Ecological Engineering, 42*, 304–315.

Mburu, N., Rousseau, D. P. L., van Bruggen, J. J. A., Thumbi, G., Llorens, E., García, J., & Lens, P. N. L. (2013). Reactive transport simulation in a tropical horizontal subsurface flow constructed wetland treating domestic wastewater. *Science of the Total Environment, 449*, 309–319.

McBride, G. B., & Tanner, C. C. (2000). Modelling biofilm nitrogen transformations in constructed wetland mesocosms with fluctuating water levels. *Ecological Engineering, 14*, 93–106.

McDonald, M., & Harbaugh, A. (1988). *A modular three-dimensional finite-difference groundwater flow model*. Reston: U.S Geological Survey.

Meyer, D. (2011). Modellierung und Simulation von Retentionsbodenfiltern zur weitergehenden Mischwasserbehandlung (Modelling and simulation of constructed wetlands for enhanced combined sewer overflow treatment). PhD thesis, Institute of Urban Water Management, Technical University of Kaiserslautern, Germany.

Meyer, D., Langergraber, G., & Dittmer, U. (2006). Simulation of sorption processes in vertical flow constructed wetlands for CSO treatment. In *Proceedings 10th international conference on wetland systems for water pollution control* (pp. 599–609). Lisbon, Portugal: MAOTDR.

Meyer, D., Molle, P., Esser, D., Troesch, S., Masi, F., & Dittmer, U. (2013). Constructed wetlands for combined sewer overflow treatment – Comparison of German, French and Italian approaches. *Water, 5*, 1–12.

Meyer, D., Chazarenc, F., Claveau-Mallet, D., Dittmer, U., Forquet, N., Molle, P., Morvannou, A., Pálfy, T., Petitjean, A., Rizzo, A., Samsó, R., Scholz, M., Soric, A., Langergraber, G. (in press). Modelling constructed wetlands: Scopes and aims – A review. *Ecological Engineering*.

Mitsch, W. J., & Wise, K. M. (1998). Water quality, fate of metals, and predictive model validation of a constructed wetland treating acid mine drainage. *Water Research, 32*(6), 1888–1900.

Morvannou, A., Forquet, N., Vanclooster, M., & Molle, P. (2013). Which hydraulic model to use in vertical flow constructed wetlands? In J. Šimůnek & R. Kodešová (Eds.), *Proceedings of the 4th international conference HYDRUS software applications to subsurface flow and contaminated transport problems* (p. 74). Prague, Czech Republic: Czech University of Life Sciences Prague.

Moutsopoulos, K. N., Poultsidis, V. G., Papaspyros, J. N. E., & Tsihrintzis, V. A. (2011). Simulation of hydrodynamics and nitrogen transformation processes in HSF constructed wetlands and porous media using the advection–dispersion-reaction equation with linear sink-source terms. *Ecological Engineering, 37*, 1407–1415.

Ojeda, E., Caldentey, J., Saaltink, M., & García, J. (2008). Evaluation of relative importance of different microbial reactions on organic matter removal in horizontal subsurface-flow constructed wetlands using a 2D simulation model. *Ecological Engineering, 34*(1), 65–75.

Pálfy, T. G., & Langergraber, G. (2013). Simulation of constructed wetland microcosms using the HYDRUS wetland module. In *Proceedings 5th international symposium on "Wetland pollutant dynamics and control WETPOL 2013"* (pp. 178–179). 13–17 Oct 2013, Nantes, France.

Parkhurst, D. L., & Appelo, C. A. J. (1999). User's guide to PHREEQC (version 2)–A computer program for speciation, batch-reaction, one-dimensional transport, and inverse geochemical calculations: U.S. Geological Survey Water-Resources Investigations Report 99–4259, 312 pp.

Reichert, P. (1998). AQUASIM 2.0—User manual computer program for the identification and simulation of aquatic systems. Swiss Federal Institute for Environmental Science and Technology (EAWAG), CH-8600 Dubendorf, Switzerland.

Rolle, M., Clement, T. P., Sethi, R., & Molfetta, A. D. (2008). A kinetic approach for simulating redox-controlled fringe and core biodegradation processes in groundwater: Model development and application to a landfill site in Piedmont, Italy. *Hydrological Processes, 4921* (September), 4905–4921.

Rousseau, D. P. L., Vanrolleghem, P. A., & De Pauw, N. (2004). Model-based design of horizontal subsurface flow constructed treatment wetlands: A review. *Water Research, 38*(6), 1484–1493.

Saaltink, M. W., Ayora, J., Stuyfzand, P. J., & Timmer, H. (2003). Analysis of a deep well recharge experiment by calibrating a reactive transport model with field data. *Journal of Contaminant Hydrology, 65*(1–2), 1–18.

Saaltink, M. W., Batlle, F., Ayora, C., Carrera, J., & Olivella, S. (2004). RETRASO, a code for modeling reactive transport in saturated and unsaturated porous media. *Geologica Acta, 2*(3), 235–251.

Samsó, R., & García, J. (2013a). BIO_PORE, a mathematical model to simulate biofilm growth and water quality improvement in porous media: Application and calibration for constructed wetlands. *Ecological Engineering, 54*, 116–127.

Samsó, R., & García, J. (2013b). Bacteria distribution and dynamics in constructed wetlands based on modelling results. *Science of the Total Environment, 461–462*, 430–440.

Šimůnek, J., Sejna, M., & van Genuchten M. Th. (1999). The HYDRUS-2D software package for simulating two-dimensional movement of water, heat, and multiple salutes in variably saturated media, version 2.0. Manual, U.S. Salinity Laboratory, USDA, ARS, Riverside, CA, USA.

Stein, O. R., Biederman, J. A., Hook, P. B., & Allen, C. (2006). Plant species and tempera- ture effects on the k-C* first-order model for COD removal in batch-loaded SSF wetlands. *Ecological Engineering, 26*(2), 100–112.

Suliman, F., French, H. K., Haugen, L. E., & Søvik, A. K. (2006). Change in flow and transport patterns in horizontal subsurface flow constructed wetlands as a result of biological growth. *Ecological Engineering, 27*, 124–133.

Toscano, A., Langergraber, G., Consoli, S., & Cirelli, G. L. (2009). Modelling pollutant removal in a pilot-scale two-stage subsurface flow constructed wetlands. *Ecological Engineering, 35*, 281–289.

van Genuchten, M. T. (1980). A closed-form equation for predicting the hydraulic conductivity of unsaturated soils. *Soil Science Society of America Journal, 44*, 892–898.

Zheng, C., & Wang, P. (1999). MT3DMS, a modular three-dimensional multi-species transport model for simulation of advection, dispersion and chemical reactions of contaminants in ground-water systems; Documentation and User's Guide. U.S. Army Engineer Research and Development Center, USA.

# Chapter 12
# Behaviour of a Two-Stage Vertical Flow Constructed Wetland with Hydraulic Peak Loads

Guenter Langergraber, Alexander Pressl, and Raimund Haberl

**Abstract** The behaviour of a two-stage vertical flow (VF) constructed wetland (CW) system with hydraulic peak loads has been investigated. The CW system was constructed for the Bärenkogelhaus which is located on top of a mountain 1,168 m above sea level. The Bärenkogelhaus has 70 seats and 16 rooms for overnight guests and is a popular site for day visits especially during weekends and public holidays. The system was designed for a hydraulic load of 2,500 L.day$^{-1}$ with a specific surface area requirement of 2.7 m$^2$ per person equivalent. It was built in fall 2009 and started operation in April 2010 when the restaurant was reopened. Samples have been taken between July 2010 and June 2013. In general, the measured effluent concentrations were low and the removal efficiencies high. Even in winter effluent BOD$_5$, COD and NH$_4$-N concentrations have been only slightly above the limit of detection for these parameters. Using the two-stage VF CW system, nitrogen elimination efficiency of more than 70 % could be achieved without recirculation. At hydraulic peak loads, i.e. at events at the Bärenkogelhaus with more than 100 visitors with a hydraulic load > 100 % of the design load, a very robust treatment performance could be observed. The final effluent concentrations of COD and NH$_4$-N did not rise during these peak loads and remained around the limit of detection.

**Keywords** Constructed wetland • Hydraulic peak load • Two-stage vertical flow

## 12.1 Introduction

Constructed wetland (CW) systems are a simple technology in construction as well as in operation and maintenance and have a high buffer capacity for hydraulic and organic load fluctuations and high robustness and process stability. CWs are therefore a suitable technological solution for small villages and single objects

G. Langergraber (✉) • A. Pressl • R. Haberl
Institute for Sanitary Engineering and Water Pollution Control, University of Natural Resources and Life Sciences, Vienna (BOKU), Muthgasse 18, Vienna A-1190, Austria
e-mail: guenter.langergraber@boku.ac.at

(Kadlec and Wallace 2009). When stringent effluent thresholds regarding nitrification have to be met, vertical flow (VF) CWs with intermittent loading have to be used (Haberl et al. 2003).

The requirements of the Austrian regulation (1.AEVkA 1996) regarding maximum effluent concentrations and minimum removal efficiencies can be met using single-stage VF CWs. According to the Austrian design standards (ÖNORM B 2505, 2009), the VF CWs can be loaded with an organic load of 20 g $COD.m^{-2}.d^{-1}$ (i.e. 4 $m^2$ per person equivalent (PE)). It was shown that a two-stage design of the VF system loaded with 40 g $COD.m^{-2}.day^{-1}$ (i.e. 2 $m^2$ per PE) fulfils the Austrian requirements regarding effluent concentrations and removal efficiencies and additionally achieves a nitrogen removal efficiency of more than 60 % at high removal rates of about 1300 g $N.m^{-2}.year^{-1}$ (Langergraber et al. 2008a, b, 2010, 2011).

Results from the first full-scale implementation of the two-stage CW system were presented. While Langergraber et al. (2013) present the overall results from the full three years investigation period (July 2010 until June 2013), this manuscript presents the special investigations carried out in 2012, i.e. a tracer study and intensified sampling campaigns to investigate the behaviour of the system during hydraulic peak loads at events.

## 12.2 Material and Methods

### 12.2.1 Site Description

The CW system is located at 1,168 m above sea level on top of the mountain Bärenkogel in Styria. The two-stage VF CW system treats the wastewater of the Bärenkogelhaus. The Bärenkogelhaus is a popular site for day visits especially during weekends and public holidays and has 70 seats in the restaurant and 16 rooms for overnight guests. The design hydraulic load was 2,500 $L.day^{-1}$ resulting in a specific surface area requirement of 2.7 $m^2$ per PE

The two beds of the VF CW system have a surface area of 50 $m^2$ each. Mechanically pretreated wastewater is loaded intermittently on stage 1 using a siphon. Also stage 2 is loaded intermittently using a siphon. The volume of a single load is about 580 L for both stages. Stage 1 consists of a 50 cm main layer of sand with a grain size distribution of 2–4 mm, a 10 cm top layer of gravel (4–8 mm) and a 20 cm drainage layer of gravel (8–16 mm) at the bottom. Stage 2 has a 50 cm main layer of sand with a grain size distribution of 0.06–4 mm, whereas the top and drainage layers are the same as for stage 1. The drainage layer of stage 1 is impounded (water level about 15 cm), whereas there is free drainage in stage 2. Both beds are planted with common reed (*Phragmites australis*). The system started its operation in April 2010 when the restaurant reopened.

In 2010, during the first year of operation, the Bärenkogelhaus was operated as a restaurant that was open continuously five days a week (closed on Monday and

Tuesday). As the lessee stopped his contract at the end of 2010 and no new lessee could be found, the Bärenkogelhaus since 2011 is open on demand only for events. The first events took place in July 2011.

#### 12.2.1.1 Experimental Programme, Online Measurements, Sampling, Analyses and Data Evaluation

Routine samples have been taken biweekly from the influent of the two-stage system after mechanical pretreatment as well as from the effluents of the first and second bed. The samples have been taken by the lessee or the owner and stored in a fridge. During sampling, a check list was completed for documentation. During investigations of peak load events in 2012, automatic samplers (ISCO) have been used to collect samples for a period of about one week before and after the event.

Collected samples have been analysed for TSS, $BOD_5$, COD, $NH_4$-N, $NO_2$-N, $NO_3$-N and TN in the lab of the Institute for Sanitary Engineering at BOKU University using standard methods. Samples from events have been analysed for COD and $NH_4$-N. For data evaluation, measured values below the limit of detection have been considered as the value of the limit of detection.

In the influent chamber of stage 1 in which the siphon is placed, the water level has been measured online using a pressure probe. The number of loadings has been calculated from the changes of the water level. Additionally, the temperatures of air, influent and effluent water and temperatures in different depths of the VF filters of stage 1 and 2 have been measured online.

For the tracer experiment, the tracer salt (KCl) was mixed in a separate 1 $m^3$ tank with tap water and an additional simulated loading of 580 L was applied on the first stage. The influent electrical conductivity (EC) during the simulated loading was 10 mS $cm^{-1}$. To guarantee subsequent loadings during the tracer experiment, a water tap was kept running at a toilet of the Bärenkogelhaus. During the experiment, EC was measured online in the influents and effluents of the beds using WTW probes and recorded using a data logger. Additionally, the volumetric effluent flows of stages 1 and 2 have been measured online using an electromagnetic flow metre (Endress + Hauser).

### 12.3 Results and Discussion

#### 12.3.1 Influent and Effluent Concentrations and Removal Efficiencies

Data from analysis of routine sampling are presented for two periods according to the operation of the Bärenkogelhaus: (1) start of the investigations until the end of

**Table 12.1** Influent concentrations (mg.L$^{-1}$) of the two-stage CW system at Bärenkogelhaus

| Parameter | Continuous operation ||||Event operation ||||
|---|---|---|---|---|---|---|---|---|
| | BOD$_5$ | COD | NH$_4$-N | TN | BOD$_5$ | COD | NH$_4$-N | TN |
| Number of samples | 10 | 10 | 10 | 10 | 32 | 39 | 39 | 32 |
| Median value | **560** | **1,015** | **50.8** | **65.3** | **149** | **346** | **56.6** | **66.0** |
| Average value | 543 | 974 | 48.6 | 62.9 | 156 | 350 | 53.4 | 64.6 |
| Standard deviation | 114 | 220 | 5.6 | 8.1 | 96 | 160 | 24.4 | 29.4 |
| 95 % confidence interval | 71 | 136 | 3.5 | 5.0 | 33 | 50 | 7.7 | 10.2 |
| Maximum | 691 | 1,230 | 53.6 | 71.0 | 363 | 720 | 103.0 | 119.2 |
| Minimum | 282 | 499 | 34.4 | 42.4 | 17 | 71 | 12.8 | 16.2 |

**Table 12.2** Final effluent concentrations (mg.L$^{-1}$) of the two-stage CW system at Bärenkogelhaus

| Parameter | Continuous operation ||||Event operation ||||
|---|---|---|---|---|---|---|---|---|
| | BOD$_5$ | COD | NH$_4$-N | TN | BOD$_5$ | COD | NH$_4$-N | TN |
| Number of samples | 10 (3*) | 10 (9*) | 10 (1*) | 10 | 32 (31*) | 39 (15*) | 39 (14*) | 32 |
| Median value | 3 | 20 | 0.06 | 19.2 | 3 | 12 | 0.03 | 16.6 |
| Average value | 3 | 22 | 0.06 | 16.5 | 3 | 14 | 0.06 | 17.4 |
| Standard deviation | 0 | 5 | 0.02 | 6.7 | 0 | 5 | 0.07 | 10.3 |
| 95 % confidence interval | 0 | 3 | 0.01 | 4.2 | 0 | 1 | 0.02 | 3.6 |
| Maximum | 4 | 35 | 0.09 | 26.1 | 4 | 26 | 0.33 | 35.2 |
| Minimum | 3 | 20 | 0.03 | 5.5 | 3 | 10 | 0.03 | 1.1 |

*Number of analysis below the limit of detection: 3 mg BOD$_5$/L, 20 mg COD/L (2010) and 10 mg COD/L, respectively (since 2011, lower limit of detection due to new analytical method) and 0.03 mg NH$_4$-N/L

2010 (five day continuous operation of the restaurant per week) and (2) since June 2011 (start of the events at Bärenkogelhaus).

Table 12.1 shows the influent concentrations for the two periods. During the "continuous operation" of the restaurant, the influent BOD$_5$ and COD concentrations to the first stage have been significantly higher compared to "event operation". The higher influent concentrations of the organic parameters can be explained by the fact that the kitchen of the Bärenkogelhaus was always operating during "continuous operation" and much less in use during "event operation". The average specific organic load was 10.3 g COD m$^{-2}$ day$^{-1}$ during "continuous operation", i.e. about only 32 % of the dimensioning value. In this period, the organic load reached about 80–90 % of the dimensioning value during weekends with many visitors. During "event operation", the average specific organic load was 1.4 g COD m$^{-2}$ day$^{-1}$, less than 5 % of the dimensioning value. In contrary to the organic parameters, the NH$_4$-N and TN influent concentrations were not lower during "event operation".

Tables 12.2 and 12.3 show the final effluent concentrations and the removal efficiencies, respectively. During "continuous operation", the maximal allowed

**Table 12.3** Removal efficiencies of the two-stage CW system at Bärenkogelhaus

|  | Continuous operation |  |  |  | Event operation |  |  |  |
|---|---|---|---|---|---|---|---|---|
| Parameter | BOD$_5$ | COD | NH$_4$-N | TN | BOD$_5$ | COD | NH$_4$-N | TN |
| Number of samples | 10 | 10 | 10 | 10 | 32 | 39 | 39 | 32 |
| Median value | **99.4 %** | **98.0 %** | **99.88 %** | **70.5 %** | **98.0 %** | **96.0 %** | **99.92 %** | **74.4 %** |
| Average value | 99.4 % | 97.7 % | 99.88 % | 74.3 % | 96.2 % | 95.1 % | 99.84 % | 71.0 % |
| Standard deviation | 0.2 % | 0.8 % | 0.1 % | 9.3 % | 4.3 % | 2.9 % | 0.25 % | 18.9 % |
| 95 % confidence interval | 0.1 % | 0.5 % | 0.0 % | 5.8 % | 1.5 % | 0.9 % | 0.08 % | 6.6 % |
| Maximum | 99.6 % | 98.4 % | 99.92 % | 87.0 % | 99.2 % | 98.1 % | 99.96 % | 97.8 % |
| Minimum | 98.9 % | 96.0 % | 99.83 % | 59.3 % | 82.4 % | 85.9 % | 98.52 % | 2.0 % |

effluent concentrations (25 mg BOD$_5$.L$^{-1}$; 90 mg COD.L$^{-1}$ and 10 mg NH$_4$-N.L$^{-1}$, respectively) have not been exceeded as well as the required minimum removal efficiencies (95 % for BOD$_5$ and 85 % for COD, respectively) have been reached.

Also during "event operation", the legally allowed maximal effluent concentrations have not been exceeded. The required minimum removal efficiency for COD (85 %) has been met during the whole period, whereas for BOD$_5$ the required minimum removal efficiency (95 %) was not reached for seven sampling dates. However, at these sampling dates, the influent concentrations have been very low and the measured effluent concentrations were below the limit of detection (3 mg BOD$_5$/L).

In both periods, the NH$_4$-N effluent concentrations have been far below the maximum allowed 10 mg/L (which has to be met when the effluent water temperature is higher than 12 °C). The maximum measured NH$_4$-N effluent concentration during winter was < 0.5 mg/L. In addition to the legal requirements, a stable nitrogen removal efficiency of > 70 % was obtained.

### 12.3.1.1 Tracer Experiment

The tracer experiment was started on 25 May 2012 at 14:06 by adding an additional simulated loading with an EC of 10 mS/cm. During the experiment, 32 loadings occurred which represent a hydraulic load of 22.6 mm/d. Figure 12.1 shows the measured volumetric effluent flows of stage 1 and stage 2 and the respective loadings and Fig. 12.2 the measured EC. In Fig. 12.1, also the effect of heavy rain on 4 June 2013 on the volumetric effluent flows can be seen. The hydraulic retention time was calculated to be one day and 8 eight hours for stage 1 and three days and two hours for the whole system, respectively (Fig. 12.3).

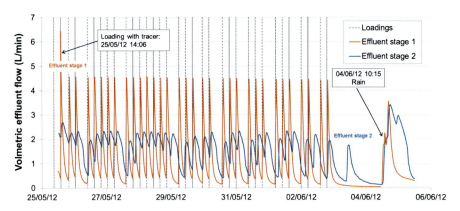

**Fig. 12.1** Volumetric effluent flows of stage 1 and 2 during the tracer experiment (25.05.–06.06.2012)

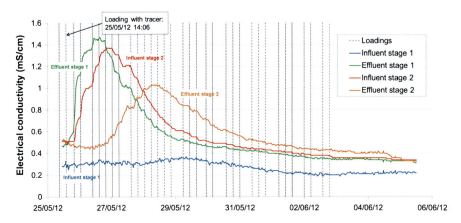

**Fig. 12.2** Electrical conductivity during the tracer experiment (25.05.–06.06.2012)

### 12.3.1.2 Sampling Campaigns at Events

Table 12.4 shows the events selected for sampling campaigns in 2012. The events have been selected together with the owner of Bärenkogelhaus in January 2012 according to expected visitors. It has been shown that the events selected have been quite representative in terms of visitors (Table 12.4).

The first event sampling campaign took place at the weekend 18/19 February 2012. To prevent freezing of the automatic sampler, insulated boxes with heating systems were installed. Figure 12.4 shows measured temperatures of the air and at the surface of the main layer of the first VF bed between January and March 2012. The period of the event sampling is highlighted. It can be clearly seen that the sampling was done during a very cold period. The constant temperature at the surface of the main layer indicated that during this period, there was a snow layer on

12 Behaviour of a Two-Stage Vertical Flow Constructed Wetland with Hydraulic... 181

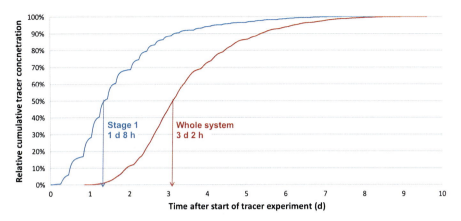

**Fig. 12.3** Relative cumulative tracer concentration (25.05.–06.06.2012)

**Table 12.4** Events selected for sampling campaigns

| Weekend | # | Event | Visitors | Overnight stays |
|---|---|---|---|---|
| 18/19 February 2012 | 1 | Event | 90 | 4 |
| 16/17 June 2012 | 2 | Concert | 100 | – |
| 23/24 June 2012 | 3 | Wedding 1 | 150 | 25 |
| 28/29 July 2012 | 4 | Wedding 2 | 90 | 20 |
| 08/09 September 2012 | 5 | Event | 500 | – |

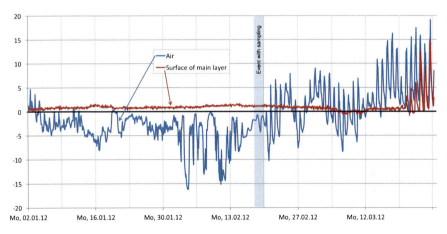

**Fig. 12.4** Air temperature and temperature at the surface of the main layer of bed 1 between January and March 2012

top of the bed. After melting of the snow (mid of March 2012), typical daily temperature patterns can also be observed at the surface of the main layer. Around the period of the event sampling, the diurnal variations of air temperature have been

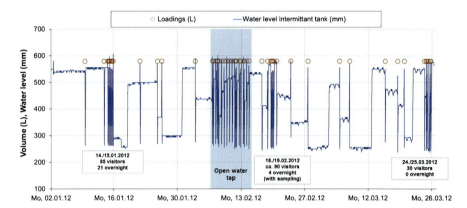

**Fig. 12.5** Water level in the influent chamber and single loadings of the first VF bed (January to March 2012)

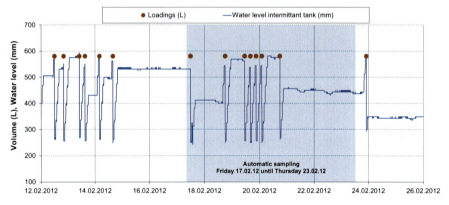

**Fig. 12.6** Water level in the influent chamber and single loadings of the first VF bed (12–26 February 2012)

lower than before and after. This was caused by the fact that during this period, a bank of snow was covering the temperature sensor. Therefore, the actual air temperature has been lower than the measured one.

Figure 12.5 shows the measured water level in the influent chamber and the loadings of the first VF bed during the period January to March 2012. Figure 12.6 shows a more detailed picture for a two weeks period around the event sampling in February 2012. In the first three months of 2012, three events have been at Bärenkogelhaus, i.e. on the weekends 14/15 January, 18/19 February and 24/25 March 2012, respectively. In Fig. 12.5, these periods can be identified due to the higher amount of loadings. The loadings with a very constant interval between 6 and 14 February was caused by an open water tap in a toilet of the Bärenkogelhaus.

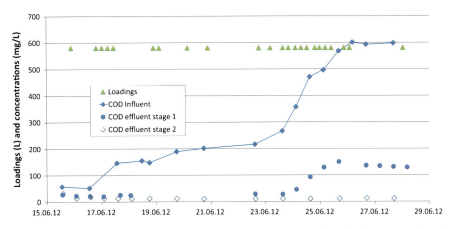

**Fig. 12.7** Loadings and COD influent and effluent concentrations during events 2 and 3 (16–28 June 2012)

Figure 12.6 shows that during the event sampling period at the weekend 18/19 February 2012, seven loadings occurred between Friday (17 February) and Thursday (23 February). Ninety visitors at the evening event on 18 February (plus four guests staying overnight) produced about 1.8 m$^3$ of wastewater. Between Sunday 19 February 2012 11:00 and Monday 20 February 2012 18:00, five loadings were recorded resulting in a hydraulic load of 94 % of the design load.

Due to the open water tap, the wastewater stored in the pretreatment of the system was strongly diluted for more than a week prior to the event resulting in influent concentrations during the event sampling of only $243 \pm 18$ mg COD.L$^{-1}$ and $21.2 \pm 1.7$ mg NH$_4$-N.L$^{-1}$ (N = 5), respectively. The effluent concentrations of the CW system have been $19 \pm 3$ mg COD.L$^{-1}$ (N = 9) and $0.03 \pm 0.01$ mg NH$_4$-N.L$^{-1}$ (N = 17, with 11 measurements below the limit of detection of 0.03 mg NH$_4$-N.L$^{-1}$). The additional loadings at the event had no influence on the final effluent concentrations of COD and NH$_4$-N. The event did not show the expected rise in the effluent concentrations that was expected for peak loads in winter, mainly caused by the low influent concentrations. However, from a technical point of view, the sampling campaign in winter worked well.

The influent concentrations of the second and third event sampling campaigns which were carried out on two successive weekends (i.e. on 16/17 and 23/24 June 2012) were influenced by tracer experiments carried out prior to these events. Figure 12.7 shows the COD influent and effluent concentrations as well as the loadings for the period 16–28 June 2012. For the second event (concert), four loadings occurred between 16 June 2012 19:28 and 19 June 2012 02:28 (hydraulic loading rate 11.14 mm/d) and, for the third event (wedding), ten loadings between 23 June 2012 15:28 and 26 June 2012 01:28 (hydraulic loading rate 24.00 mm/d). The influent COD concentration was still low at the beginning of the measurements and increased as wastewater was produced during the events. After the third event, "normal" COD influent concentrations (around 600 mg L$^{-1}$) were reached again.

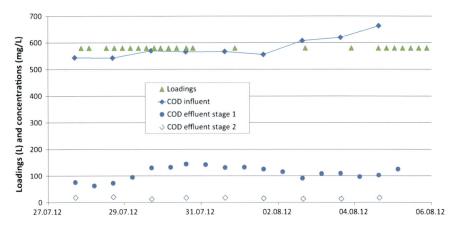

**Fig. 12.8** Loadings and COD influent and effluent concentrations during event 4 (27 July–5 August 2012)

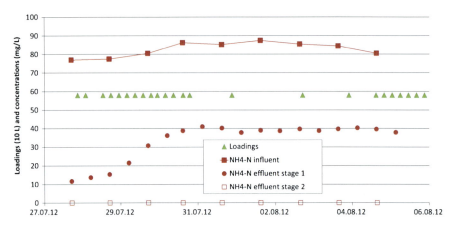

**Fig. 12.9** Loadings and NH$_4$-N influent and effluent concentrations during event 4 (27 July–5 August 2012)

The effluent concentration of stage 1 increased to about 130 mg L$^{-1}$ during the third event. The effluent concentration, however, remained around the limit of detection. A similar pattern was measured for NH$_4$-N concentrations (not shown); after in the third event, the influent concentrations were around 80 mg L$^{-1}$ and the effluent concentrations of stage 1 around 10 mg L$^{-1}$. Again, this increase had no impact on the final effluent concentrations.

The sampling for event 4 (wedding, 28/29 July 2012) took place between 27 July and 5 August 2012. There was an additional event on the following weekend. Between 27 July 2012 21:03 and 30 July 2012 19:03, 13 loadings were recorded; the hydraulic loading rate was 25.85 mm/d. Figure 12.8 shows the loadings and COD influent and effluent concentrations and Fig. 12.9 the loadings and NH$_4$-N influent and effluent concentrations. The influent concentrations (N = 9) during this period were $582 \pm 40$ mg COD L$^{-1}$ and $82.7 \pm 3.9$ mg NH$_4$-N L$^{-1}$, respectively. The COD effluent concentrations of stage 1 and 2 were $111 \pm 25$ mg L$^{-1}$ (N = 18) and $17 \pm 3$ mg L$^{-1}$ (N = 9), respectively. During the event, an increase in the NH$_4$-N effluent concentrations of stage 1 from 10 to 40 mg NH$_4$-N L$^{-1}$ could be observed. However, there was no increase of the final NH$_4$-N effluent concentration.

For event 5, the sampling started on 8 September and lasted until 14 September 2012. From 8 to 10 September 2012, ten loadings were recorded resulting in a hydraulic loading rate of 19.20 mm/d. The results were similar compared to the previous event. The measured influent concentrations were $604 \pm 97$ mg COD L$^{-1}$ and $65.1 \pm 2.5$ mg NH$_4$-N L$^{-1}$ (N = 6), respectively. Average COD effluent concentrations of stage 1 and 2 were $93 \pm 26$ mg L$^{-1}$ (N = 11) and $22 \pm 6$ mg L$^{-1}$ (N = 6), respectively, and average NH$_4$-N effluent concentrations $39.6 \pm 3.5$ mg L$^{-1}$ (N = 12) and $0.05 \pm 0.02$ mg L$^{-1}$ (N = 6), respectively.

Table 12.5 summarises measured COD and NH$_4$-N concentrations during the sampled events. For all events, the final effluent concentrations were far below the legal requirements. The final effluent concentrations were not affected by the hydraulic peak loads.

### 12.3.1.3 Summary of Events in 2012

Table 12.6 shows the overview of the events held in the Bärenkogelhaus in 2012. The five events for which sampling campaigns have been carried out are shaded grey. For each event (when recorded), the number of visitors (V) and overnight stays (O) is given. The hydraulic load of the CW system was calculated based on the recorded loadings (L) during the event duration (D). As for some of the events, the measured COD influent concentrations were very low (see above); the organic load of the CW system was calculated using the median value of the measured COD influent concentrations in 2012 (i.e. 556 mg l$^{-1}$). The calculated organic loading rate during the high hydraulic load events was a maximum of 15 g COD m$^{-2}$ day$^{-1}$ and thus only about 50 % of the design organic load. The average hydraulic load in 2012 was 16 % of the design load (291 loadings were recorded when subtracting the loadings caused by the open water tap and the additional loadings due to the tracer experiment); the calculated organic load was 2.3 g COD m$^{-2}$ day$^{-1}$, about 8 % of the design organic load.

**Table 12.5** Summary of measured COD and NH$_4$-N concentrations during sampled events (average values ± standard deviation in mg L$^{-1}$, number of samples in brackets)

|   | COD |   |   | NH$_4$-N |   |   |
|---|---|---|---|---|---|---|
|   | Influent | Effluent stage 1 | Effluent stage 2 | Influent | Effluent stage 1 | Effluent stage 2 |
| 1 | 243 ± 18 (5) | – | 19 ± 3 (9) | 21.2 ± 1.7 (5) | – | 0.03 ± 0.01 (17, 11*) |
| 2 | 145 ± 62 (8) | 24 ± 3 (8) | 14 ± 6 (10, 2*) | 16.9 ± 6.9 (8) | 2.8 ± 0.8 (9) | 0.06 ± 0.03 (10) |
| 3 | 483 ± 124 (8) | 118 ± 34 (8) | 11 ± 1 (5, 2*) | 66.1 ± 17.1 (8) | 22.5 ± 8.3 (11) | 0.04 ± 0.01 (5) |
| 4 | 582 ± 40 (9) | 111 ± 25 (18) | 17 ± 3 (9) | 82.7 ± 3.9 (9) | 33.4 ± 10.2 (18) | 0.05 ± 0.01 (9, 1*) |
| 5 | 604 ± 97 (6) | 93 ± 26 (11) | 22 ± 6 (6) | 65.1 ± 2.5 (6) | 39.6 ± 3.5 (12) | 0.05 ± 0.02 (6) |

*Number of analysis below the limit of detection: 10 mg COD/L and 0.03 mg NH$_4$-N/L, respectively

**Table 12.6** Overview of events at Bärenkogelhaus and loading of the CW system in 2012

| Date | Event | V | O | D | L | Hydraulic load | Organic load (g COD·m$^{-2}$·d$^{-1}$) measured | calculated * |
|---|---|---|---|---|---|---|---|---|
| 14.01.2012 | Birthday party | 85 | 21 | 2.09 | 8 | 78% | – | 11.0 |
| 19.02.2012 | **Event** | **90** | **4** | **1.29** | **5** | **72%** | **4.5** | **10.2** |
| 24.03.2012 | Birthday party | 30 | – | 2.54 | 8 | 64% | – | 9.0 |
| 05.05.2012 | Event | 65 | 5 | 2.71 | 7 | 51% | – | 7.2 |
| 30.05.2012 | Christening | 40 | – | – | – | – | – | – |
| 16.06.2012 | **Concert** | **100** | **–** | **1.75** | **6** | **66%** | **2.5** | **9.3** |
| 23.06.2012 | **Wedding** | **150** | **25** | **3.29** | **13** | **85%** | **10.6** | **11.9** |
| 28.07.2012 | **Wedding** | **90** | **20** | **2.92** | **14** | **103%** | **14.9** | **14.6** |
| 04.08.2012 | Wedding | 80 | 20 | 4.00 | 13 | 70% | – | 9.8 |
| 25.08.2012 | Wedding | 80 | 25 | 3.08 | 12 | 83% | – | 11.7 |
| 31.08.2012 | Event | 48 | – | 1.67 | 6 | 70% | – | 9.8 |
| 09.09.2012 | **Event** | **500** | **–** | **2.42** | **9** | **77%** | **11.7** | **10.8** |
| 15.09.2012 | Birthday party | 85 | 9 | 2.42 | 10 | 86% | – | 12.2 |
| 21.09.2012 | Concert | 80 | 0 | 0.50 | 3 | 93% | – | 13.1 |
| 13.10.2012 | Event | 30 | 2 | 1.17 | 4 | 56% | – | 7.8 |
| 21.10.2012 | Concert | 90 | 4 | 1.81 | 6 | 60% | – | 8.4 |
| 14.12.2012 | Christmas party | 66 | – | 0.63 | 4 | 104% | – | 14.6 |
| 27.12.2012 | Christmas party | 50 | – | 0.21 | 2 | 104% | – | 14.6 |
| 2012 | Including "additional" loadings ** | | | | 363 | 19 % | | 2.6 |
| | Without "additional" loadings** | | | | 291 | 16 % | | 2.3 |

V ... Visitors; O ... Overnight stays; D ... Duration (d); L ... Number of loadings
*Calculated with 556 mg/L the median value of the COD influent concentration on 2012
**"Additional" loadings due to open water tap in February and during tracer experiments in May and June 2012

## 12.4  Conclusions

It can be concluded that the two-stage VF CW system:

- Fulfils the requirements of the Austrian regulations regarding effluent concentration and removal efficiencies
- Additionally enables achieving nitrogen elimination efficiencies of more than 70 % without recirculation
- Shows a very robust treatment performance at high fluctuations of flow and concentrations
- Is therefore suitable for the application at objects with high hydraulic peak loads, e.g. restaurants with event operation, even at an altitude of above 1,100 m above sea level

**Acknowledgements**  The experiments were carried out in the course of the research project *"Begleitende Untersuchungen zur praktischen Anwendung eines 2-stufigen bepflanzten Bodenfilters beim Gasthaus Bärenkogel"* funded by the Austrian Ministry for Agriculture,

Forestry, Environment and Water Management. The authors are grateful for the support and especially thank the Hammer family, the owner of the Bärenkogelhaus, for their support onsite.

# References

1.AEVkA (1996). 1. Abwasseremissionsverordnung für kommunales Abwasser (Austrian regulation for emissions from domestic wastewater). BGBl.210/1996, Vienna, Austria [*in German*].

Haberl, R., Grego, S., Langergraber, G., Kadlec, R. H., Cicalini, A. R., Martins Dias, S., et al. (2003). Constructed wetlands for the treatment of organic pollutants. *The Journal of Soils and Sediments, 3*(2), 109–124.

Kadlec, R., & Wallace, S. (2009). *Treatment wetlands* (2nd ed.). Boca Raton: CRC Press.

Langergraber, G., Leroch, K., Pressl, A., Rohrhofer, R., & Haberl, R. (2008a). A two-stage subsurface vertical flow constructed wetland for high-rate nitrogen removal. *Water Science and Technology, 57*(12), 1881–1887.

Langergraber, G., Prandtstetten, C., Pressl, A., Sleytr, K., Leroch, K., Rohrhofer, R., & Haberl, R. (2008b). Investigations on nutrient removal in subsurface vertical flow constructed wetlands. In J. Vymazal (Ed.), *Wastewater treatment, plant dynamics and management in constructed and natural wetlands* (p. 199). Dordrecht: Springer.

Langergraber, G., Pressl, A., Leroch, K., Rohrhofer, R., & Haberl, R. (2010). Comparison of the behaviour of one- and two-stage vertical flow constructed wetlands for different load scenarios. *Water Science and Technology, 61*(5), 1341–1348.

Langergraber, G., Pressl, A., Leroch, K., Rohrhofer, R., & Haberl, R. (2011). Long-term behaviour of a two-stage CW system regarding nitrogen removal. *Water Science and Technology, 64*(5), 1137–1141.

Langergraber, G., Pressl, A., & Haberl, R. (2013). Experiences from the full-scale implementation of a new 2-stage vertical flow constructed wetland design. *Water Science Technology*, submitted.

ÖNORM B 2505. (2009). Bepflanzte Bodenfilter (Pflanzenkläranlagen) – Anwendung, Bemessung, Bau und Betrieb (Subsurface-flow constructed wetlands – Application, dimensioning, installation and operation). Österreichisches Normungsinstitut, Vienna, Austria [in German].

# Chapter 13
# A New Concept of Multistage Treatment Wetland for Winery Wastewater Treatment: Long-Term Evaluation of Performances

Fabio Masi, Riccardo Bresciani, and Miria Bracali

**Abstract** The wastewater produced by a Tuscan winery has been treated with a multistage constructed wetland (CW or MSTW) treatment system since 2001. In recent years, the production at the winery has increased greatly and consequently the treated flows ranged from 35 $m^3 d^{-1}$ to ~100 $m^3 d^{-1}$. A prolonged overload (for about 2 years) has resulted in clogging of the 1st stage HF CW bed in the former configuration, a 2-stage CW treatment plant: horizontal subsurface flow (HF) + free water system (FWS). In order to solve this problem, an upgrade of the existing plant was implemented in 2009. The upgraded configuration consists of four stages: a "French style" CW for raw wastewater followed by an HF system, formed by the existing restored bed plus three new beds, all in parallel, and finally by the existing FWS followed by a small sand filter. This is an innovative design for CW treatment systems, showing a unique and powerful combination of different typologies with the insertion of the "French style" reed beds as first stage and the consequent elimination of the former Imhoff tank. This paper will present the results obtained in the first 3 years of operation of the upgraded CW treatment system comparing them with the historical data of the last 10 years. Results indicate that a COD reduction of up to 96 % has been achieved even at peak loads. The results from this study confirm the potential of CWs as a suitable technology for treating winery wastewater and the efficiency of French systems as primary treatment for raw winery wastewater.

**Keywords** COD removal • French CW systems • Multistage constructed wetland • Winery wastewater

---

F. Masi (✉) • R. Bresciani
IRIDRA, Srl – Via La Marmora51, 50121 Florence, Italy
e-mail: masi@iridra.it

M. Bracali
Casa Vinicola Luigi Cecchi e Figli – Loc, Casina dei Ponti, 56, 53011 Castellina in Chianti, Siena, Italy

© Springer International Publishing Switzerland 2015
J. Vymazal (ed.), *The Role of Natural and Constructed Wetlands in Nutrient Cycling and Retention on the Landscape*, DOI 10.1007/978-3-319-08177-9_13

## 13.1 Introduction

Wine production generates large quantities of relevant fluxes of wastewater, produced during the grapes processing (vintage and racking) period and the following months where bottling and cleaning of containers are almost continuous operations. These fluxes of winery wastewater are often an environmental problem of great concern in wine-producing countries (Serrano et al. 2011). This particular wastewater is characterized by fluctuations in terms of quality and quantity during the whole year, and that depend on several factors such as the adopted industrial process chain and its seasonality or the kind of produced wine (Fernández et al. 2007; Grismer et al. 2001).

The winery wastewater produced by the Casa Vinicola Luigi Cecchi and Sons (Castellina in Chianti – Siena) has been treated with a constructed wetland treatment system (CWTP) since 2001. The system consisted of a primary treatment phase with an Imhoff septic tank, followed by a single stage horizontal subsurface flow system (HF) ($A = 480$ m$^2$) and then by a free water system (FWS) ($A = 850$ m$^2$). The system was designed to treat flows of 35 m$^3$ day$^{-1}$. In recent years, the production at the winery has greatly increased, and consequently, flows to the CW have increased up to 70 m$^3$ d$^{-1}$. A prolonged overload resulted in a severe clogging of the HF bed. In order to solve this problem, an upgrade of the existing plant was designed by IRIDRA srl, Florence, and its realization completed in 2009 (Fig. 13.1, Table 13.1). The upgraded configuration consists of four stages: a "French style" CW for raw wastewater followed by an HF system, formed by the existing restored bed plus three new beds, all in parallel, and finally by the existing FWS followed by an optional small sand filter to be used before the discharge the effluents in fresh water (Gena river) in case of algal blooming into the FWS itself or

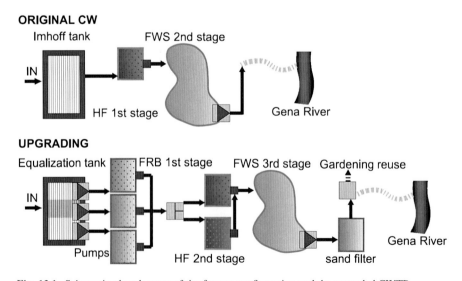

**Fig. 13.1** Schematic plant layouts of the former configuration and the upgraded CWTP

**Table 13.1** Main features of Cecchi CW treatment plant

| Load (pe) | 1,500 |
|---|---|
| Flow ($m^3\ d^{-1}$) | 100 |
| Surface area 1st stage FRB ($m^2$) | 1,200 |
| Surface area 2nd stage HF ($m^2$) | 960 |
| Surface area 3rd stage FWS ($m^2$) | 850 |
| Surface area final sand filter ($m^2$) | 50 |
| FRB beds depth (m) | 0.85 |
| FRB filling media (m) | Top 40 cm gravel ⌀ 2–8 mm |
|  | 20 cm gravel ⌀ 5–20 mm |
|  | Bottom 25 cm gravel ⌀ 40 mm |
| HF beds depth (m) | 0.80 |
| HF beds water level (m) | 0.70 |
| HF gravel size (mm) | 8–12 |

in general whenever there is still a too high content of TSS. The multistage CW system is designed to treat flows up to 100 $m^3\ d^{-1}$ with an average organic content of about 3,800 mg COD $L^{-1}$. This particular winery wastewater is characterized by a quite constantly high organic matter content, low level of nutrients, light acidity and large variations during the usual daily operations that produce effluents. This winery, in fact, produces wine elsewhere or buys it from local producers, and the industrial complex where the CWTP is located only bottles and ages the wine in cellars. So the effluents are created by the washing procedures of bottles, tanks, silos, ground floors, pipes, etc., operations that are planned with an industrial approach and well distributed in the production cycle throughout the year. Various studies (Shepherd et al. 2001; Masi et al. 2002; Müller et al. 2002; Grismer et al. 2003; Mulidzi 2007; Serrano et al. 2011; Vymazal 2009; de la Varga et al. 2013) demonstrated the efficiency of CWs as a low cost, low maintenance and energy-saving technology for the treatment of wineries wastewater. Several of these experiences have also shown lessons to be learnt, such as some limits in the tolerance of the HF and VF classic CWs to the strength of the wineries wastewater, especially in the 1st stage of multistage systems. Therefore, the consequent choice of a French system for raw wastewater as the 1st stage of a hybrid CW system represented an innovation for treatment of wineries wastewater in Italy and more generally at a worldwide level. The use of this technology enhances the sustainability of the treatment plant by the reduction of sludge production and sludge cycle management costs and also provides more robustness to the treatment train, minimizing a big part of the problems observed in the above-cited experiences and at the same site with the older "Imhoff + HF + FWS" configuration. The French CW system has been successfully tested with more than 600 plants by Cemagref (nowadays IRSTEA) and some very active private companies in France (Molle et al. 2005; Boutin et al. 1997) for treating raw domestic and municipal wastewater. The FRB system consists of a batch fed filtering bed, mostly comparable to unsaturated vertical flow constructed wetlands. The main characteristic of this system is that it receives raw wastewater directly without a primary settlement stage. The settled materials remain on the surface of the basins as a sludge layer, which is kept in a highly aerated condition by the presence of reeds growing in it. In the specific case, the installation of the 1st stage FRB has resulted in dismissing the old Imhoff tank, which also created some problems in the HF CW for frequent events of exceptionally high flows and linked wash-out events from the digester, when relevant amounts of primary sludge reached the inlet section of the HF, surely contributing to its deterioration at the time.

## 13.2 Methods

The dimensioning tools utilized for the design of the system were based on the following guidelines and publications: IWA Technical Report n.8 (Kadlec et al. 2000) and the French literature data (Molle et al. 2005). The required treatment area was estimated on the basis of required outfalls, average and peak hydraulic and organic loads. The total hydraulic retention time of the system is about 6–7 d (depending on the flow which varies from 40 to 100 $m^3\ d^{-1}$). The 1st stage FRB system is divided in three lines working in parallel. Each line is intermittently loaded by a set of three independent submerged pumps placed in the equalization tank and connected to a different bed: a timer gives the start eight times per day, with 3 h intervals, to one pump after the other so that every bed is fed every 9 h. The resting period of each line (a minimum of 9 h, then almost 2 days every week during the weekend including all the holiday periods and almost all of August) ensures a good oxygen transfer and mineralization of the surface layer, preventing odour diffusion. Every single load is about 10 $m^3$. There are two floating sensors in the equalization tank that turn off the power to the pumps when the tank is empty and gives a start-up command when the peak flows reach the maximum top level, still maintaining the progressive order of the loadings by the three pumps.

The hydraulic loading rate of each line averaged 0.083 $m^3\ m^{-2}\ d^{-1}$, and the organic loading rate is on average 314 g COD $m^{-2}\ d^{-1}$ for the RBF beds. Molle et al. (2005) recommended a hydraulic load of 0.12 $m^3\ m^{-2}\ d^{-1}$ (maximum 0.37 $m^3\ m^{-2}\ d^{-1}$) and an organic load of 100 gCOD $m^{-2}\ d^{-1}$ (maximum 300 g COD $m^{-2}\ d^{-1}$). The effluent of the FRB system is fed to the HF CWs by gravity and then it flows to the FWS. The final sand filter is optionally fed by a centrifugal submerged pump. Finally, the treated wastewater is discharged into the Gena River by gravity or reused for gardening or irrigation.

The monitoring of each stage of the plant has taken place continuously since the start-up in 2001 under the supervision of the Regional Environmental Protection Agency (ARPAT). The outlet standards for industrial wastewater discharging into superficial water bodies were established in the Italian National Law D. Lgs. 152 of 2006 (Table 13.2).

All the series of samplings have been performed by appropriate internal personnel of the winery, and a large part of the chemical analyses have been performed

**Table 13.2** Required outlet concentrations for discharge in freshwater – D.L.152/2006

| Parameters | Max effluent acceptable concentration (mg $L^{-1}$) |
|---|---|
| BOD$_5$ | 40 |
| COD | 160 |
| TSS | 80 |
| NH$_4^+$ | 15 |
| P total | 10 |
| Sulphides | 1 |

directly in their certified laboratory; the samples were mainly grab samples taken at least twice per month during the peak seasons or every month outside the peak season, with the exclusion of the periods when there is no production of wastewater. All analytical parameters were determined by methods in accordance with Standard Methods (APHA 1992).

## 13.3 Results and Discussion

The average concentrations for the main parameters and pH monitored at the input and at the output of the plant are reported in Table 13.3. In the beginning of the operational period of the first treatment plant configuration, the daily flow during about 220 working days in the year averaged 30–40 m$^3$ d$^{-1}$. In the last two monitored years, the flow averaged about 70 m$^3$ d$^{-1}$. The treated wastewater exhibited quite constant high organic content, while acidity and concentrations of nutrients were low. During all the 11 years of operation, the development and the health of the *Phragmites* community and several other macrophytes in the FWS seemed to be never affected by phytotoxic effects generated by the wastewater with the exception of a short period during which the reeds of the only HF bed of the 1st configuration had a massive deployment. During this time, only the borders of the bed were vegetated with all the central part showing superficial runoff as a consequence of severe clogging (Fig. 13.2). The FRBs were planted with *Phragmites* with the density of four shoots per m$^2$ and have shown a normal growth and regular density after 1 year (Fig. 13.3).

**Table 13.3** Average inlet and outlet pH and COD, Ptot, Ntot, anionic surfactants (MBAS), sulphides concentrations at the Cecchi – Castellina constructed wetland

|  | IN | OUT | n° samples |
|---|---|---|---|
| pH | 6.54 ± 0.7 | 7.72 ± 0.3 | 13 |
| COD (mg $L^{-1}$) | 3,777 ± 3,056 | 67 ± 99 | 24 in/125 out |
| Ptot (mg $L^{-1}$) | 10.98 ± 4.75 | 3.70 ± 1.14 | 12 |
| Ntot (mg $L^{-1}$) | 2.21 ± 1.94 | 0.13 ± 0.08 | 12 |
| MBAS (mg $L^{-1}$) | 0,53 ± 2.19 | 0.27 ± 0.30 | 12 |
| S$^{2-}$ (mg $L^{-1}$) | 1.22 ± 2.60 | 0.05 ± 0.17 | 12 |

**Fig. 13.2** The initial 500 m$^2$ HF CW before (*left*) and after (*right*) the occurrence of severe clogging generated by progressively increasing overloading of the system

**Fig. 13.3** RBF at the plantation time (*top*) and after 1 year of operation (*bottom*)

The use of an FRB as the first stage of treatment resulted in high treatment efficiency, with a removal rate of 51 % for COD. The COD inlet concentration of the 2nd HF stage ranges from 603 to 6,320 mg L$^{-1}$ (except peak loads). The second HF stage has a HRT of 2.2 days and achieved a mean COD removal of 90 %, reaching an outlet concentration in the range 33–570 mg L$^{-1}$. The FWS provides a long residence time (about 5 days), ensuring a quite constant outlet quality. The mean outlet COD concentration was 67 mg L$^{-1}$. All the FWS outlet samples respected the required limits for discharge into freshwaters. The final sand filter further enhanced the effluent quality, removing the organic matter and the solids that can be generated in the FWS due to sediment resuspension and algal blooms. This stage, when activated, achieves a residual COD average removal efficiency of 40 %. The whole multistage CW system has a final effluent with a mean concentration of less than 70 mg L$^{-1}$ COD (average removal efficiency 98 %), 4.5 mg L$^{-1}$ P (average removal efficiency 50 %) and 0.25 mg S L$^{-1}$ sulphides.

Figure 13.4 shows the whole series of available data for both the treatment plant configurations that are shown here in the same graphic; during the 11 years, the daily flow and the treatment surface have both increased, and the obtained results are quite constant for the whole period as functions of the improvements made. During the critical period, when the daily flow was increasing and consequently the organic load to the 1st stage HF and the clogging of the HF were definitely diagnosed, still the 2nd stage FWS offered a valuable buffering action, keeping the outlet concentrations within the law limits with only few and very limited exceptions.

The only exceptional values in the observed period were two particular episodes of very high inlet concentrations (9,850 mg L$^{-1}$ COD and 6,678 mg L$^{-1}$ COD) due to accidental spill off of wine. However, the CWTP provided a high removal efficiency even under these peak loads (99.6 % and 98.9 %) producing an effluent with COD concentrations, respectively, of 43 mg L$^{-1}$ COD and 72.6 mg L$^{-1}$ COD.

Another relevant aspect of this type of CW treatment plant is the high environmental sustainability: the introduction of FRBs as the first stage of treatment allows

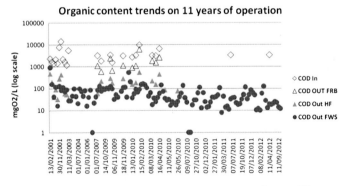

**Fig. 13.4** Historical trends of inlet and outlet COD concentrations for the different stages of the CWTP Cecchi – Castellina in Chianti

for absence of the primary sedimentation system. The plant does not produce sludge which needs to be removed and transported to the treatment facility. The sludge layer on the surface of the basins grew 1.5–2.0 cm per year and can be removed after 10–14 years. The French experience has shown that the sludge removed is well mineralized with a dry matter content of 21–28 % and an organic matter content of 34–50 %. These characteristics allow for reuse as agricultural compost directly by the wine producers, with optimized overall energy consumption if this approach is compared to the traditional treatment scheme by constructed wetlands and a primary treatment producing primary sludge that has usually to be transported elsewhere for its treatment.

In Fig. 13.5, outflow concentrations of Ntot, Ptot, MBAS and sulphides during 2012 are presented together with the discharge limits for all parameters.

The comparison of this specific system SRRs (surface removal rates), OLR (surface organic loading rate) and %R (removal efficiency) with the values obtained in Spain (de la Varga et al. 2013; Serrano et al. 2011), Slovenia (Zupancic Justin et al. 2009) and South Africa (Mulidzi 2010) brings the following observations:

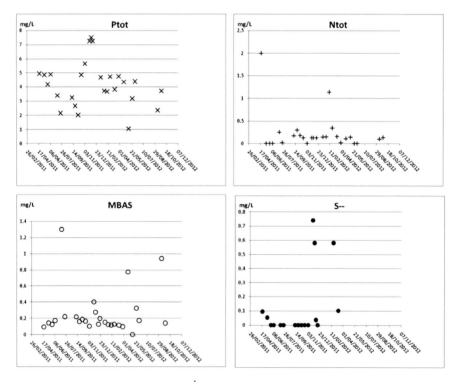

**Fig. 13.5** Outlet concentrations (mg L$^{-1}$) for Ptot (10), Ntot (20), MBAS (2) and sulphides (1) in 2012; in *brackets*, the discharge limits are shown (mg L$^{-1}$)

1. Figure 13.6 shows the data obtained for COD concentrations observed at every step of the treatment though the year 2012, by monthly sampling events and the related SRRs. The observed trends show relevant influence of the load on the FRBs removal rates, with the 1st stage buffering the inlet concentrations for all following stages, which are therefore less affected by the raw wastewater quality. In this present configuration, the multistage CW system is not affected by COD inlet concentrations >600 mg L$^{-1}$ as reported previously by de la Varga et al. (2013). Also, in the 2nd stage HF beds, the higher values of SRR were linked to the highest concentrations of about 1,000 mg L$^{-1}$ of COD. This represents the inlet SRR of 75 g COD m$^{-2}$ d$^{-1}$. This value is much higher than that of 15–20 g COD m$^{-2}$ d$^{-1}$ reported by de la Varga et al. (2013) in Spain.
2. The overall performances of the whole system seem more stable and slightly better as compared to Slovenian and South African systems, even though in both cases the inlet pH of the wastewater was lower by about 1–1.5 units than the monitored system in Italy.
3. Figure 13.7 shows the data obtained for Ntot concentrations observed at each step of the treatment system during the year 2012 and the related SRRs. The data indicate very good treatment efficiency, especially for the FRBs.

In the CWTP, the energy consumption is given by:

- FRB feeding pumps
- Mixer in the equalization tanks
- Pump in the FWS for feeding the sand filter and/or discharging into the river

Assuming the values described in Table 13.4, a total consumption per day of 29.6 kWh has been estimated and consequently a total annual energy consumption of 7,800 kWh (263 working days per year).

Making use of a primary sedimentation unit (i.e. Imhoff tank) and assuming a BOD$_5$ concentration value of 1,300 mg L$^{-1}$ with a daily flow of 70 m$^3$, the production of sludge would be 12 t year$^{-1}$ of dry matter. Activated sludge WTP for the 1,500 pe could generate about 22 t year$^{-1}$ dry matter, corresponding to 440 m$^3$ year$^{-1}$ of sludge with 5 % solids content. The power consumption for other alternatives such as ASTP or SBRs ranges from 55 to 80 kWh pe$^{-1}$ year$^{-1}$. Considering the organic load, the system can treat 1,500 pe, so the energy consumption can approximately be estimated at 82,500 kWh year$^{-1}$ (Table 13.5).

Moreover, the whole CWTP does not produce any waste and the harvested reeds (5–7 t year$^{-1}$) are used inside the farm as soil conditioner. The treatment by this CW configuration reduces the total CO$_2$ emissions in comparison with the alternatives listed above, with an approximate reduction of about 43 t year$^{-1}$ of CO$_2$ emissions and 91 % of saved energy (Table 13.5). For the first 3 years of operation, the measured sludge grow rate in FRBs was 0.8 cm year$^{-1}$.

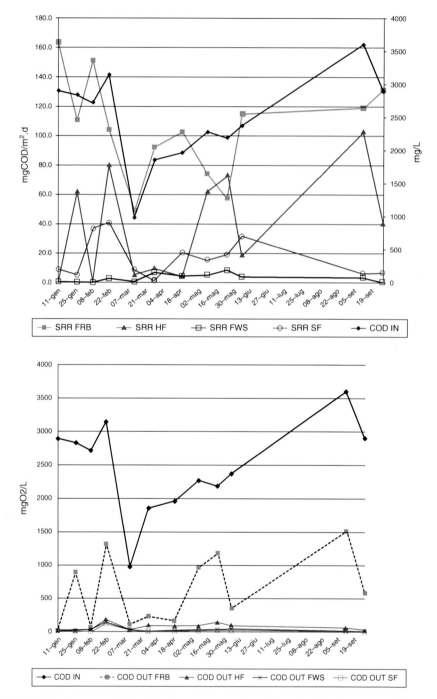

**Fig. 13.6** Surface removal rates (*Top*) and concentrations (*Bottom*) of the four stages of the new configuration compared to the COD inlet values in the year 2012

13 A New Concept of Multistage Treatment Wetland for Winery Wastewater... 199

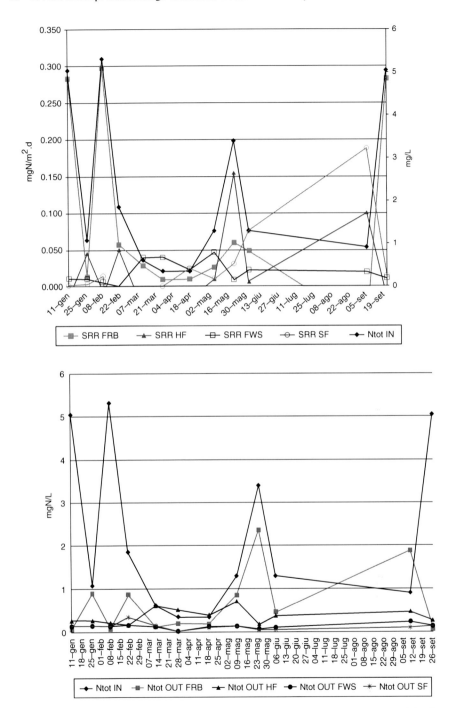

**Fig. 13.7** Inlet and outlet surface removal rates (*Top*) and concentrations (*Bottom*) of the four stages of the new configuration in the year 2012

**Table 13.4** Energy consumption in the winery CWTP

| Component | Nominal power (kW) | Working time (h d$^{-1}$) | Consumed energy (kWh d$^{-1}$) |
|---|---|---|---|
| **RBF pumps** | 7 | 1.6 | 11.2 |
| **Mixer** | 1.5 | 1.6 | 2.4 |
| **Effluent pump** | 2 | 8 | 16 |

**Table 13.5** Comparison of performances with a conventional WWTP

| | Activated sludge | Cecchi CWTP |
|---|---|---|
| Energy consumption (k Wh year$^{-1}$) | 82,500 | 7,800 |
| Residual sludge (t year$^{-1}$) | 22 | 0 |
| $CO_2$ emissions (t year$^{-1}$) | 47.5 | 4.5 |

## 13.4 Conclusions

The results from this study confirm the potential of multistage CWs as a suitable technology for treating winery wastewater and the efficiency of French systems as primary treatment for raw wastewater.

The introduction of FRBs improves the overall performances and the environmental benefits: the lower amount of sludge production reduces the $CO_2$ emissions due to sludge transport and treatment. The removal of more than half of the carbon content by aerobic reactors in the 1st stage such as FRBs substantially reduced the emission of $CH_4$ by the HF bed used as 2nd stage of the treatment. In this system, the water depth in the HF and in the configuration that had been in operation for the first 8 years before the upgrading averaged 0.7 m and a relevant part of the C removal took place by anaerobic processes.

The average OLR to the RBFs ranged in 2012 from 56 to 205 gCOD m$^{-2}$ d$^{-1}$, while the average COD removal was 75 %. For the following HF unit, the average OLR ranged from 2 to 107 gCOD m$^{-2}$ d$^{-1}$, while COD average removal was 87 %. Surface removal rates up to 232 gCOD m$^{-2}$ d$^{-1}$ were reached by the overall (FRB + HF + FWS) system. Overall system removals ranged from 96 to 99 % of COD.

The results from this study confirm the potential of CWs as a suitable technology for treating winery wastewater. The French system (FRB) applied as 1st stage has shown good performances in the treatment of raw winery wastewater, reaching removal efficiency up to 75 % and eliminating the clogging problems of the 2nd stage HF CWs.

## References

APHA. (1992). *Standard methods for the examination of water and wastewater analysis* (19th ed.). Washington, DC: American Public Health Association AWWA and WPCF.

Boutin, C., Liénard, A., & Esser, D. (1997). Development of a new generation of reed-bed filters in France: First results. *Water Science and Technology, 35*(5), 315–322.

de la Varga, D., Ruiz, I., & Soto, M. (2013). Winery wastewater treatment in subsurface constructed wetlands with different bed depths. *Water, Air, & Soil Pollution, 224*(4), 1–13.

Fernández, B., Seijo, I., Ruiz-Filippi, G., Roca, E., Tarenzi, L., & Lema, J. M. (2007). Characterization, management and treatment of wastewater from white wine production. *Water Science and Technology, 56*(2), 121–128.

Grismer, M. E., Ross, C. C., Valentine, G. E., Smith, B. M., & Walsh, J. L. (2001). Literature review: Food processing wastes. *Water Environment Research, 73*(5), CD-Rom.

Grismer, M. E., Carr, M. A., & Shepherd, H. L. (2003). Evaluation of constructed wetland treatment performance for winery wastewater. *Water Environmental Research, 75*(5), 412–421.

Kadlec, R. H., Knight, R. H., Vymazal, J., Brix, H., Cooper, P. F., & Haberl, R. (2000). *Constructed wetlands for pollution control: Processes, performances, design and operation* (IWA scientific and technical report no. 8). London, UK: IWA Publishing.

Masi, F., Conte, G., Martinuzzi, N., & Pucci, B. (2002). Winery high organic content wastewater treated by constructed wetland in Mediterranean climate. In *Proceedings of the 8th international conference on wetland system for water pollution control* (pp. 274–282). Arusha: IWA Publishing and University of Dar El Salaam.

Molle, P., Liénard, A., Boutin, C., Merlin, G., & Iwema, A. (2005). How to treat raw sewage with constructed wetlands: an overview of the French systems. *Water Science and Technology, 57*(9), 11–21.

Mulidzi, A. R. (2007). Winery wastewater treatment by constructed wetlands and the use of treated wastewater for cash crop production. *Water Science and Technology, 56*(2), 103–109.

Mulidzi, A. R. (2010). Winery and distillery wastewater treatment by constructed wetland with shorter retention time. *Water Science and Technology, 61*(10), 2611–2615.

Müller, D. H., Dobelmann, J. K., Hahn, H., Pollatz, T., Romanski, K., & Coppik, L. (2002). The application of constructed wetlands to effluent purification in wineries. In *Proceedings of the 8th international conference on wetland systems for water pollution control* (pp. 599–605). Arusha: IWA Publishing and University of Dar El Salaam.

Serrano, L., de la Varga, D., Ruiz, I., & Soto, M. (2011). Winery wastewater treatment in a hybrid constructed wetland. *Ecological Engineering, 37*, 744–753.

Shepherd, H. L., Grismer, M. E., & Tchobanoglous, G. (2001). Treatment of high-strength winery wastewater using a subsurface flow constructed wetland. *Water Environmental Research, 73*(4), 394–403.

Vymazal, J. (2009). The use constructed wetlands with horizontal sub-surface flow for various types of wastewater. *Ecological Engineering, 35*, 1–17.

Zupancic Justin, M., Vrhovsek, D., Stuhlbacher, A., & Griessler Bulc, T. (2009). Treatment of wastewater in hybrid constructed wetland from the production of vinegar and packaging of detergents. *Desalination, 247*, 101–110.

# Chapter 14
# Polishing of Real Electroplating Wastewater in Microcosm Fill-and-Drain Constructed Wetlands

Adam Sochacki, Olivier Faure, Bernard Guy, and Joanna Surmacz-Górska

**Abstract** This chapter presents a study of the performance of metals removal mechanisms in microcosm fill-and-drain constructed wetlands used for polishing of real-life electroplating wastewater. Two types of columns were used in the experiment: with compost or sand bed media. All the columns were planted with common reed. The main design goal of the experimental system was to promote precipitation of metals with biogenic sulfides. The feed of the system contained mainly metals (Al, Cu, Ni, Zn), B, and cyanides. The substrate from the columns was sampled after cessation of the experiment (56 weeks) and was analyzed using a sequential extraction procedure and a scanning electron microscopy–energy-dispersive X-ray spectroscopy (SEM-EDS) method. The studied columns were found to remove Al, Cu, and cyanides to a high extent, but were less efficient for other metals and B. The obtained results showed that the intended process for metals removal, which was the precipitation of metal sulfides, was responsible for binding only a minor fraction of metals. The major portion of metals was present as exchangeable or reducible fraction.

**Keywords** Constructed wetlands • Cyanide • Electroplating wastewater • Fill-and-drain • Metals

A. Sochacki (✉)
Environmental Biotechnology Department, Faculty of Power and Environmental Engineering, Silesian University of Technology, ul. Akademicka 2, PL-44 100 Gliwice, Poland

GéoSciences & Environnement Département, Ecole Nationale Supérieure des Mines, CNRS: UMR 5600, EVS, 158, cours Fauriel, F-42023 Saint-Etienne Cedex 2, France
e-mail: adam.sochacki@polsl.pl

O. Faure • B. Guy
GéoSciences & Environnement Département, Ecole Nationale Supérieure des Mines, CNRS: UMR 5600, EVS, 158, cours Fauriel, F-42023 Saint-Etienne Cedex 2, France

J. Surmacz-Górska
Environmental Biotechnology Department, Faculty of Power and Environmental Engineering, Silesian University of Technology, ul. Akademicka 2, PL-44 100 Gliwice, Poland

© Springer International Publishing Switzerland 2015
J. Vymazal (ed.), *The Role of Natural and Constructed Wetlands in Nutrient Cycling and Retention on the Landscape*, DOI 10.1007/978-3-319-08177-9_14

## 14.1 Introduction

The occurrence of metals in industrial wastewater is a matter of concern because these contaminants are often present at significant levels and, if discharged directly into the environment, can have severe effects on the environment and public health. Moreover, metals at toxic levels may inhibit biological processes in activated sludge systems (Kiptoo et al. 2004). The composition of electroplating wastewater depends on the type of coatings applied and the coated objects used in electroplating workshops, but various metals, cyanides, and sulfates are inherently present in electroplating wastewater. High concentrations of metals and cyanides, often in hundreds of mg L$^{-1}$ (Bartkiewicz 2006), prevent direct application of constructed wetlands (CWs) for treatment of electroplating wastewater. Pretreated electroplating wastewater could be polished in CWs; however, carbon deficiency of this wastewater makes this treatment a significant challenge. This is because the precipitation of metals as sulfides, which appear to be the most promising sink for metals, is dependent on heterotrophic bacteria, namely, sulfate-reducing bacteria (SRB) (Liamleam and Annachhatre 2007). The process of metal sulfide precipitation is possible in CWs when the requirements for the growth of SRB are met, i.e., the presence of reducing conditions, sulfates, and simple organic compounds (such as acetates or lactates). When only complex organic matter is available (e.g., peat or compost), SRB are dependent on the other groups of bacteria such as fermentative bacteria. The volatile fatty acids utilized by SRB are generated by fermentative bacteria as a product of their metabolism (Sturman et al. 2008). The fermentative bacteria utilize simple organic compounds obtained by enzymatic hydrolysis of detrital matter composed of cellulose, hemicellulose, proteins, lipids, waxes, and lignin (Reddy and DeLaune 2008). On the other hand, SRB may also compete with other groups of bacteria. Methanogens and SRB share many ecological and physiological similarities and they often coexist in reducing zones of CWs. When the source of an electron donor is not limited, both methanogens and SRB can thrive because competition is greatly reduced. There are several factors that govern this competition, e.g., pH, COD/SO$_4^{2-}$ ratio, substrate type, or temperature (Khanal 2008). Removal of metals in sulfate-reducing wetlands was reported to follow two mechanisms: the adsorption of metals onto organic matter and the formation of metal-sulfide precipitates. Adsorption onto organic matter appears to occur much more rapidly than sulfide precipitation. Metals adsorbed onto organic matter appear to convert slowly to sulfide precipitates (Wallace and Knight 2006). Also, the SRB biofilm is able to attract metals and bind them by several mechanisms (Reddy and DeLaune 2008). It is noteworthy that sulfide precipitation itself is a multistage process, which initially consists in deposition of non-stoichiometric phases that subsequently undergo diagenetic reorganization to crystalline forms. Sulfides are (meta)stable under anaerobic conditions but undergo decomposition at the oxic--anoxic interface, which leads to release of covalently bound metals and co-precipitated and adsorbed metals or metalloids (Lens et al. 2007). In general, metal sulfides are less soluble than their carbonate or hydroxide counterparts, achieving more complete precipitation and stability over a broader pH range (Blais et al. 2008). It can be stated that the solubility of metal sulfides is extremely low providing that anaerobic conditions prevail. In a CW or a reactor with suspended SRB biomass, it can be observed that the production of $H_2S$ exceeds its consumption by metal precipitation. Excessive amounts of $H_2S$ (or $HS^-$) may adversely affect the environment (toxicity to aquatic live and foul smell of gaseous form) and increase metal mobility in the system (Lewis 2010).

The two approaches mostly employed and used for the study of the spent bed media are mineralogical analysis and chemical extractions (Neculita et al. 2008b). Direct determination of the chemical form of metal in solid-phase environmental samples can be achieved by means of various instrumental techniques. Only a few methods are directly applicable to the mineralogical analysis of wetland bed media due to the poor crystallinity of the precipitates and poor detection limits (Neculita et al. 2008b). These methods include scanning electron microscopy equipped for backscattered electron imaging (SEM-BSE), which was the most reliable tool, whereas X-ray diffraction (XRD) or Mossbauer analyses have been of lesser applicability in detecting amorphous metal sulfides (Machemer et al. 1993). The SEM approach coupled with energy-dispersive X-ray spectroscopy (SEM-EDS) proved successful for identifying sulfides in reactive mixtures from CWs (Machemer et al. 1993; Neculita et al. 2008b). Neculita et al. (2008b) used the SEM-EDS and XRD techniques to evaluate mineralogy of spent reactive mixtures withdrawn from bench-scale sulfate-reducing column reactors. Gammons and Frandsen (2001) examined the solids from anaerobic CWs using SEM and XRD. The second approach includes a sequential extraction procedure (SEP), which consists in application of a series of reagents added to the same sample to subdivide the total metal content. Each step of the procedure tends to be more vigorous than the previous one, starting with initial mild conditions (e.g., shaking with water, a salt solution or dilute acetic acid) to end with much harsher reagents (e.g., hot mineral acid). Generally, the potentially toxic elements extracted early during the procedure are those most weakly bound to the solid phase (Bacon and Davidson 2008). The applicability of chemical analysis for anoxic sediments has been widely discussed in the literature. It was suggested that commonly applied sequential extraction protocols may underestimate the amount of sulfides. The chemical analyses of anoxic sediments should be evaluated using X-ray absorption spectroscopy or SEM-EDS (Peltier et al. 2005). The distribution, mobility, and bioavailability of metals in the environment depend on their concentration and also (to an even larger extent) on the form in which they are bound to a solid phase. The short-term or long-term changes of the chemical−physical conditions in the environment may increase the mobility of metals causing contamination. The trait of being bioavailable that is characteristic for metals is associated with many variables such as characteristics of the particle surface, on the type of binding, and properties of the solution in contact with the solid phase (Filgueiras et al. 2002). Another crucial factor deciding whether an element will be incorporated into an organism is its physiology. As stated by Bacon and Davidson (2008), functional speciation depends strongly on the context and aim of the experiment so that the same metal pool could be considered as "bioavailable" when plant uptake is of interest but "mobile" or "labile" in leaching studies.

The goal of the experiment was to study the feasibility of polishing electroplating wastewater in the fill-and-drain (FaD) microcosm CWs and to elucidate the metals removal mechanism based on the speciation of metals in the microcosm systems studied.

## 14.2 Methods and Materials

### 14.2.1 Experimental System

The experimental system comprised two types of FaD columns distinguished by the type of bed media used: mineral or organic. The columns used in the experiment were made of PVC and were 80 cm high and 20 cm in diameter. Each type of a column was duplicated, thus four columns were used in the experiment. The columns were filled up to 63.5 cm with silica quartz filtration sand (1–2.5 mm, more than 5 % of free silica) or equivolume compost-sand mixture (hereafter referred to as "compost"). The compost which was used was composed of composted manure, algae and bark, and also brown and white peat. The other properties of the compost were organic matter content 50 % in DM, conductivity 45 mS m$^{-1}$, pH 6.8, and water-holding capacity 700 mL L$^{-1}$. The bed media were inoculated with SRB by adding cattle manure obtained from a local agricultural farm. All the columns were planted with potted seedlings (three per column) of *Phragmites australis* (Cav.) Trin. ex Steud (Common reed). The columns with compost are hereafter denoted as FaD-CP (**f**ill-**a**nd-**d**rain columns with **c**ompost and **p**lants), and the columns with sand are denoted as FaD-SP (**f**ill-**a**nd-**d**rain columns with **s**and and **p**lants). The columns were fed manually from the top and drained from the bottom by opening a valve. A cycle in this type of batch columns consisted of filling phase (1 h), holding phase (2 weeks), and draining phase (1 h). During the draining phase, air was sucked into the bed media, thereby promoting aerobic transformation until onset of anaerobic conditions as soon as oxygen is depleted. The height of water layer above the bed media was 6.5 cm, thus total water height during holding phase was also 70 cm. The system was situated at the premises of an electroplating plant in the region of Auvergne, France, and has been operated for 56 weeks. The system was operated under the outdoor conditions from June to October 2011 and April to June 2012. For the autumn–winter season, the system was moved into the building of a WWTP of the company.

The system was fed with real wastewater originating from an electroplating plant and containing Al, B, Cu, Ni, Fe, Zn, cyanides, and S as the contaminants of concern. The system was fed with final effluent wastewater (feed A) and pretreated wastewater (feed B), which were fed to the system from one-cubic-meter containers. Feed A was fed from June 2011 to June 2012, and feed B was fed during

May and June 2012 to the columns which were previously fed with feed A. In May and June 2012, columns were fed in parallel by both types of influent (feed A and feed B); therefore, they were not replicated.

The main goal of the system design was to promote bacterial sulfate reduction and subsequent precipitation of metals with biogenic sulfides. To this end, organic bed medium (compost) was used in selected columns to provide organic carbon for SRB and the operation of columns was favoring anaerobic conditions.

### 14.2.2 Wastewater Analysis

Metals (Al, Cu, Fe, Mn, Ni, Zn), B, and S were analyzed with inductively coupled plasma atomic emission spectrometry (ICP-AES; Horiba Jobin-Yvon JY138 Ultrace apparatus) after acidifying filtered (0.22 μm) samples with 65 % $HNO_3$ to pH < 2. The detection limits for metal analysis by ICP-AES were 0.001, 0.01, 0.001, 0.002, 0.001, 0.002, 0.2, and 0.001 mg/L for Al, B, Cu, Fe, Mn, Ni, S, and Zn.

The concentration of free cyanide in filtered samples was determined photometrically at 588 nm with Hach Lange cuvette test LCK 315 (measuring range 0.01–0.60 mg/L $CN^-$) using Hach Lange DR2800 spectrophotometer. Chemical oxygen demand (COD) in filtered samples was determined by Hach Lange cuvette test LCI 500 according to ISO method 15705 (2002) using Hach Lange DR2800 spectrophotometer. Oxidation-reduction potential (ORP) was measured (with a redox electrode) in the water samples from the bottom, top, and water layer of the columns directly after withdrawal. pH was measured in the influent and effluent using an electrode connected to a pH-meter WTW 330.

### 14.2.3 Substrate Analysis

#### 14.2.3.1 Sequential Extraction Procedure

The manner in which metals (and S) were bound to substrate in the experimental system was assessed by the SEP based on the BCR2 protocol (Rauret et al. 1999). This procedure consists of three steps devised to extract three operationally defined fractions. In the first step, exchangeable, water- and acid-soluble fraction was targeted and 0.11 M acetic acid was used. The second step was to extract reducible phase by the use of 0.5 M hydroxylammonium hydrochloride at pH 1.5. The target phase in step 3 was reducible fraction and reagents used were 8.8 M $H_2O_2$ at 85 °C and then 1 M ammonium acetate at pH 2.0. As an internal check on the procedure, the residue from step 3 was digested in *aqua regia* (step 4), and the amount of metals extracted (i.e., sum of step 1, step 2, step 3, and residue) was compared with that obtained by independent *aqua regia* digestion of a separate sample of the sediment. The SEP was conducted with approx. 1 g of substrate accurately weighed in 50 mL polypropylene centrifuge tubes, to which extracting solutions were added in a volume twice as small as in the BCR2 protocol. Substrate samples were taken from the bottom and top of a column. It was assumed that these zones were represented by 150 mm layer from the bottom or top surface. Each step (1–3) involved shaking the samples with extracting solution for 16 h on the end-over-end shaker. The extract from the solid residue was separated by centrifugation at 3000 g for 20 min, and the supernatant was decanted into a PE container, filtered, and stored at 4 °C for ICP-AES analyses (Horiba Jobin-Yvon JY138 Ultrace apparatus). The residue was washed by adding 20 mL of distilled water, shaking for 15 min, centrifuging for 20 min, and then discarding the supernatant. Reaction with $H_2O_2$ in step 3 and *aqua regia* digestion were performed in 30 mL PTFE vials. The *aqua regia* digestion was based on the protocols of ISO 11466 (1995) and Sastre et al. (2002). In this procedure 1 g of sample was weighed into 30 mL PTFE vials and 12 mL of *aqua regia* was added (9 mL of 37 % HCl and 3 mL of 65 % $HNO_3$) and digested at room temperature for 16 h. Afterward the suspension was digested at 130 °C for 2 h in closed vials. The obtained suspension was filtered through an ashless filter and then diluted to 32 mL with distilled water and stored in PE bottles at 4 °C for analyses. The amount of metals extracted (and S) in the independent *aqua regia* digestion was assumed to represent pseudo-total concentration of metals in the substrate. For the evaluation of measurement precision and accuracy, the certified reference material (soil) NCS ZC73006 (China National Analysis Centre for Iron and Steel) was used. The reproducibility expressed as a relative standard deviation of the quality control samples were less than 10 % (n = 7). Average recoveries between analyzed and certified reference material values, for steps 1–4 of the BCR protocol, were (n = 8): Al 39.2 %, Cu 65.52 %, Fe 57.5 %, Mn 83.4 %, Ni 65.7 %, S 137.4 %, and Zn 67.4 %. The recoveries between the sum of steps 1–4 and single-step *aqua regia* digestion of substrate samples were (n = 27) Al 82.12 %, Cu 105.41 %, Fe 70.19 %, Mn 94.10 %, Ni 108.82 %, S 85.87 %, and Zn 180.46 %.

The low recovery for some of the elements in comparison with the certified reference material is not unusual. *Aqua regia* digestion, which is recommended in the BCR2 protocol to extract the residual fraction, is not sufficient to achieve complete dissolution of siliceous and refractory materials (Larner et al. 2006). Recoveries as low as 39 % for Al and 57 % for Cr were obtained in the *aqua regia* digestion of a certified reference material by Larner et al. (2006). The discrepancies between the amount of metals determined by steps 1–4 of the BCR protocol and the independent *aqua regia* digestion may result from the method but also from the properties of the sample. The former aspect is widely discussed in the literature (Filgueiras et al. 2002; Bacon and Davidson 2008), whereas the latter may be related to poor homogeneity of the samples and the presence of fine organic debris that could be lost from the extracting solution. Li et al. (2010) found recoveries between 87 and 124 % for Zn, Cu, Pb, and Cd to be indicating "good agreement" between the sum of steps 1–4 and the independent *aqua regia* digestion. Hang et al. (2009) extracted metals from river sediments with recoveries between 82 and 119 %.

### 14.2.3.2 Scanning Electron Microscopy-Energy Dispersive Spectroscopy

Surface morphology of the samples was performed using a scanning electron microscope (SEM) HITACHI S-3400 N with backscattered electron detector (BSE). The analysis of chemical composition of the samples was performed using energy-dispersive X-ray spectroscopy (EDS) Thermo Noran System (System Six). EDS analysis is a semiquantitative method and the limit of detection is ca. 0.2 atomic % depending on element. SEM imaging and chemical microanalysis were performed in a low vacuum (50 Pa) and at accelerating voltage of 15 kV. Similarly to the procedure for SEP substrate samples were taken from the bottom and top of a column. It was assumed that these zones were represented by 150 mm layer from the bottom or top surface. The samples analyzed by SEM-EDS were subdivided from the samples taken for the SEP analysis and then placed on titanium sample mounts covered with carbon tape.

## 14.3 Statistical Analysis

Most of the subsets of the performance data had non-normal distribution; therefore, nonparametric statistical tests were used to compare these subsets. Median and the median absolute deviation (MAD) were used as descriptors of the central tendency and dispersion of data distributions for non-normally distributed data. The Shapiro–Wilk W test was employed to test for normality of the data. Dependent groups were analyzed using the Wilcoxon matched pair test, and independent groups were analyzed using the Mann–Whitney $U$ test. Differences were considered statistically significant if $p < 0.05$. Statistical testing was performed using the STATISTICA 10 software (StatSoft, Inc 2011).

## 14.4 Results and Discussion

### 14.4.1 Wastewater Polishing Efficiency

Influent and effluent quality and the corresponding treatment efficiency in the FaD columns are presented in Tables 14.1 and 14.2, for feed A and feed B, respectively. Both types of influents had similar qualitative composition with the exception of cyanides, which were present only in feed B. The other contaminants were metals: Al, Cu, Ni, Zn, Fe, and Mn; B, S, and organics measured as COD. Feed A was fed to the columns throughout all of the experiment and it was the treated effluent that was discharged by the electroplating plant into a stream. Concentration of B, S, Mn, Ni, and Zn was higher in this type of wastewater compared to feed B. Feed B was

**Table 14.1** Overall inlet and effluent concentrations (median ± MAD, mg L$^{-1}$) and removal efficiencies (median ± MAD, %) of contaminants in the FaD columns fed with feed A, (n = 6), weeks 1–56

| Parameter | Influent (feed A) | Effluent from FaD-CP CWs Concentration | Effluent from FaD-CP CWs Removal[a] | Effluent from FaD-SP CWs Concentration | Effluent from FaD-SP CWs Removal[a] |
|---|---|---|---|---|---|
| Al | 0.264 ± 0.638 | 0.063 ± 0.040 | 79.1 ± 10.7 | 0.064 ± 0.048 | 84.4 ± 13.5 |
| Cu | 0.083 ± 0.016 | 0.003 ± 0.002 | 97.2 ± 1.4 | 0.001 ± 0.001 | 98.9 ± 0.3 |
| Fe | 0.012 ± 0.007 | 0.258 ± 0.170 | −1912.5 ± 1954.1 | 0.648 ± 0.531 | −8117.1 ± 7999.5 |
| Mn | 0.003 ± 0.001 | 0.848 ± 0.125 | −19842.8 ± 5469.3 | 0.199 ± 0.044 | −7017.2 ± 4114.6 |
| Ni | 0.058 ± 0.013 | 0.013 ± 0.004 | 77.9 ± 7.4 | 0.013 ± 0.004 | 80.3 ± 5.4 |
| Zn | 0.022 ± 0.015 | 0.054 ± 0.030 | −167.6 ± 89.3 | 0.046 ± 0.015 | −296.0 ± 277.0 |
| B | 5.43 ± 0.14 | 3.47 ± 0.46 | 37.7 ± 28.5 | 5.34 ± 0.15 | 1.4 ± 11.0 |
| COD | 38.0 | 108.7 | −186.6 | 84.6 | −123.1 |
| S | 538.9 ± 4.7 | 375.9 ± 85.4 | 18.3 ± 3.4 | 515.7 ± 43.1 | 4.3 ± 8.9 |
| pH | 8.5 | 6.74 | | 6.59 | |

[a]Negative values indicate release of a given element

**Table 14.2** Overall influent and effluent concentration (median ± MAD, mg L$^{-1}$) and reduction of contaminants (median ± MAD, %) in columns fed with feed B, (n = 4), weeks 45–56

| Parameter | Influent (feed B) | Effluent from FaD-CP CWs Concentration | Removal[a] | Effluent from FaD-SP CWs Concentration | Removal[a] |
|---|---|---|---|---|---|
| Al | 2.768 ± 0.111 | 0.242 ± 0.094 | 91.4 ± 3.0 | 0.270 ± 0.097 | 90.4 ± 3.6 |
| Cu | 2.315 ± 0.251 | 0.003 ± 0.003 | 99.9 ± 0.1 | 0.005 ± 0.003 | 99.9 ± 0.1 |
| Fe | 0.033 ± 0.001 | 0.059 ± 0.008 | −140.8 ± 76.5 | 0.270 ± 0.026 | −724.6 ± 73.6 |
| Mn | 0.002 ± 0.001 | 0.439 ± 0.010 | −24008.5 ± 2053.0 | 0.052 ± 0.004 | −2547.2 ± 486.1 |
| Ni | 0.037 ± 0.012 | 0.032 ± 0.006 | 28.2 ± 17.3 | 0.044 ± 0.004 | −47.0 ± 55.4 |
| Zn | 0.015 ± 0.009 | 0.040 ± 0.004 | −304.2 ± 320.4 | 0.032 ± 0.023 | −332.3 ± 394.4 |
| B | 4.03 ± 0.03 | 2.21 ± 0.38 | 46.8 ± 7.4 | 4.36 ± 0.50 | −1.9 ± 13.5 |
| free CN | 1.835 ± 0.090 | 0.255 ± 0.059 | 86.0 ± 4.0 | 0.246 ± 0.033 | 86.9 ± 2.6 |
| COD | 40.2 | 244.5 | −488.3 | 145.5 | −299.3 |
| S | 393.9 ± 0.1 | 293.5 ± 30.6 | 21.4 ± 7.6 | 350.0 ± 10.9 | 7.9 ± 3.4 |
| pH | 9.05 | 6.88 | | 7.22 | |

[a]Negative values indicate release of a given element

characterized by elevated concentration of Al, Cu, and detectable amounts of cyanide. This influent was the pretreated wastewater taken from a metal-precipitation step of the treatment in the WWTP of the electroplating plant.

The removal efficiency for metals was in the decreasing order of Cu > Al > Ni. The removal of Al and Cu was invariably very high, > 79 % and >90 %, respectively. The removal efficiency for Al and Cu was higher in the columns fed with feed B ($p < 0.05$). In contrast, the removal of Ni was lower in the columns fed with feed B. Ni was even released from the FaD-SP column fed with this influent in the last 12 weeks of the experiment. Fe, Mn, and Zn were released from both types of the FaD columns. Interestingly, the Ni concentration was lower in feed B, however, not statistically different from feed A. Thus, it can be assumed that the decreased removal of Ni can be attributed to the effect of the other pollutants present in feed B in higher concentrations, namely, Al, Cu, and cyanide, or pH of this influent. The experimental system was also very efficient in terms of cyanide removal amounting to $\geq 86$ % in both FaD columns fed with feed B. Statistically significant differences between the treatment efficiency in the FaD-CP and FaD-SP columns were observed for B, S, and Mn for both types of influent. Reduction of both B and S was higher in the columns filled with compost. The removal of B was markedly higher in FaD-CP columns for both types of wastewater. Enhanced removal of B in the FaD-CP columns may be attributed to its sorption to organic matter within the bed of the columns. Higher removal efficiency of B was recorded for the FaD-CP columns fed with feed B, in which B concentration was lower by 1.4 mg/L. The FaD-SP columns were inefficient in the removal of B with almost no reduction or slight release. The removal of S was much higher in the FaD-CP columns, which might be related to the stimulative effect of internal carbon source on SRB. This, however, had no effect on the removal of metals. It was observed that the COD value of the treated wastewater increased by one- to twofold. This was caused by the presence of manure in the FaD-SP CWs and both manure and compost in the FaD-CP CWs, in which the release was higher. The release of Mn was higher in the FaD-CP columns by an order of magnitude. This might be related to the fact that the Mn concentration in compost was tenfold higher compared to sand (see Sect. 15.3.2). Apart from Mn, also Fe and Zn were released from both FaD columns. The release of Mn was higher for the FaD-CP column and Fe for FaD-SP column and was ten- to hundredfold higher as compared to the influent concentration. The release of Fe and Mn can probably be attributed to dissolution of Fe and Mn oxides and hydroxides, which is promoted under anaerobic conditions. In contrast to Fe and Mn, the release of Zn was only one- to threefold higher than the influent concentration. The provenance of Zn is probably different, as Fe and Mn were probably present in the raw bed media, and Zn was mainly desorbed after the sorption capacity of the media was exhausted. The composition of raw bed media was analyzed and is presented and discussed in Sect. 14.4.2.

As shown in Tables 14.1 and 14.2, it is noteworthy that electroplating wastewater is a poor source of carbon for stimulation of the SRB activity. COD as a lumped parameter does not determine whether organic matter could be utilized by heterotrophic bacteria. Speciation analyses were not performed but it is likely that organic matter in the real electroplating wastewater could not be readily available for bacteria. The COD of the wastewater could be related to residual organic auxiliary compounds used in electroplating, which can be slowly biodegradable or biorecalcitrant. Considering the potential for bacterial sulfate reduction, it should be mentioned that S content was high enough for this process to occur; thus, carbon was the limiting substrate.

### 14.4.2 Metal Removal Mechanisms

#### 14.4.2.1 Sequential Extraction Procedure

The SEP was performed for the substrates sampled from the top and bottom of the columns after cessation of the experiment. The presented results are for the FaD columns fed with feed B, for the FaD-CP column in Fig. 14.1 and for the FaD-SP column in Fig. 14.2.

Aluminum was bound mostly within the residual fraction, which ensures its stability in substrate. However, most of the residual fraction could originate from the raw material either sand or compost. It was observed that the reducible fraction was higher in the top of the substrate, probably due to the effect of rhizosphere. Copper was bound to more stable fractions, with the exception of the bottom of the FaD-SP column where most of Cu is bound with exchangeable fraction. This "strong binding" of Al and Cu corresponds to their high removal. The results presented in Figs. 14.1 and 14.2 show that most of Mn, Zn, and Ni is present in exchangeable fraction which can be easily released from the columns filled with either organic and mineral media. These findings are in agreement with the results presented in Table 14.2, which shows that these elements were released from all the FaD columns fed with feed B. It should be noted that the exchangeable fraction of metals and S is generally higher in the substrate of the FaD-SP column. The exception is the fractionation of Zn and Mn. This may stem from the fact that the exchangeable fraction of Mn and Zn was higher in compost than in sand as shown in Fig. 14.3.

When the fractionation of metals and S in the substrate from the FaD-CP column and the raw compost were compared, it was observed that: (i) the exchangeable fraction of Al increased slightly and also that the residual fraction increased; (ii) for Cu the reducible fraction increased and residual fraction diminished; (iii) the residual fraction of Fe diminished markedly in the bottom part (by ca. 50 %) and slightly in the top part (by ca. 10 %), and the reducible fraction (less than 10 % in the raw material) increased (by 50 %) in the bottom layer and by cc. 20 % in the top layer; (iv) the Mn exchangeable fraction increased by ca. 15 % in the bottom part, and decreased by 10 % for the residual fraction; (v) the Ni residual fraction decreased by more than 20 %, and the exchangeable fraction increased by more than 30 %; (vi) for S the decrease of the oxidizable fraction was observed, and the increase of the exchangeable fraction by more than 30 % in the bottom part and less

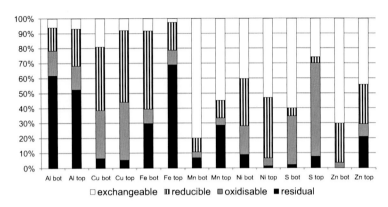

**Fig. 14.1** Metal and S fractions (by SEP) in *top* and *bottom* (bot.) substrate layers of the FaD-CP column fed with feed B

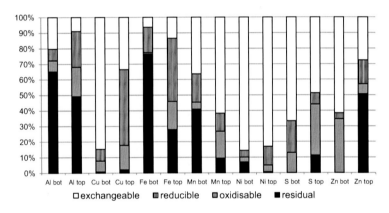

**Fig. 14.2** Metal and S fractions (by SEP) in *top* and *bottom* (bot.) substrate layers of the FaD-SP column fed with feed B

than 5 % in the top part; and (vii) the increase of the residual fraction of Zn in the top part by less than 5 % was observed and the increase of the exchangeable fraction by less than 20 %, in the bottom part of the column the exchangeable fraction of Zn increased by almost 50 % and the residual fraction was approx. 0 %.

When the fractionation of metals and S in the substrate from the FaD-SP column and the raw compost were compared, it was observed that: (i) the exchangeable fraction of Al increased slightly by about 15 % in the bottom part and by less than 5 % in the top part of the column, the residual fraction decreased by about 10 % and 25 % for the bottom and the top, respectively, the residual fraction is less than 5 % in both top and bottom of the column, and the exchangeable fraction increased by about 85 % in the bottom part and by about 35 % in the top part; (ii) for Fe and Mn the residual fraction decreased more in the top parts; (iii) for Ni the residual fraction decreased by about 50 %; (iv) the exchangeable fraction of S increased by at least about 20 % and the reducible fraction diminished proportionally in the bottom part,

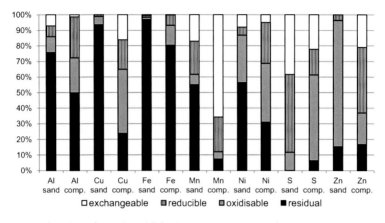

**Fig. 14.3** Fractionation of metals and S in the raw substrates: sand and compost (comp.)

**Table 14.3** Pseudo-total concentration of metals (mg kg$^{-1}$ DW) in the bottom and top layers of substrate sampled from the FaD columns fed with feed B and in the raw materials (sand and compost)

| Column | Sample location | Al | Cu | Fe | Mn | Ni | S | Zn |
|---|---|---|---|---|---|---|---|---|
| FaD-CP | Bottom | 111.2 | 1.13 | 260.8 | 6.02 | 0.63 | 293.2 | 4.09 |
|  | Top | 846.5 | 70.96 | 4578 | 33.7 | 23.1 | 865.4 | 15.43 |
| FaD-SP | Bottom | 28.6 | 2.81 | 129.5 | 0.92 | 1.13 | 88.98 | 1.47 |
|  | Top | 155.8 | 58.16 | 2852.2 | 7.12 | 17.6 | 358.2 | 7.50 |
| Compost[a] | Raw material | 1113.1 | 2.45 | 1186.9 | 42.8 | 2.26 | 211.1 | 9.2 |
| Sand | Raw material | 148.3 | 0.018 | 1464.7 | 4.83 | 1.25 | 17.36 | 4.57 |

[a]Compost–sand mixture

whereas in the top part the residual and oxidizable fraction increased markedly and the exchangeable fraction increased by 10 %; and (v) the residual fraction of Zn in the top part of the column increased by about 35 % and the exchangeable fraction increased by ca. 25 % and in the bottom part the exchangeable fraction of Zn increased by 60 % and the oxidizable fraction decreased proportionally, so that the residual fraction was practically completely washed out.

The amount of metals and S extracted in the single-step *aqua regia* digestion was assumed to represent a pseudo-total concentration of these elements in the substrates (Table 14.3). *Aqua regia* is not able to liberate the metals that are silicate-bound, but the use of this reagent to give a "pseudo-total" analysis of the substrate is acceptable in this case as the silicate fraction was not of interest in this study (Dean 2003).

Generally, the concentration of metals and S is higher in the substrate of the FaD-CP column than in the FaD-SP column. This is because the initial concentration of these elements was higher in the compost than in the sand. This observation is not viable for all the metals, for example, Fe, Ni, and Cu. The initial and the final concentrations of the elements allow for observation whether they were released or stored in the substrate. The release of the element might be attributed to its mobilization and subsequent fate such as washout from the system, uptake by plants, or removal in the downstream parts of the column. In Table 14.3 it can be seen that all the elements, save S and Cu, were leached to some extent from the bottom part of the FaD columns. Most of the elements (in terms of concentration) were accumulated in the top part of the FaD columns. The exceptions were Al and Mn in the FaD-CP, which were leached from the top part of this column. These differences between the top and bottom parts of the FaD columns may result probably from the fact that the influent was fed from the top; therefore, major particulate fraction of the elements was entrapped in the top part of the FaD columns. The dense system of rhizomes and roots in the upper part of the columns was presumably playing a major role in filtering out the contaminants and further removal in the oxidized root zone. The soluble fraction of contaminants was probably adsorbed on the upper part of the column, either by the substrate or the plant biomass, and could be precipitated as sulfides or oxy(hydroxides), depending on the ORP.

### 14.4.2.2 SEM-EDS Analysis of the Substrate

The substrate from the FaD-CP column fed with feed B was analyzed by the SEM-EDS method, taking into account the location of sample: bottom and top of the column. The results of the SEM-EDS analysis of the substrate sampled from the bottom of the column are shown in Fig. 14.4 and in Table 14.4. Figure 14.4 presents two SEM-BSE images of the sample with numbered areas analyzed by EDS. These areas are indicated by white or black rectangles and corresponding numbers, which were used interchangeably depending on the background shade in order to increase clarity. Table 14.4 shows the results of the EDS analyses of the areas indicated in Fig. 14.4. The samples in Table 14.4 are named according to the name of a sample

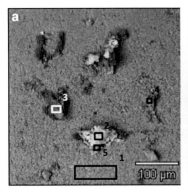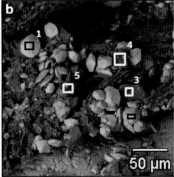

**Fig. 14.4** SEM-BSE images of wetland substrates sampled from the bottom (bot) of the FaD-CP column fed with feed B (**a**) Sample FaD-CP-bot(1), (**b**) Sample FaD-CP-bot(2)

14 Polishing of Real Electroplating Wastewater in Microcosm Fill-and-Drain...          217

**Table 14.4** Elemental (atomic %) composition of the areas (ar#) indicated in the images in Fig. 14.4 determined using EDS

| Sample_area | Atomic %[a] |||||||||||
|---|---|---|---|---|---|---|---|---|---|---|---|
| | Na | Mg | Al | Si | P | S | Cl | K | Ca | Ti | Mn | Fe |
| FaD-CP-bot(1)_ar1 | – | – | 1.9 | 98.1 | – | – | – | – | – | – | – | – |
| FaD-CP-bot(1)_ar2 | – | – | 2.1 | 49.6 | 0.4 | 0.7 | – | 1.1 | 0.5 | – | 42.2 | 3.3 |
| FaD-CP-bot(1)_ar3 | 1.7 | 0.7 | 7.6 | 75.7 | 2.0 | 2.7 | – | 1.3 | 0.7 | 0.5 | | 7.2 |
| FaD-CP-bot(1)_ar4 | 2.5 | – | 8.5 | 58.0 | 1.7 | 2.7 | – | 1.1 | 0.8 | – | 17.8 | 6.9 |
| FaD-CP-bot(1)_ar5 | 3.4 | 0.7 | 5.6 | 73.0 | 1.1 | 3.5 | – | 1.3 | 1.8 | – | 5.3 | 4.3 |
| FaD-CP-bot(2)_ar1 | 1.2 | 0.2 | 1.0 | 11.8 | – | 44.1 | – | | 41.0 | – | – | 0.7 |
| FaD-CP-bot(2)_ar2 | 0.8 | – | 0.6 | 14.4 | – | 42.4 | – | 0.3 | 41.0 | – | – | 0.6 |
| FaD-CP-bot(2)_ar3 | 9.7 | 1.9 | 4.2 | 21.3 | – | 31.8 | 0.9 | 2.7 | 24.6 | 0.4 | – | 2.4 |
| FaD-CP-bot(2)_ar4 | 12.8 | 2.3 | 5.2 | 34.1 | 0.9 | 24.6 | – | 3.2 | 11.6 | 1.2 | – | 4.0 |
| FaD-CP-bot(2)_ar5 | 11.9 | 2.5 | 9.9 | 40.0 | 2.1 | 14.8 | 0.6 | 3.4 | 5.5 | 0.8 | – | 8.5 |

[a]Compositions normalized to 100 %

written below the images in Fig. 14.4. The names of the samples include the name of a column and sampling spot (top or bottom). In Table 14.4 the areas for the left-hand image (Fig. 14.4a) were listed first.

The FaD-CP-bot(1) sample contains mainly detrital matter associated with Al silicate and probably Fe-Mn oxy(hydroxides). The composition of area 1 of this sample corresponds to the composition of the sand used in this study (Table 14.8). The image of sample FaD-CP-bot(2) presents a cluster consisting of several phases: calcium sulfates probably with detrital grains of quartz and other silica minerals. Also the presence of Al hydroxysulfate and carbonates cannot be excluded.

The SEM-BSE images of the samples taken from the top part of the FaD-CP column are presented in Fig. 14.5, and the elemental composition of the selected areas are listed in Table 14.5.

The FaD-CP-top(1) sample contained small particles of Fe (areas 1, 2 and 5) bound with various elements disallowing determination of the form. It can be only presumed that Fe was in a form of (oxy)hydroxides or phosphates. The increased amount of Fe could be attributed to the oxidizing effect of the plants. The FaD-CP-top(2) had similar composition but small particles were aggregated in a larger cluster containing organic material.

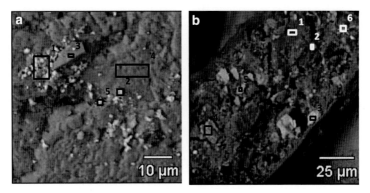

**Fig. 14.5** SEM images of wetland substrates sampled from the *top* of the FaD-CP column fed with feed B (**a**) Sample FaD-CP-top(1), (**b**) Sample FaD-CP-top(2)

In all the four samples of the substrate from the FaD-CP column sulfidic forms of metals were observed. In fact the metals of concern Ni, Cu, and Zn, save Al, were not detected.

### 14.4.2.3 SEM-EDS Analysis of the Substrate from the FaD-SP Column

The SEM-BSE images of the samples taken from the bottom part of the FaD-SP column presented in Fig. 14.6, and the elemental composition of the selected areas are listed in Table 14.6.

The composition of the FaD-SP-bot(1) sample indicates the presence of calcium sulfate (area 1) and titanium Ti as a native component of the medium. The FaD-SP-bot(2) sample contained certain forms of Fe, probably hydroxides or associated with calcium sulfate.

The SEM-BSE images of the samples taken from the top part of the FaD-SP column are presented in Fig. 14.7, and the elemental composition of the selected areas are listed in Table 14.7.

Both samples of the substrate taken from the top part of the FaD-SP column shown in Fig. 14.7 differ from the sample of the bottom part of the column in that they contained plant detrital matter and the metals of concern (Ni and Cu). These metals were present in small particles containing Fe and S, which may suggest the presence of sulfides, but also the presence of oxidized forms of metals cannot be excluded. Sample FaD-SP-top(1) contained calcium sulfate (area 4) and probably elemental Sn (area 6).

These findings are in agreement with the chemical analysis of the substrate from the FaD-SP column. As it was shown in Fig. 14.2 and Table 14.3, the concentration of Cu and Ni was markedly higher in the top part of the column and was mainly present in the exchangeable and reducible fraction.

**Table 14.5** Elemental (atomic %) composition of the areas (ar#) indicated in the images in Fig. 14.5 determined using EDS

| Sample_area | Atomic %[a] Na | Mg | Al | Si | P | S | K | Ca | Mn | Fe | Ag | La | Ce | Nd |
|---|---|---|---|---|---|---|---|---|---|---|---|---|---|---|
| FaD-CP-top(1)_ar1 | 2.0 | 0.5 | 2.1 | 57.2 | 3.3 | 3.9 | 0.5 | 4.9 | – | 25.5 | – | – | – | – |
| FaD-CP-top(1)_ar2 | 1.7 | 0.5 | 1.7 | 60.6 | 3.7 | 1.4 | 0.5 | 2.5 | – | 27.3 | – | – | – | – |
| FaD-CP-top(1)_ar3 | – | – | – | 99.6 | – | – | – | 0.4 | – | – | – | – | – | – |
| FaD-CP-top(1)_ar4 | – | – | 1.1 | 98.9 | – | – | – | – | – | – | – | – | – | – |
| FaD-CP-top(1)_ar5 | 1.8 | 0.5 | 3.5 | 50.2 | 4.8 | 0.8 | 0.3 | 2.1 | – | 36.1 | – | – | – | – |
| FaD-CP-top(2)_ar1 | 3.3 | 1.3 | 15.4 | 51.6 | 1.1 | 7.3 | 5.3 | 11.4 | – | 3.2 | – | – | – | – |
| FaD-CP-top(2)_ar2 | 1.9 | 0.8 | 9.0 | 73.7 | 1.2 | 3.1 | 2.0 | 5.9 | – | 2.3 | – | – | – | – |
| FaD-CP-top(2)_ar3 | – | 11.5 | 19.6 | 46.1 | – | 0.7 | 2.3 | 1.6 | 0.4 | 17.7 | – | – | – | – |
| FaD-CP-top(2)_ar4 | 8.3 | – | 16.5 | 71.3 | – | 0.9 | 0.6 | 2.4 | – | – | – | – | – | – |
| FaD-CP-top(2)_ar5 | 1.0 | 1.5 | 24.5 | 56.0 | – | 2.3 | 7.7 | 3.8 | – | 3.0 | – | – | – | – |
| FaD-CP-top(2)_ar6 | – | 0.7 | 10.3 | 49.1 | 10.6 | 5.9 | 2.2 | 10.9 | – | 2.4 | 0.7 | 1.8 | 4.1 | 1.3 |

[a]Compositions normalized to 100 %

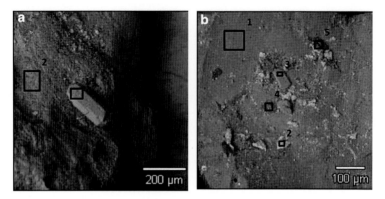

**Fig. 14.6** SEM images of wetland substrates sampled from the *bottom* (bot) of the FaD-SP column fed with feed B (**a**) Sample FaD-SP-bot(1), (**b**) Sample FaD-SP-bot(2)

**Table 14.6** Elemental (atomic %) composition of the areas (ar#) indicated in the images in Fig. 14.6 determined using EDS

| Sample_area | Atomic %[a] | | | | | | | | | |
|---|---|---|---|---|---|---|---|---|---|---|
| | Na | Al | Si | P | S | Cl | K | Ca | Ti | Fe |
| **FaD-SP-bot(1)_ar1** | 1.3 | 2.8 | 36.0 | – | 29.8 | – | 0.4 | 26.6 | 1.1 | 2.0 |
| **FaD-SP-bot(1)_ar2** | 1.6 | 6.0 | 85.7 | – | 1.8 | – | 0.5 | 0.6 | 1.4 | 2.5 |
| **FaD-SP-bot(2)_ar1** | – | 0.7 | 97.5 | – | – | – | – | – | 1.8 | – |
| **FaD-SP-bot(2)_ar2** | 1.5 | 4.2 | 52.3 | 1.1 | 2.3 | – | – | 0.3 | 1.5 | 36.8 |
| **FaD-SP-bot(2)_ar3** | 0.4 | 1.2 | 35.5 | – | 30.8 | – | 0.2 | 27.5 | 1.9 | 2.5 |
| **FaD-SP-bot(2)_ar4** | 1.6 | 7.5 | 78.8 | – | 3.1 | – | 0.8 | 0.5 | 1.9 | 5.9 |
| **FaD-SP-bot(2)_ar5** | 2.0 | 5.4 | 65.6 | 1.8 | 3.6 | 0.4 | 1.1 | – | 2.2 | 17.8 |

[a]Compositions normalized to 100 %

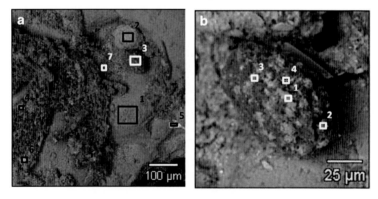

**Fig. 14.7** SEM images of wetland substrates sampled from the *top* of the FaD-SP column fed with feed B influent (**a**) Sample FaD-SP-top(1), (**b**) Sample FaD-SP-top(2)

Table 14.7 Elemental (atomic %) composition of the areas (ar#) indicated in the images in Fig. 14.7 determined using EDS

| Sample_area | Atomic %[a] | | | | | | | | | | | | | | |
|---|---|---|---|---|---|---|---|---|---|---|---|---|---|---|---|
| | Na | Mg | Al | Si | P | S | Cl | K | Ca | Ti | Fe | Ni | Cu | Zr | Sn |
| FaD-SP-top(1)_ar1 | – | – | 1.7 | 95.9 | 1.5 | – | – | – | 0.9 | – | – | – | – | – | – |
| FaD-SP-top(1)_ar2 | 1.4 | 0.9 | 18.0 | 57.5 | 2.0 | 1.7 | – | 4.0 | 1.6 | 1.2 | 11.0 | – | 0.7 | – | – |
| FaD-SP-top(1)_ar3 | 5.5 | 1.6 | 10.2 | 52.1 | 4.1 | 9.6 | – | 3.3 | 7.2 | – | 2.9 | – | 3.6 | – | – |
| FaD-SP-top(1)_ar4 | 0.9 | 0.2 | 2.2 | 22.1 | 1.3 | 36.3 | – | 0.4 | 34.5 | 0.3 | 0.9 | – | 0.8 | – | – |
| FaD-SP-top(1)_ar5 | 1.3 | – | 3.5 | 75.6 | – | 2.7 | – | 1.2 | 2.8 | 0.6 | 3.6 | – | – | 8.8 | – |
| FaD-SP-top(1)_ar6 | 5.0 | 1.3 | 7.0 | 41.5 | 4.6 | 3.0 | – | – | – | – | 2.5 | 1.1 | 3.3 | – | 30.7 |
| FaD-SP-top(1)_ar7 | 1.2 | 0.5 | 4.0 | 86.3 | 3.4 | – | – | 0.4 | 1.3 | – | 1.5 | 0.7 | 0.6 | – | – |
| FaD-SP-top(2)_ar1 | 2.0 | 0.8 | 22.8 | 53.4 | 1.9 | 5.7 | – | 6.2 | 4.4 | – | 1.7 | – | 1.0 | – | – |
| FaD-SP-top(2)_ar2 | 8.0 | 2.1 | 8.5 | 52.2 | 3.1 | 7.2 | 0.8 | 2.7 | 6.0 | 0.3 | 2.6 | 1.3 | 5.3 | – | – |
| FaD-SP-top(2)_ar3 | 6.0 | 1.5 | 12.1 | 36.1 | 12.9 | 9.7 | – | 3.1 | 8.3 | 0.4 | 3.6 | 0.8 | 3.4 | – | 2.1 |
| FaD-SP-top(2)_ar4 | 8.7 | 2.1 | 6.0 | 54.8 | 2.3 | 8.2 | 0.6 | 2.4 | 6.0 | – | 2.0 | 1.1 | 5.6 | – | – |

[a]Compositions normalized to 100 %

### 14.4.2.4 SEM-EDS Analysis of the Raw Sand

The above-presented results of the SEM-EDS of the spent substrate contained several areas in which the elemental composition was much different from the other analyzed areas. In this situation it was assumed that the detected elements were of native origin. This means that they were present in the media before filling the columns. There was certain evidence, however indirect, that the bed media contained some contaminants that were leached into the wastewater flowing through the columns. This was especially a case for Mn and Fe. The chemical analysis confirmed these assumptions for sand and compost used in the experiment. The SEM-EDS analysis was performed to obtain more information on the forms of native contaminants in the raw media.

The SEM-BSE images of the samples of sand used in the experiment are presented in Fig. 14.8, and the elemental composition of the selected areas are listed in Table 14.8.

The sand used in the experimental system contained invariably Si and much lower amounts of Al. The other elements that were detected were Ca, Ti, Fe, and Zr. Iron was present probably in the oxidized form and could be leached from the sand under anaerobic conditions.

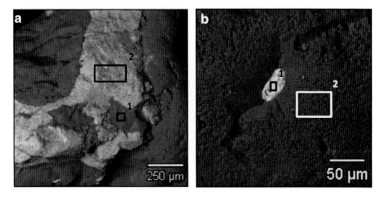

**Fig. 14.8** SEM images of the raw sand used as a bed medium (**a**) Sample Sand(1), (**b**) Sample Sand(2)

**Table 14.8** Elemental (atomic %) composition of the areas (ar#) indicated in the images in Fig. 14.8 determined using EDS

| Sample_area | Atomic %[a] | | | | | |
|---|---|---|---|---|---|---|
| | Al | Si | Ca | Ti | Fe | Zr |
| **Sand(1)_ar1** | 1.9 | 88.0 | – | 0.6 | 9.5 | – |
| **Sand(1)_ar2** | 4.2 | 29.6 | 0.8 | 0.5 | 64.7 | 0.3 |
| **Sand(2)_ar1** | 2.0 | 68.9 | 1.8 | – | – | 27.3 |
| **Sand(2)_ar2** | 2.6 | 96.8 | 0.6 | – | – | – |

[a]Compositions normalized to 100 %

## 14.4.2.5 Discussion on the Sequential Extraction Procedure and SEM-EDS Analyses of the Substrates

It was observed that the intended removal of metals in the form of sulfide precipitates was a minor metal removal mechanism in the studied microcosm system. It was also found that the observed, however rarely occurring, clusters of sulfides contained Fe more frequently than the other metals of concern. In general, metals studied in this work were bound mostly to exchangeable and reducible fraction. The exchangeable fraction contains weakly adsorbed metals bound to the solid surface by relatively weak electrostatic forces, metals that can be released by ion-exchange processes and metals that can co-precipitate with carbonates. Remobilization of metals belonging to this fraction may occur upon changes of the ionic composition (affecting adsorption–desorption equilibrium) or lowering of pH (Filgueiras et al. 2002). The reducible fraction contains mostly hydrous Fe-Mn oxides, which exist in large proportion in soil and sediments as nodules, concretions, cement between particles, or coating on particles. Scavenging of trace metals by these oxides can occur by any or a combination of the following mechanisms: co-precipitation, adsorption, surface complex formation, ion exchange, and penetration of the lattice. They have high affinity for trace metals and are thermodynamically unstable under anoxic conditions and are attacked by benthic organisms (Filgueiras et al. 2002). Interestingly, the observations made for the reported experiments are to some extent similar to the findings for other types of wastewater treated in CWs. The findings of Neculita et al. (2008b) based on the experiment in a microcosm downflow columns indicated that only 15 % of total metals were removed as sulfides. In that experiment 3.5 L columns were fed with simulated acid mine drainage with Ni and Zn concentration in the range of 13–15 mg $L^{-1}$, 4 g $L^{-1}$ of sulfate and acidic pH. Their findings suggested that the major removal of metals was attributed to adsorption or complexation with organic matter and co-precipitation as (oxy)hydroxide minerals. The duration of their experiment (44 weeks) (Neculita et al. 2008a) was shorter than the experiments described in this work. Similarly to the study presented in this paper Neculita et al. (2008b) used SEP and the SEM-EDS to analyze the substrates sampled from the columns. The additional method that was applied by these authors was determination of acid volatile sulfide-extractable metals, which yields the ratio between the metals bound with sulfides and the volatile sulfides. The ratio between extractable metals and acid-volatile sulfides determined by these authors was much greater than 1 indicating high mobility of metals in acidic conditions. This analysis was not applied in this study but the similarity of the other results allows for hypothesis that the mobility of metals in the investigated systems should be a matter of concern. Song (2003) found that Zn and Pb sulfides were sparse and randomly scattered in the substrate of microcosm CW fed with simulated mine drainage water containing 50 μg $L^{-1}$ Pb, 0.5 mg $L^{-1}$ Zn, 0.23 mg $L^{-1}$ Fe, 34.2 mg $L^{-1}$ $SO_4^{2-}$, and with pH approx. 8. This system was operated for 4 years. Moreover, Song (2003) stated that SEM-EDS analysis might not be sufficient to observe metal sulfides at low

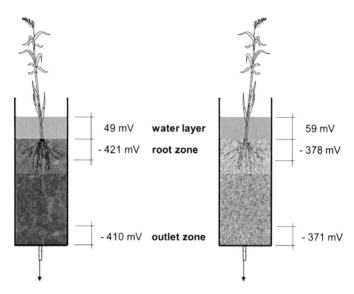

**Fig. 14.9** Oxidation-reduction potential in the substrate and water layer of the FaD-CP (*left*), FaD-SP (*right*)

concentrations present in the investigated CW. The increased affinity of Fe for sulfides was also reported by Song (2003) who elucidated that binding of Fe with sulfides is more thermodynamically favored as compared to Pb or Zn, for example.

### 14.4.3 Oxidation-Reduction Potential

The oxidation-reduction potential (ORP) was measured several times in the end of the experiment. The mean values are given for the water layer and the 10-cm-thick top and bottom parts of the substrate in the FaD columns (Fig. 14.9).

The conditions in the substrate of the FaD columns were strongly reducing and the ORP was below the threshold allowing for methanogenesis ($-200$ mV). The columns with organic bed media had slightly lower ORP of the substrate and water layer. It was also observed that the ORP in the root zone was slightly lower than in the outlet zone suggesting that the oxygen transfer from the above ground parts was negligible; however, the measurements were not made in close proximity to the roots. In these redox conditions, SRB and methanogens may compete for electron donors. Interestingly, approximately 14 % of methanogens may use acetic acid, which is used by SRB but also produced by the lactate-oxidizing SRB. Both groups of bacteria may also compete for dimethyl sulfide, which is abundant in anoxic wetlands (Paul and Clark 1996; Vymazal and Kröpfelová 2008). The emission of methane and other gases from the system was not quantified; however, this negative aspect should be considered when planning full-scale implementation.

## 14.5 Conclusions

This article presents, to our best knowledge, the first results for the polishing of electroplating wastewater in subsurface flow CWs. The studied columns were found to remove Al and Cu to high extent, and also Ni, but only in the columns fed with influent containing no detectable concentration of cyanides. Precipitation of metals with biogenic sulfide was found not to be a major sink for metals in the studied microcosm columns. It may stem from the fact that the organic carbon source for SRB was poor, either in terms of wastewater quality or the availability of carbon present in bed media. Regarding the first aspect, it was observed that the electroplating wastewater used in the experiment was carbon-deficient. Regarding the second aspect, it was found that compost did not enhance the removal efficiency of metals. It was observed that the presence of compost stimulated bacterial sulfate reduction, but this had no effect on removal of metals when compared to gravel bed medium, with the exception of Ni removal in columns fed with feed B. Apart from the effect of carbon deficiency of wastewater and bed media, it is probable that the cycling aerobic and anaerobic conditions in the columns may not be advantageous for bacterial sulfate reduction. The major portion of metals was bound to exchangeable or reducible fractions, which allows conclusion that adsorption and oxic removal processes are responsible for the removal of metals in a short-time experiment. These removal processes are not recommended because metals can be easily leached from these phases upon change of process conditions. For the above-mentioned reasons, the application of FaD CWs for the precipitation of metal sulfides should be treated with caution. The improvement of treatment efficiency of metals removal could be improved by providing readily biodegradable carbon source for SRB and constant anaerobic conditions in the system by applying saturated flow in the system.

FaD columns can efficiently remove cyanides from pretreated electroplating wastewater. The application of compost as organic bed medium enhances the removal of B in comparison to mineral bed media, in which B was a conservative element.

**Acknowledgments** The project was supported by Grant N N523 561938 from the Polish Ministry of Science and Higher Education. The lead author was granted scholarship in the Project "SWIFT (Scholarships Supporting Innovative Forum of Technologies)" POKL.08.02.01-24-005/10 cofinanced by the European Union under the European Social Fund.

## References

Bacon, J. R., & Davidson, C. M. (2008). Is there a future for sequential chemical extraction? *Analyst, 133*, 25–46.

Bartkiewicz, B. (2006). *Oczyszczanie ścieków przemysłowych* (Industrial wastewater treatment). Warszawa: Wydawnictwo Naukowe PWN. [in Polish]

Blais, J. F., Djedidi, Z., Cheikh, R. B., Tyagi, R. D., & Mercier, G. (2008). Metals precipitation from effluents: Review. *Practice Periodical of Hazardous, Toxic, and Radioactive Waste Management, 12*, 135–149.

Dean, J. R. (2003). *Methods for environmental trace analysis. Analytical techniques in the sciences*. Chichester: John Wiley & Sons, Ltd.

Filgueiras, A. V., Lavilla, I., & Bendicho, C. (2002). Chemical sequential extraction for metal partitioning in environmental solid samples. *Journal Environmental Monitoring, 4*(2002), 823–857.

Gammons, C., & Frandsen, A. (2001). Fate and transport of metals in $H_2S$-rich waters at a treatment wetland. *Geochemical Transactions, 2*, 1–15.

Hang, X., Wang, H., Zhou, J., Du, C., & Chen, X. (2009). Characteristics and accumulation of heavy metals in sediments originated from an electroplating plant. *Journal of Hazardous Materials, 163*, 922–930.

ISO 11466. (1995). *Soil quality – Extraction of trace elements soluble in aqua regia*. Geneve: ISO.

Khanal, S. (2008). *Anaerobic biotechnology for bioenergy production: Principles and applications*. New York: John Wiley & Sons.

Kiptoo, J. K., Ngila, J. C., & Sawula, G. M. (2004). Speciation studies of nickel and chromium in wastewater from an electroplating plant. *Talanta, 64*, 54–59.

Larner, B. L., Seen, A. J., & Townsend, A. T. (2006). Comparative study of optimised BCR sequential extraction scheme and acid leaching of elements in the certified reference material NIST 2711. *Analitica Chimica Acta, 556*, 444–449.

Lens, P. N. L., Vallero, M., & Esposito, G. (2007). Bioprocess engineering of sulphate reduction for environmental technology. In L. L. Barton & W. A. Hamilton (Eds.), *Sulphate-reducing bacteria: Environmental and engineered systems* (pp. 383–404). Cambridge, NY: University Press.

Lewis, A. E. (2010). Review of metal sulphide precipitation. *Hydrometallurgy, 104*, 222–234.

Li, J., Lu, Y., Shim, H., Deng, X., Lian, J., Jia, Z., & Li, J. (2010). Use of the BCR sequential extraction procedure for the study of metal availability to plants. *Journal of Environmental Monitoring, 12*, 466–471.

Liamleam, W., & Annachhatre, A. P. (2007). Electron donors for biological sulfate reduction. *Biotechnology Advances, 25*, 452–463.

Machemer, S. D., Reynolds, J. S., Laudon, S. L., & Wildeman, T. R. (1993). Balance of S in a constructed wetland built to treat acid mine drainage, Idaho Springs, Colorado, USA. *Applied Geochemistry, 8*, 587–603.

Neculita, C. M., Zagury, G. J., & Bussière, B. (2008a). Effectiveness of sulfate-reducing passive bioreactors for treating highly contaminated acid mine drainage: I. Effect of hydraulic retention time. *Applied Geochemistry, 23*, 3442–3451.

Neculita, C. M., Zagury, G. J., & Bussière, B. (2008b). Effectiveness of sulfate-reducing passive bioreactors for treating highly contaminated acid mine drainage: II. Metal removal mechanisms and potential mobility. *Applied Geochemistry, 23*, 3545–3560.

Paul, E. A., & Clark, F. E. (1996). *Soil microbiology and biochemistry* (2nd ed.). San Diego, CA: Academic.

Peltier, E., Dahl, A. L., & Gaillard, J. F. (2005). Metal speciation in anoxic sediments: When sulfides can be construed as oxides. *Environmental Science and Technology, 39*, 311–316. doi:10.1021/es049212c.

Rauret, G., López-Sánchez, J. F., Sahuquillo, A., Rubio, R., Davidson, C., Ure, A., & Quevauviller, P. (1999). Improvement of the BCR three step sequential extraction procedure prior to the certification of new sediment and soil reference materials. *Journal of Environmental Monitoring, 1*, 57–61.

Reddy, K. R., & DeLaune, R. (2008). *Biogeochemistry of wetlands: Science and applications*. Boca Raton, FL: CRC Press.

Sastre, J., Sahuquillo, A., Vidal, M., & Rauret, G. (2002). Determination of Cd, Cu, Pb and Zn in environmental samples: Microwave-assisted total digestion versus aqua regia and nitric acid extraction. *Analytica Chimica Acta, 462*, 59–72.

Song, Y. (2003). *Mechanisms of lead and zinc removal from lead mine drainage in constructed wetland*. PhD dissertation, Civil Engineering Department, Faculty of Graduate School, Univ. Missouri-Rolla, Rolla, MO., USA.

StatSoft, Inc. (2011). STATISTICA (data analysis software system) (Version 10). www.statsoft.com

Sturman, P. J., Stein, O. R., Vymazal, J., & Kröpfelová, L. (2008). Sulfur cycling in constructed wetlands. In J. Vymazal (Ed.), *Wastewater treatment, plant dynamics and management in constructed and natural wetlands* (pp. 329–344). Dordrecht: Springer.

Vymazal, J., & Kröpfelová, L. (2008). *Wastewater treatment in constructed wetlands with horizontal sub-surface flow*. Dordrecht: Springer.

Wallace, S. D., & Knight, R. L. (2006). *Small-scale constructed wetland treatment systems: Feasibility, design criteria, and O&M requirements*. Alexandria: Water Environment Research Foundation.

# Chapter 15
# Relationship Between Filtering Material and Nitrification in Constructed Wetlands Treating Raw Wastewater

**Georges Reeb and Etienne Dantan**

**Abstract** Treatment performance measured in young constructed wetlands related to the particle sizes of the upper layers which receive the raw wastewater is compared with long-term hydraulic security. The objective of the study is to identify treatment limitations related to particle size of the upper mineral layer. In general, a site performs better with age, thanks to the accumulation/evolution of organic deposits on the surface of stage 1. The treatment performance in terms COD and TKN in the cases presented was >80 %. Regarding grading of gravel and sand, performance >90 % was reached even with $d_{10}$ between 4 and 5 mm, despite an unstable performance tendency observed when the $d_{10}$ is over 4 mm if at the same time the $D_{60}$ is over 7 mm. An alternative to granular mineral substrate, whose extraction is a worrying environmental problem, is considered.

**Keywords** Constructed wetlands • Gravel • Sand • Mineral aggregates • Raw wastewater • Hydraulics • Clogging

## 15.1 Introduction

Atelier Reeb has been designing reed bed water treatment plants for 20 years, primarily for rural communities. Long-term reliability is one of our priorities: we envisage treatment sites being fully functional for at least 30 years. For any potential problems during a site's lifetime, wooded buffer zones or vegetated discharge areas appear to provide an effective and inexpensive means of protecting the surrounding environment.

The lifetime of a filter depends primarily on the development of the mineralisation of organic matter. At present, we do not know how to evaluate the effect of the rhizosphere on the longevity of our planted filters, and the question of potential clogging over time remains open.

---

G. Reeb (✉) • E. Dantan
Atelier Reeb, 2 rue de Genève, F 67000 Strasbourg, France
e-mail: georges.reeb@atelier-reeb.fr

The flooding and drying sequence of planted helophyte filters (feed and rest) is linked to the build-up and mineralisation phases of organic matter. We believe that these phases of mineralisation are the major factor of the life span of our installations. And yet, pollutant loads being equal, the dynamics and effectiveness of this mineralisation are linked to the particle size of the mineral layer above.

For dual-stage treatment plants with a vertical flow receiving raw untreated water (the main type of installation in France), the design guidelines (Groupe macrophytes et traitement des eaux 2005; Ministère de la pêche et de l'agriculture 2007) recommended the use of 2–8-mm gravel, at least 30 cm deep, for the upper layer of the first filtering stage, and 0–4-mm sand, also 30 cm deep, for the upper layer of the second filtering stage. To date, Atelier Reeb has data on the functioning of numerous treatment plants, some of which have implemented this recommendation, while others have not. In fact, in order to promote the effectiveness of the mineralisation phases in the long term, Atelier Reeb recommends a particle size between 4 and 10 mm for the filtering layer of stage 1. These derogations are often the subject of discussion, even disagreements, with the various consultant engineers with whom we have been in contact. Two arguments are most commonly put forward: treatment performance would be inferior and the risk of organic matter leaching into stage 2 would be greater.

It is for this reason that we are beginning a long-term study of the performance of our treatment plants, with focus on the following three features: treatment performance, dynamics of the mineralisation of organic matter on the surface of the filtering bed of stage 1 and the development of the surface of the filtering bed of stage 2.

This chapter will set out and discuss the treatment performance measured at certain "young" treatment plants over the last 5 years, relative to the particle sizes of the upper layers which receive the wastewater.

## 15.2 Methods

In Table 15.1, basic design parameters of nine monitored constructed wetlands are presented. The samples were taken over 24-h periods and analysed for COD and TKN. As regards the grading of the gravel and sand, the value of $d_{10}$ refers to the diameter in millimetres of the mesh which lets through 10 % of the aggregate by weight, while $D_{60}$ refers to the diameter in millimetres of the mesh which lets through 60 % of the aggregate by weight.

# 15 Relationship Between Filtering Material and Nitrification in Constructed...

**Table 15.1** Basic design parameters of monitored constructed wetlands

| Site | PE | Operation since | Upper layer stage 1 ||||| Upper layer stage 2 |||
|---|---|---|---|---|---|---|---|---|---|
| | | | Material | Fraction (mm) | $d_{10}$; $D_{60}$ (mm) | Depth (cm) | Material | Fraction (mm) | Depth (cm) |
| A | 800 | 2006 | Gravel | 4–11.2 | 4.3; 7.4 | 30 | Sand | 0–4 | 30 |
| B | 300 | 2007 | Gravel | 2–4 | 1.4; 3.2 | 30 | Sand | 0–4 | 30 |
| C | 1200 | 2010 | Gravel | 4–10 | 4.6; 8.0 | 30 | Sand | 2–4 | 30 |
| D | 1000 | 2010 | Gravel | 4–10 | 4.6; 8.0 | 50 | Sand | 0–4 | 5–10 |
| | Additions in 2011 |||||||||
| E | 600 | 2012 | Gravel | 2–6.3 | 2.2; 4.9 | 10 | Sand | 0–4 | 40 |
| F | 260 | 2010 | Gravel | 2–4 | 2.0; 3.2 | 20 | Sand | 0–4 | 50 |
| G | 1650 | 2011 | Gravel | 2–6.3 | 2.1; 4.6 | 10 | Sand | 0–4 | 30 |
| H | 300 | 2010 | Gravel | 2–5.6 | 2.5; 3.8 | 5 | Sand | 0.5–4 | 30 |
| I | 730 | 2008 | Gravel | 2–6.3 | 2.2; 4.6 | 40 | Sand | 0–4 | 40 |
| | | | Gravel | 4–6.3 | 4.1; 5.5 | 40 | | | |

## 15.3 Results and Discussion

### 15.3.1 Results Applicable to All Sites

Treatment performance of two-stage systems was the subject of an extensive study carried out on existing treatment sites, the results of which shall be used as a point of reference. At the outlet of the constructed wetland, it is therefore expected to see a reduction in COD of about 90 % and a reduction in TKN of about 85 % (Molle et al. 2005).

The treatment performance for COD and TKN in monitored sites is mostly >80 % (Figs. 15.1 and 15.2, Table 15.2). Performance >90 % was reached even with a $d_{10}$ between 4 and 5 mm. Nevertheless, a tendency can be observed for the performance to become unstable when the $d_{10}$ is over 4 mm if at the same time the $D_{60}$ is over 7 mm. This tendency needs to be confirmed by studying larger datasets, particularly for medium aggregate of 4–6 mm or 4–8 mm ($d_{10}$ over 4 mm and $D_{60}$ less than 6 mm).

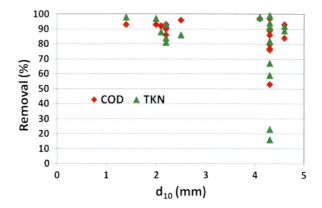

**Fig. 15.1** *Graph representing the treatment performance observed according to the $d_{10}$ of the first filtration layer of stage 1 (the values from site D taken in December 2010 and April 2011 have not been included on account of the type of 2–4-mm sand instead of 0–4-mm sand—used in the second stage; see 15.3.2)*

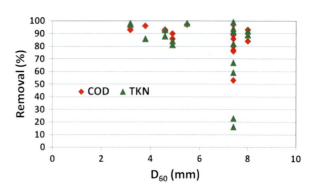

**Fig. 15.2** *Graph representing the treatment performance observed according to the $D_{60}$ of the first filtration layer. For details on site D, see Fig. 15.1*

**Table 15.2** Removal efficiency for COD and TKN and $d_{10}/D_{60}$ values at all monitored sites

| Treatment site | d10 upper layer stage 1 | D60 upper layer stage 1 | Removal of COD (%) | Removal of TKN (%) | Sampling time |
|---|---|---|---|---|---|
| A | 4.3 | 7.4 | 86 | 91 | 1/2009 |
|   |     |     | 97 | 99 | 10/2009 |
|   |     |     | 90 | 59 | 4/2010 |
|   |     |     | 76 | 23 | 11/2010 |
|   |     |     | 89 | 82 | 5/2011 |
|   |     |     | 77 | 94 | 10/2011 |
|   |     |     | 53 | 67 | 4/2012 |
|   |     |     | 80 | 16 | 9/2012 |
| B | 1.4 | 3.2 | 93 | 98 | 7/2012 |
| C | 4.6 | 8.0 | 84 | 89 | 6/2010 |
|   |     |     | 93 | 92 | 9/2011 |
| D | 4.6 | 8.0 | 88 | −83 | 12/2010 |
|   |     |     | 84 | 51 | 4/2011 |
|   | 2.2 | 4.9 | 90 | 84 | 3/2012 |
|   |     |     | 86 | 81 | 4/2012 |
| E | 2   | 3.2 | 93 | 97 | 7/2012 |
| F | 2.1 | 4.6 | 92 | 88 | 4/2012 |
| G | 2.5 | 3.8 | 96 | 86 | 2/2011 |
| H | 2.2 | 4.6 | 93 | 93 | 8/2011 |
| I | 4.1 | 5.5 | 97 | 98 | 4/2011 |

## 15.3.2 Results Specific to Site D

This treatment plant was constructed using aggregate 4–10 mm for the filtering layer of stage 1 and a coarse sand (2–4 mm) for the filtering layer of stage 2.

The expected performance was not achieved after 2 years in operation. The upper layers were modified during the second year as follows: a 10-cm layer of 2–6-mm gravel was added to the surface of stage 1 and a 5–10-cm layer of 0–4-mm sand was added to the surface of stage 2. These additions resulted in an increase of treatment efficiency from 80 to 88 % for COD and from 39 to 81 % for TKN. This performance improvement illustrates the considerable impact (which is, for that matter, already known) of the first 10 cm of the upper layer of each stage. This fact is supported by the results from site G (removal of 96 % for COD and 86 % for TKN) where the filtering layer of stage 1 composed of gravel 2–6 mm is only 5 cm thick but provides similar results to those sites constructed with at least 30 cm of 2–6-mm gravel.

## 15.4 Discussion

In order to focus our attention on the parameters connected to the particle size of the filtering layer of stage 1, we have voluntarily concealed in this study the following parameters:

- Age of the site
- Type of aggregate (crushed, semi-crushed or smooth)
- Hydraulic and organic surface load
- Density of helophytes (aerial stems, rhizomes, roots)
- Density of other plant species, where necessary

However, it will later be necessary to reveal a potential variability connected with the age of the site. Indeed, one can reckon that the system treatment performance improves with the age of the system thanks to the accumulation and/or evolution of organic deposits on the surface of stage 1.

In fact, we know that the organic layer of stage 1 forms gradually by the mineralisation/humification of the matter in the raw wastewater, providing more robust and flexible filtration in the long term (Chazarenc and Merlin 2005; Liey and Reeb 2010). Nevertheless, this is not supported by the data presented here for site A after 6 years of operation. It is possible that the accumulation of deposits on the surface of stage 1 is very slow thanks to excellent mineralisation associated notably with a 4–11-mm particle size (better oxygenation and drying out than with 2–6-mm fraction). It is also possible that the colonisation of the filtering bed by the reed rhizomes opens up preferential passages which reduce the overall residence time of the water in the filtration beds.

It will also be necessary to monitor a possible variability in the results according to the type of aggregate such as calcareous, siliceous, etc.

Lastly, the evaluation of the treatment sites according to their surface load (hydraulic, organic and nitrogen loads) would also allow the observed results to be qualified, where necessary.

## 15.5 Conclusion

This study is the beginning of a long-term observation of the functioning of our treatment sites, with the intent to obtain data and reliable readings in order to refine our construction recommendations regarding the objectives of each project. In fact, since the beginning of our work with constructed wetlands, we have been convinced that it is generally appropriate to prioritise long-term hydraulic security (e.g. with a $d_{10}$ of more than 4 mm for the filtering layer of stage 1) over short-term treatment efficiency (easily achieved with a $d_{10}$ of less than 3 mm and a $D_{60}$ of less than 5 mm for the filtering layer of stage 1), especially since treatment efficiency can increase over time. We should therefore be able to identify the treatment limitations (especially regarding organic compounds and TKN) of our sites over which hydraulic security has been given priority and confirm where necessary that they are not incompatible with the protection requirements of the receiving environment.

Visual observation campaigns of the filtering media in operation will complement our performance data. For example, during the summer of 2013, we carried out an audit of an 8-year-old treatment site in Brittany. Stage 1, consisting of a surface layer of 2–6-mm aggregate to a depth of 60 cm, above a transition layer of 6–10 mm and a drainage layer of 20–40 cm, had been severely clogged since November 2012. We created a test gravel pit in July 2013 and observed signs of anoxia from a depth of about 30 cm (black area where the gravel had the appearance of being covered in bitumen). We also found that the 2–6-mm gravel was coated throughout with a translucent substance, brown and sticky but not viscous, which was causing the particles to cluster together. These observations suggest a shift in the initial porosity at the expense of effective re-oxygenation. The type of gravel (2–6 mm) and the thickness of the layer are therefore called into question in this case.

In addition to this study of the mineral aggregates at our treatment sites, we are beginning a study of the treatment capabilities of granular organic materials (wood chippings), which are easy to reintroduce in the organic matter cycle and renewable. They could be used as an alternative to granular mineral substrate, deposits of which are running low and whose extraction is a worrying environmental problem in many areas.

Lastly, whatever the type and particle size of the filtering layer, good maintenance of the plant beds contributes enormously to their ability to deal with the water properly and therefore their role as a filter. That is to say, the development of the layer of deposits at stage 1 must be monitored, as must be plant growth: prompt removal of tree seedlings, control of plants that would compete with the helophytes, etc.

## References

Chazarenc, F., & Merlin, G. (2005). Influence of surface-layer on hydrology and biology of gravel bed vertical flow constructed wetlands. *Water Science and Technology, 51*(9), 91–97.

Groupe macrophytes et traitement des eaux, Agence de l'Eau Rhône Méditerranée Corse. (2005). *Epuration des eaux usées domestiques sur lits plantés de macrophytes: Prescriptions et recommandations pour la conception et la réalisation.* Lyon: Agence de l'Eau Rhone Méditerranée Corse.

Liey, S., & Reeb, G. (2010). Role, characteristics and importance of deposits at stage 1 of vertical flow constructed wetlands receiving raw sewage. In Masi, F., & Nivala, J. (Eds.), *Proc. 12th internat. conf. on wetland systems for water pollution control* (pp. 1366–1371). Padova: IWA, IRIDRA Srl and PAN Srl.

Ministère de la pêche et de l'agriculture (2007). *Cadre guide pour un cahier des clauses techniques particulières Filtres plantés de roseaux.* Lyon: Agence de l'Eau Rhone Méditerranée Corse.

Molle, P., Liénard, A., Boutin, C., Merlin, G., & Iwema, A. (2005). How to treat raw sewage with constructed wetlands: An overview of the French systems. *Water Science and Technology, 51*(9), 11–21.

# Chapter 16
# Single-Family Treatment Wetlands Progress in Poland

Hanna Obarska-Pempkowiak, Magdalena Gajewska, Ewa Wojciechowska, and Arkadiusz Ostojski

**Abstract** Long distances from one farm to another as well as unfavourable terrain configuration in rural areas in Poland result in high investment and operation costs of sewerage systems. Treatment wetlands (TWs) can be an alternative solution to sewage treatment in rural communities, schools, at campsites and for single houses. In this paper the results of 4 years monitoring of single-family treatment plants (SFTWs) in Kaszuby Lake District are presented. Based on experience with the TWs in Poland and in Europe, three configurations of TWs were proposed: two with vertical flow beds and the third one with a horizontal flow bed preceded by a pre-filter. In the first 2 years of operation, good treatment effectiveness was observed: $BOD_5$ 64.0–92.0 %, TN 44.0–77.0 %, TP 24.0–66.0 %. After 3 years of operation, the removal efficiency of organic matter and TN increased by 12–20 %. Comparison of three configurations of SFTWs indicated that facilities with VSSF beds were more effective in pollutants removal. Configuration I, consisting of a single VSSF bed with a larger unit area, provided more effective nitrification, while configuration II, with two sequential VSSF beds, ensured better organic matter removal. The final pollutants concentrations in effluent from these two configurations were quite similar and fulfilled the requirements of the Regulation of Environmental Ministry from 24 July 2006.

**Keywords** Single-family treatment plant • Treatment wetlands • Vertical flow • Nutrients • Organics

## 16.1 Introduction

Long distances from one farm to another as well as unfavourable terrain configuration in rural areas in Poland result in high investment and operation costs of sewerage systems. According to the European Committee, building of sewerage systems is not recommended in cases when the length of sewerage pipelines is

---

H. Obarska-Pempkowiak • M. Gajewska • E. Wojciechowska (✉) • A. Ostojski
Faculty of Civil and Environmental Engineering, Gdansk University of Technology, Narutowicza 11/12, 80-233 Gdansk, Poland
e-mail: esien@pg.gda.pl

higher than 10.0 m/pe. In Poland, there are many areas where this value is exceeded and varies from 10.0 to 19.0 m/pe. Thus simple, reliable and cost-saving solutions in the field of sewage management in rural areas are needed. Treatment wetlands (TWs) can be an alternative solution to sewage treatment in rural communities, schools, at campsites and for single houses. It is estimated that in the coming years there will be a need for construction of over 600,000 single-family treatment plants (SFTP) to ensure good quality of surface water according to the requirements of RWD/2000/60/EU.

Although TWs in Europe are usually designed to serve up to 500 inhabitants, most of the existing facilities receive sewage from less than 50 inhabitants or even from individual houses. TWs for more than 1,000 inhabitants are quite rare (Cooper 1998; Vymazal 1998; Masi et al. 2013; Kadlec and Wallace 2009). Moreover, the newest publications proved that TWs technology could be recommended as the best available technology for wastewater treatment for single families (Steer et al. 2002; Paruch et al. 2011; Jóźwiakowski 2012). In many European countries, this method of treatment is recommended and simple design guidelines are defined (Molle et al. 2005; Brix and Arias 2005; Arbeitsblatt DWA-A 262 2006; ÖNORM B 2505 (2009); Paruch et al. 2011). After 2000 the Ministry of Environment in Denmark, adapting to the requirements of the Water Framework Directive 2000/60/EC, published a set of guidelines concerning wastewater treatment at rural areas. Among others, it is recommended that settlements with a population of less than 30 should use the wetlands in the form of a single stage vertical subsurface bed (VSSF). The minimum unit area of 3.2 $m^2$/pe is recommended, but the minimum bed area for a single household should be above 16 $m^2$ (Brix 2005).

In Poland, since the late 1980s, several hundreds of treatment wetlands (TWs) used in the second stage of domestic sewage treatment have been built. During this time only one stage TWs with a horizontal subsurface flow (HSSF) have been constructed. The monitoring results of the single-family treatment wetlands indicated that the HSSF facilities designed at the second stage of sewage treatment provided effective removal of $BOD_5$ and COD as well as TSS. The effectiveness of BOD removal varied from 25.6 to 99.1 % (average 62.4 %) for the loadings from 11.2 to 115 kg $ha^{-1}d^{-1}$. However, the removal effectiveness of the total nitrogen was lower and varied from 22.4 to 84.2 % (average 44.5 %), for the loadings from 8.5 to 34.0 kg $ha^{-1}d^{-1}$ (Obarska–Pempkowiak et al. 2010). Since 2004 the vertical subsurface flow (VSSF) TWs have become more popular in Poland. Over 1,000 of such type facilities have been built in Poland, most of them located in Podlasie region. The treatment wetlands in Podlasie are being monitored by several research institutions; however, the monitoring results are contradictory. According to Halicki (2009), the average outflow concentrations were as follows: $BOD_5 < 13$ mg $L^{-1}$, COD $< 93$ mg $L^{-1}$, and TSS $< 42$ mg $L^{-1}$. The analyses performed by Obarska-Pempkowiak et al. (2010) confirmed the low effectiveness of total phosphorus and TSS removal. The share of organic suspended solids in the total suspended solids at the effluent varied from 49 to 95 % and was also the reason for high concentrations of organic matter ($BOD_5$ and COD). As a consequence the effluent from three out of five analysed SFTWs exceeded the admissible pollutants concentrations according to Regulation of Environmental Ministry from 24 July 2006.

In this chapter the results of 4 years monitoring of SFTWs in Kaszuby Lake District are presented. Within the research project *Innovative Solutions for Wastewater Management in Rural Areas* (financed by Polish Ministry of Science and Higher Education E033/P01/2008/02 and EOG Financial Mechanism and Norwegian Financial Mechanism PL0271), the conception of sewage treatment and sewage sludge utilisation at the TWs for individual households in a rural area was developed and nine TWs were built in 2009.

## 16.2 Methods

### 16.2.1 Study SFTWs Description

The project was launched in the catchment area of the river Borucinka in the Municipality of Stezyca, Pomerania Region. The idea was to prepare a ready-to-implement solution for community in a rural area through using TWs for individual households. Local terrain configuration and dispersed development in the selected area made it useless to build a conventional sewerage system. Domestic sewage used to be collected in cesspools, sometimes leaking to groundwater. Based on the experience with the TWs in Poland and in Europe, three configurations of TWs were proposed: two with vertical flow beds and the third one with a horizontal flow bed preceded by a pre-filter (Fig. 16.1). The construction of TWs was carried out by the farmers themselves.

The proposal of an innovative sanitary system was based on an idea of a closed cycle of organic matter and nutrients in the environment. The nutrient substances such as N, P, and K present in sewage should be used for soil fertilisation. Primary treatment of sewage takes place in septic tanks. TWs do not produce excessive secondary sludge. The sludge from septic tanks is periodically removed and

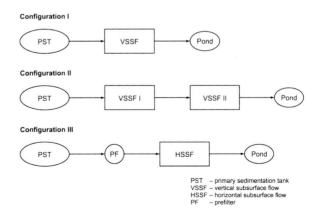

**Fig. 16.1** Schemes of applied configurations in SFTWs in Kaszuby Lake District

PST – primary sedimentation tank
VSSF – vertical subsurface flow
HSSF – horizontal subsurface flow
PF – prefilter

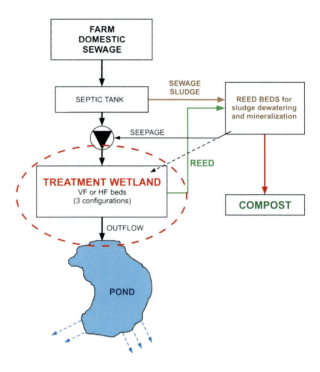

**Fig. 16.2** The concept of sewage and sludge management for SFTW

discharged to reed drying beds for sludge dewatering and mineralization processes. The seepage generated in the dewatering process is recirculated to the TWs and treated together with wastewater (Fig. 16.2). The dewatered and stabilised sewage sludge becomes a valuable humus substance that can be used as soil fertiliser at the farm lands. In this way the matter cycle is closed. The sewage treatment technology as well as sludge processing is simple to operate.

Configuration I consisted of a single VSSF bed with a unit area of 4 $m^2$/pe. In configuration II two sequentially connected VSSF beds were applied with total unit area of 2.5 $m^2$/pe. The configurations I and II were tested to investigate how double contact time affects the removal efficiency. Configuration III was designed in accordance with the Norwegian recommendations: pre-filter followed by HSSF bed. Instead of LECA, locally available material Pollytag was used (Table 16.1).

A major aspect and a distinctive feature of the project was that all nine SFTWs working in three configurations were constructed and put to operation in summer 2009. Since then they have been working as full-scale systems under real operational conditions. This enables comparison of efficiency of applied configurations in sewage treatment and makes the outcomes as well as the data on the removal of wastewater pollutants representative.

**Table 16.1** Characteristics of the SFTWs

| Configuration | HRT in primary tank [day] | Unit area per pe m$^2$ | Filling material of the beds | Pond area [m$^2$][a] |
|---|---|---|---|---|
| I – with single VSSF | Max 3 | 3.0 | 20 cm drainage layer 16–32 mm | Min 1 m$^2$/pe with the deep max 1.2 m |
| | | | 40 cm –filtration layer 2–8 mm | |
| | | | Up to 10 cm – mixture of peat and wood bark | |
| II – with two VSSFs | From 5 to 7 days | I st VSSF – 1.5 | 20 cm drainage layer 16–32 mm | Min 1 m$^2$/pe with the deep max 1.2 m |
| | | II nd VSSF- 1.0 | 20 cm – filtration layer 8–16 mm | |
| | | | 20 cm – filtration layer 2–8 mm | |
| | | | Up to 10 cm – mixture of peat and wood bark | |
| | | | [b] | |
| III – with pre-filter and HSSF | From 3 to 7 days | 5.0 | Distribution and collection zone 16–32 mm | Min 1 m$^2$/pe with the deep max 1.2 m |
| | | | Main filtration layer 2–8 mm | |

[a]Minimal designed (recommended) dimension – all farmers decided to build ponds as recipient of treated sewage
[b]The same layers for the I and the II stage beds

## 16.2.2 Sampling and Laboratory Analyses

The sampling period was divided into two sub-periods: I, years 2010 and 2011 with seven sampling events, and II, from October 2012 till May 2013 with three sampling events. The first monitoring period took place in the vegetation season and the second period should be regarded as a post-vegetation season. The grab samples of wastewater inflowing and outflowing (from the pond) from single-family TWs were collected. Collected samples were analysed for COD, TSS, total nitrogen (TN) and total phosphorus (TP). Analytical procedures recommended by Hach Chemical Company and Dr Lange GmbH were applied. The analyses were performed according to Polish Norms and guidelines given in Polish Environmental Ministry Regulation of 24 July 2006.

## 16.3 Results and Discussion

### 16.3.1 Suspended Solids and Organic Matter

According to the Regulation of the Minister of Environment from 24 July 2006 (Dz. U. 137/984 year 2006), domestic sewage from single houses in the amount $< 5.0$ m$^3$ d$^{-1}$ are obliged to only reduce the loads of BOD$_5$ (at least 20 % of load) and TSS (at least 50 %) if the treated effluent is discharged in the area of settlements. In Figs. 16.3 and 16.4, the concentrations of total suspended solids (TSS) and COD in inflowing and outflowing wastewater are presented.

The quality of wastewater fed to SFTWs after mechanical treatment in septic tanks differed substantially between the investigated TWs. The highest concentrations of COD were measured in facilities working in configuration III, both in vegetation and post-vegetation periods. The concentrations of COD varied from 900 to 1,300 mg L$^{-1}$ and were substantially higher in comparison with COD concentrations discharged to the TWs working in configuration I and II in both periods (<800 mg L$^{-1}$).

The concentration of pollutants inflowing to the investigated TWs was much higher than reported by Vymazal (2005), Heistad et al. (2006), Steer et al. (2002) and Brix and Jenssen (1999). So high inflow concentrations are quite typical for sewage in rural areas in Poland and could be caused either by improper maintenance and operation of septic tanks or by the inflow of high strength wastewater (manure, run-off from the fields or leakages from farmyard) as well as by low water consumption equal to 90–100 L pe$^{-1}$d$^{-1}$ (Gajewska et al. 2011; Jóźwiakowski 2012).

The high concentrations of COD were not related to particulate organic matter form since the concentrations of TSS were substantially lower and varied from

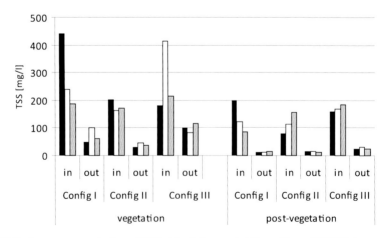

**Fig. 16.3** Average concentration of TSS at the inflow and at the outflow of SFTWs in vegetation and post-vegetation seasons

# 16 Single-Family Treatment Wetlands Progress in Poland

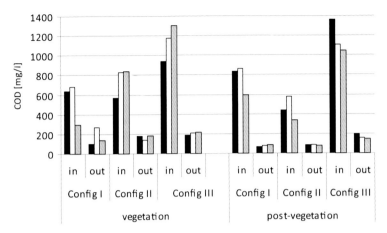

**Fig. 16.4** Average concentration of COD at the inflow and at the outflow of SFTWs in vegetation and post-vegetation seasons

160 to 450 mg L$^{-1}$. This indicates that most organic matter incoming into SFTWs was in soluble or colloidal forms. Moreover, in post-vegetation period, the concentration of TSS decreased substantially, most probably due to better operation of septic tanks, while the COD concentrations remained high.

The concentration of pollutants decreased; however, the sewage samples collected from the last stage of treatment (the pond) in many cases did not fulfil the requirements of the Regulation of Environmental Ministry from 24 July 2006 during the vegetative monitoring period. This was, however, during the period of the first 2 years of SFTWs operation and the facilities did not reach the full operation effectiveness yet. Relatively low quality of the effluent in this period seemed to result from short period of TWs operation (not complete development of roots or rhizomes and biofilms) as well as short wastewater retention in the pond. During post-vegetation period (after 2 years of operations of the systems), the quality of the effluent improved significantly. Effluents from SFTWs working in configurations I and II met all the requirements.

At most SF TWs, high efficiency of suspended solids and organic matter removal has been observed despite of high pollutants concentrations at the inflow (Table 16.2).

The efficiency of TSS and COD removal was improving in configuration I, consisting of a single VSSF bed, with the lowest loading rate in comparison to the TWs working in configurations II and III.

In the vegetation period TSS were generally more effectively removed in facilities with vertical flow beds. In the post-vegetation period, the removal efficiency for TSS in the facilities in configuration III improved significantly and was similar to others.

In the first 2 years of operation (vegetation period), configuration III was the most effective in COD removal, while after 2 years of operation, COD was removed with the highest efficiency in configuration I (Table 16.2, Fig. 16.4).

**Table 16.2** The TSS and COD removal efficiency in single-family TWs in both monitoring periods

| SF TWs | TSS | | | COD | | |
|---|---|---|---|---|---|---|
| | I | II | III (Configuration) | I | II | III |
| Vegetation period | | | | | | |
| 1 | 89.1 | 85.5 | 43.6 | 84.1 | 68.6 | 79.9 |
| 2 | 58.3 | 72.3 | 79.9 | 60.0 | 83.1 | 82.5 |
| 3 | 67.1 | 78.0 | 45.8 | 53.1 | 78.6 | 83.3 |
| Mean | **71.5** | **78.6** | **56.4** | **65.7** | **76.8** | **81.9** |
| Post-vegetation period | | | | | | |
| 1 | 93.5 | 79.9 | 83.7 | 91.5 | 81.5 | 85.5 |
| 2 | 90.7 | 86.8 | 81.2 | 91.2 | 84.7 | 85.8 |
| 3 | 82.0 | 91.3 | 86.9 | 86.3 | 76.5 | 85.9 |
| Mean | **88.8** | **86.0** | **83.9** | **89.6** | **80.9** | **85.7** |

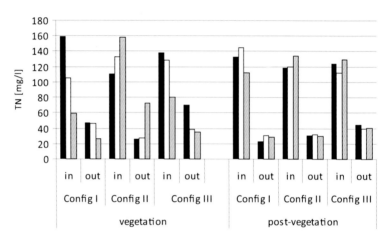

**Fig. 16.5** Average concentrations of TN at the inflow and outflow of SFTWs in vegetation and post-vegetation seasons

## 16.3.2 Nutrient Compounds Removal

According to Regulation of the Ministry of Environment from 24 July 2006 (Dz.U. 137/984/2006), there are no requirements concerning nitrogen and phosphorus concentrations if wastewater flow is below 5 $m^3\ d^{-1}$. However, due to the urgent need for protection of the Baltic Sea against eutrophication, even small amounts of wastewater should be treated with regard to low levels of discharged nutrients (TN below 30 mg $L^{-1}$ and TP below 5 mg $L^{-1}$).

In Figs. 16.5 and 16.6, the concentrations of total nitrogen (TN) and total phosphorus (TP) in inflowing and outflowing sewage are presented.

Wastewater discharged to SFTWs contained high concentrations of TN, varying from 60 to 160 mg $L^{-1}$ in both periods (Fig. 16.5). These concentrations were

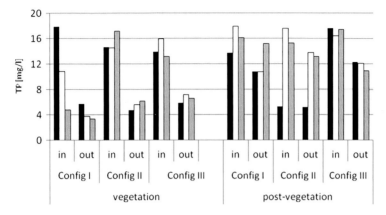

**Fig. 16.6** Average concentrations of at the inflow and outflow of SFTWs in vegetation and post-vegetation seasons

**Table 16.3** Nutrients (TN and TP) removal efficiency in single-family TWs in both monitoring periods

|  | Total nitrogen ||| Total phosphorus |||
|---|---|---|---|---|---|---|
|  | Configuration ||||||
| SF TWs | I | II | III | I | II | III |
| Vegetation period |||||||
| 1 | 70.6 | 75.9 | 49.7 | 68.2 | 67.8 | 57.6 |
| 2 | 55.9 | 79.0 | 69.9 | 65.4 | 61.4 | 54.7 |
| 3 | 55.6 | 54.3 | 55.8 | 29.1 | 64.0 | 50.0 |
| Mean | **60.7** | **69.7** | **58.5** | **54.3** | **64.4** | **54.1** |
| Post-vegetation period |||||||
| 1 | 82.3 | 73.8 | 63.6 | 21.2 | 1.9 | 30.1 |
| 2 | 78.5 | 73.5 | 64.7 | 39.7 | 21.6 | 26.2 |
| 3 | 75.0 | 77.8 | 68.8 | 5.6 | 13.7 | 37.4 |
| Mean | **78.6** | **75.0** | **65.7** | **22.1** | **12.4** | **31.2** |

considerably higher than the range of 36.3–77.5 mg L$^{-1}$ reported by Vymazal (2005) and Heistad et al. (2006). Similar high concentrations of nitrogen were present in the septic tank effluent in Podlasie Region (Poland) and Lublin Region (Poland) (Obarska-Pempkowiak et al. 2009; Gajewska et al. 2011; Jóźwiakowski 2012). The relatively high TN concentrations observed in sewage from Polish rural areas indicate that in some cases liquid wastes from animal breeding or other farm activities may be discharged together with the domestic sewage.

The highest fluctuations of TP concentrations were observed at the inflow in facilities operating in configuration I – from 5 to 18 mg L$^{-1}$ in vegetation period and for configuration II during the post-vegetation period. In other cases, the TP concentrations at the inflow were more constant and varied in the range from 14.0 to 17.5 mg L$^{-1}$ (Fig. 16.6).

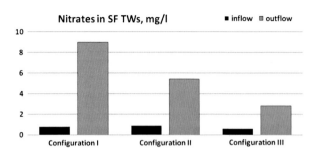

Fig. 16.7 Average concentration of nitrates in inflow and outflow from SFTWs in the third year of operations (during post-vegetation period)

Although the TN removal efficiency in configuration III increased from 58.5 % during the first 2 years of operation (vegetation period) to 65.7 % in post-vegetation period (after 2 years of operation), it was still lower than in case of configurations I and II (Table 16.3). The TP removal was quite effective in the first 2 years of operation due to high sorption capacity of the substrate. Then, after the initial sorption capacity was exhausted, the only process responsible for TP removal was bioaccumulation, resulting in the removal efficiency not exceeding 30 % (Vymazal 2010).

In Fig. 16.7, the concentrations of nitrates in wastewater from SFTWs are presented.

The high nitrates concentration could be an indicator of efficient nitrification process which is fundamental for effective nitrogen removal. In all configurations the incoming wastewater was characterised with similar low concentration of $N-NO_3^-$ which is typical for the septic tank effluent. In the third year of operation (during post-vegetation period), higher concentrations of nitrates were observed in configurations with VSSF beds (configurations I and II), especially in configuration I (with single VSSF bed) which had a higher unit area in comparison to two sequentially working VSSF beds (configuration II). This confirms that lower loadings of pollutants in case of higher unit area (configuration I) create better conditions for nitrification process in comparison to double contact time in configuration II which in turn is good for organic matter removal (Figs. 16.5 and 16.7).

## 16.4 Conclusions

The application of treatment wetlands for single-family wastewater is an effective and sustainable solution for wastewater treatment in rural areas. In spite of the fact that the TWs operated in Poland receive much higher concentration of pollutants in comparison to the TWs operated in Europe and the USA, in the first 2 years of operation good treatment effectiveness was observed: $BOD_5$ 64.0–92.0 %, TN 44.0–77.0 %, and TP 24.0–66.0 %. After 3 years of operation, the removal efficiency of organic matter and TN increased by 12–20 %. The comparison of three configurations of SFTWs indicated that facilities with VSSF beds were more effective in pollutants removal. Configuration I, consisting of a single VSSF bed with larger unit area, provided more effective nitrification, while configuration II, with two sequential VSSF beds, ensured better organic matter removal. The final pollutants concentrations in effluent from these two configurations were quite similar and fulfilled the requirements of the Regulation of Environmental Ministry from 24 July 2006. However, since there is no requirement for TN removal in case of single-household WWTPS, effective organic matter removal seems to be the key process. Thus, it can be concluded that prolongation of the contact time had more positive impact on treatment results than application of larger unit area of a single VSSF bed. This also follows the latest tendency to minimise the unit area of TWs but applying two or even more treatment stages (Steer et al. 2002; Molle et al. 2008; Vymazal 2009; Ghrabi et al. 2011).

## References

Arbeitsblatt DWA-A 262. (2006). *Grundsätze für Bemessung, Bau und Betrieb von Pflanzenkläranlagen mit bepflanzten Bodenfiltern zur biologischen Reinigung kommunalen Abwassers*. Hennef: Deutsche Vereinigung für Wasserwirtschaft Abwasser und Abfall e. V.

Brix, H., & Arias, C. A. (2005). The use of vertical flow constructed wetlands for on-site treatment of domestic wastewater: New Danish guidelines. *Ecological Engineering, 25*, 491–500.

Brix, H., & Johansen, N. H. (1999). Treatment of domestic sewage in a two-stage constructed wetland design principles. In J. Vymazal (Ed.), *Nutrient cycling and retention in natural and constructed wetland* (pp. 155–165). Leiden: Backhuys Publishers.

Cooper, P. (1998). Chapter IV: A review of the design and performance of vertical flow and hybrid reed bed treatment systems. In *Materials from 6th international conference on wetland system for water pollution control* (pp. 229–242). Brazil: Design of Wetland Systems.

*Environment Ministry Regulation according limits for discharged sewage and environmental protection* from 24 July 2006 (Dz.U.no 137 item 984).

Gajewska, M., Kopeć, Ł., & Obarska-Pempkowiak, H. (2011). Operation of small wastewater treatment facilities in a scattered settlement. *Annual Set of Environment Protection, 13*(1), 207–225.

Ghrabi, A., Bousselmi, L., Masi, F., & Regelsberger, M. (2011). Constructed wetland as a low cost and sustainable solution for wastewater treatment adapted to rural settlements: The Chorfech wastewater treatment pilot plant. *Water Science & Technology, 63*(12), 3006–3012.

Halicki, W. (2009). The schedule of natural wastewater treatment system implementation in Podlasie region. Five years of experience. In D. Boruszko & W. Dąbrowski (Eds.), *Water, wastewater and solid waste. Low-cost wastewater and sludge treatment systems* (pp. 31–39). Białystok: Technical University of Bialystok (in Polish).

Heistad, A., Paruch, A. M., Vråle, L., Adam, K., & Jenssen, P. D. (2006). A high-performance compact filter system treating domestic wastewater. *Ecological Engineering, 28*(4), 374–379.

Jóźwiakowski, K. (2012). Studies on the efficiency of sewage treatment in chosen constructed wetland systems. *Infrastructure and Ecology of Rural Areas*, 1/2012, 232.

Kadlec, R. H., & Wallace, S. D. (2009). *Treatment wetlands* (2nd ed.). Boca Raton/London/New York: CRC Press Taylor & Francis Group.

Masi, F., Caffaz, S., & Ghrabi, A. (2013). Multi-stage constructed wetlands systems for municipal wastewater treatment. *Water Science and Technology, 67*(7), 1590–1598.

Molle, P., Lienard, A., Boutin, C., Merlin, G., & Iwema, A. (2005). How to treat raw sewage with constructed wetlands: An overview of the French systems. *Water Science and Technology, 51* (9), 11–21.

Molle, P., Prost-Boucle, S., & Lienard, A. (2008). Potential for total nitrogen removal by combining vertical flow and horizontal flow constructed wetlands: A full-scale experiment study. *Ecological Engineering, 34*, 23–29.

Obarska-Pempkowiak H., Gajewska M., Wojciechowska E. (2010). Hydrofitowe oczyszczanie wód i ścieków. Wydawnictwo Naukowe PWN, 307 s.

ÖNORM B 2505. (2009). Subsurface flow constructed wetlands – Application, dimensioning, installation, and operation (update to 1997 publication). *Bepflanzte Bodenfilter (Pflanzenklaranlagen) – Anwendung, Bemessung, Bau und Betrieb*. Vienna: Osterreichisches Normungsinstitut.

Paruch, A. M., Mæhlum, T., Obarska-Pempkowiak, H., Gajewska, M., Wojciechowska, E., & Ostojski, A. (2011). Rural domestic wastewater treatment in Norway and Poland: Experiences, cooperation and concepts on the improvement of constructed wetland technology. *Water Science and Technology, 63*, 776–781.

Steer, D., Fraser, L., Boddy, J., & Seifert, B. (2002). Efficiency of small constructed wetlands for subsurface treatment of single-family domestic effluent. *Ecological Engineering, 18*, 429–440.

Vymazal, J. (1998). Czech Republic. In J. Vymazal, H. Brix, P. F. Cooper, M. B. Green, & R. Haberl (Eds.), *Constructed wetlands for wastewater treatment in Europe*. Leiden: Backhuys Publishers.

Vymazal, J. (2005). Horizontal sub-surface flow and hybrid constructed wetlands systems for wastewater treatment. *Ecological Engineering, 25*(5), 478–490.

Vymazal, J. (2009). The use constructed wetlands with horizontal sub-surface flow for various types of wastewater. *Ecological Engineering, 35*, 1–17.

Vymazal, J. (2010). Constructed wetlands for wastewater treatment. *Water, 2*, c530–c549.

# Chapter 17
# Treatment Wetland for Overflow Stormwater Treatment: The Impact of Pollutant Particles Size

Magdalena Gajewska, Marzena Stosik, Ewa Wojciechowska, and Hanna Obarska-Pempkowiak

**Abstract** The problem of stormwater treatment in urban areas has become increasingly crucial. It has been widely recognized that both mechanical and biological treatment of stormwater is necessary to protect surface water against pollution. Moreover, technology must be applied to ensure effective treatment in changing hydraulic conditions and to serve, to the extent possible, as a retention volume. Treatment wetland (TW) could be one of the best solutions for stormwater treatment. In this paper, we describe the problem of removing suspended solids in a two-stage TW (pond and horizontal subsurface flow bed). We also discuss the result of particle size distribution in compartments of the analyzed TW during different weather conditions.

**Keywords** Stormwater treatment • Suspended solids • Particle size distribution • Treatment wetland

## 17.1 Introduction

Adapting to EU requirements, Polish Environment Protection National Strategy assumes reduction of the pollutants load discharged to the Baltic Sea of 80 % by 2020 (EC, 2000). Nevertheless, achieving this goal will not be possible without limiting the load of stormwater pollutants treated insufficiently. Many studies conducted in different countries have pointed out that stormwater runoff from urban areas is highly contaminated and can have a negative impact on receiving water bodies (Sansalone and Tribouillard 1999; Revitt et al. 2004; Shutes et al. 1999). The type and character of the catchment area determines the concentration of organic and inorganic substances with different dispersion degrees in the runoff from urbanized area. The characteristic feature of stormwater runoff is

M. Gajewska (✉) • M. Stosik • E. Wojciechowska • H. Obarska-Pempkowiak
Civil and Environmental Engineering, Gdańsk University of Technology, Narutowicza st. 11/12, 80-233 Gdańsk, Poland
e-mail: mgaj@pg.gda.pl

fluctuating quantity and composition dependence, among others, on intensity and frequency of storm events and the type of catchment area. Stormwater coming from streets and parking lots contains hydrocarbons, oils and grease, lead compounds from engines, and salts used for snow and ice removal from the streets (mainly sodium and calcium chlorides).

The most contaminated is the first flush (about 20 min) of storm events which may be even more polluted than municipal wastewater (Welker 2007; Gasperi et al. 2011). Suspended solids in stormwater are the most relevant pollutant due to quite large loads of pollutants (organic matter, nutrients, heavy metals, etc.) adsorbed on the particles' surfaces. Particles of different sizes carry different types and quantities of pollutants. Worn-off particles of car tires mixed with the surface of roads, street rubbish, remains of soil erosion, incombustible leavings of fuels, as well as dusts and aerosols washed off with rain are the sources of total suspended solids in stormwater. It is estimated that 665 kg of total suspended solids outflow from 1 ha of impermeable surface. Mineral suspended solids are predominant in stormwater, reaching 90–99 % of total suspended solids (Carleton et al. 2001; Królikowski and Królikowska 2009). The share of volatile suspended solids increases in autumn when leaves fall; it is also higher when marketplaces or the green areas are located in the catchment area (Welker 2007).

Also, other pollutants are adsorbed on the suspended solids' particles, like organic compounds, heavy metals, greases, mineral oils, and phosphates (Welker 2007; Gasperi et al. 2011). According to Królikowski and Królikowska (2009), up to 92 % of organic substances (COD and $BOD_5$), up to 80 % of nitrogen, and up to 99 % of hydrocarbons and lead are associated with the TSS. In consequence, the runoff contains both mineral and organic compounds, and thus both mechanical and biochemical treatment should be applied. Suspended solids removal in sedimentation processes allows for reduction of the concentration of other pollutants as well. Thus, particle size distribution is of importance in water and wastewater treatment. Up to now, it has been proved that treatment wetlands (TWs) efficiently remove TSS (Kadlec and Knight 1996; Kadlec and Wallace 2009; Obarska-Pempkowiak et al. 2011). Sedimentation is one of the basic mechanisms of removing pollutants in treatment wetlands. The following factors may have influence on the efficiency of pollutant sedimentation in stormwater treatment wetlands: flow conditions, origin and particle size of pollutants, retention time, and weather conditions (air temperature, frequency, and amount of precipitation) (Li et al. 2007). Treatment wetland can be the stormwater best management practice since it provides retention as well as both mechanical and biological treatment.

The objective of the investigation is to evaluate the impact of TSS particle size distribution during treatment of stormwater runoff in a TW. This could be helpful in dimensioning of such systems for effective retention of fine-grained particles, which are most harmful for aquatic environment.

## 17.2 Methods

### 17.2.1 Study Facility

The investigations were carried out at a facility (pond + horizontal subsurface flow bed planted with reed) located at the Swelina Stream at the border of the towns of Gdynia and Sopot. Swelina is a short stream (total length of 2.630 km) with its springs located in the Tricity Landscape Park area, outflowing directly to the Gulf of Gdańsk, in a well-known tourist area in Sopot. The catchment area of the Swelina Stream covers 316 ha, including 237 ha of forests, 24.9 ha of urban area, and 54.7 ha of open green area. The stream receives waters from 12 stormwater drainage outlets, as well as the surface runoff from adjacent green areas. In 1994, a treatment system, consisting of a retention pond (500 m$^3$) and a horizontal subsurface flow bed (960 m$^2$, 1 m depth, providing 2 h retention time for designed flow), was constructed (Fig. 17.1). The treatment system was designed to protect waters of Swelina and was part of a coastal water protection and restoration program. The treated water is collected by a drainage system and delivered to the control well (sampling point 3, Fig. 17.1) and then discharged to waters of the stream. The flow in the dry period varies from 12 to 33 l s$^{-1}$. During intensive rainfall, the first most polluted flush is directed through overflow grid and sand trap, then kept in the retention pond, and finally treated in an HSSF bed.

In Fig. 17.1, sampling points are marked: SP-1 at the inflow to the retention tank; SP-2 after the tank, at the overflow to the HSSF bed; and SP-3 after the HSSF bed (control well).

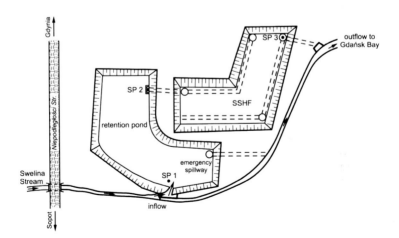

**Fig. 17.1** The scheme of the treatment system at Swelina Stream

## 17.2.2 Sampling and Measurements

The sampling events during 3 years were divided into periods with rain or snowmelt and dry weather periods in order to evaluate the effectiveness of the treatment system during different flow conditions. In collected samples, the measurements of organic matter (COD, $BOD_5$), total suspended solids (TSS), and volatile suspended solids (VSS) were carried out according to the Polish standards and guidelines given in the Regulation of Minister of Environment dated July 24 (2006). The measurement procedures are consistent with APHA (2005). A Malvern Instruments Ltd. 2000 laser granulometer was used for the measurement of suspended solid particles size distribution in the unit volume. The presence of suspended solid particles in waste samples causes laser light dispersion (diffraction), and the angle of refraction of light is inversely proportional to the size of the particles. The device allows us to measure particles of equivalent diameter from 0.02 to 2,000 μm. The measurement results can be printed in the form of histograms or curves of size composition, which makes the evaluation of granulometric composition of suspended solids possible.

## 17.3 Results and Discussion

### 17.3.1 Quality of Influent to TW Facility During Different Weather Conditions

In Fig. 17.2, the concentrations of $BOD_5$, COD, and TN in treated water during dry weather, rainy weather, and snowmelt are presented. In Table 17.1, the removal efficiency of analyzed pollutants is presented.

The highest inflow concentrations were observed during snowmelt even three times higher in comparison to dry weather. The concentrations during rainy weather also exceeded the dry weather ones. COD, $BOD_5$, and TN concentrations were not reduced in the first stage of treatment (retention tank), indicating that sedimentation at this stage

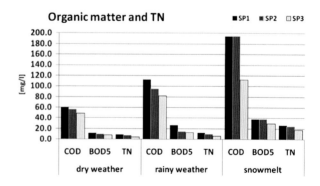

**Fig. 17.2** The fluctuation of organic matter and total nitrogen concentrations in various treatment stages of TW Swelina during different weather conditions

# 17 Treatment Wetland for Overflow Stormwater Treatment: The Impact of...

**Table 17.1** The removal efficiency (%) of organic matter, total nitrogen, and suspended solids (total, mineral, and volatile) in TW Swelina

| Weather/parameter | BOD$_5$ | COD | TN | TSS | MSS | VSS |
|---|---|---|---|---|---|---|
| Dry | 19.5 | 33.3 | 47.7 | 5.8 | −35.0 | 14.1 |
| Rainy | 27.3 | 53.4 | 45.2 | 36.4 | 63.0 | −10.1 |
| Snowmelt | 42.1 | 21.1 | 33.2 | 17.5 | 13.7 | 21.3 |

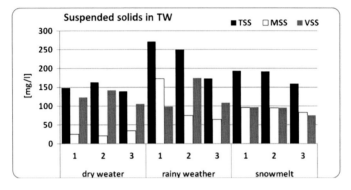

**Fig. 17.3** Concentrations of TSS, MSS, and VSS after subsequent treatment stages of TW Swelina during different weather conditions

of treatment failed in reducing the organic matter concentrations. Also, VSS concentrations were not decreased in the retention tank (Fig. 17.3). Figure 17.3 also shows that mineral suspended solids (MSS) were removed at this stage of treatment only during rainy weather periods. However, during rainy weather, VSS concentrations increased after a retention pond, indicating resuspension of fine particles of organic solids (Table 17.1). This was associated with a remarkable increase in COD concentration after this stage of treatment during rainy weather (Fig. 17.2).

COD, BOD$_5$, and TN concentrations decreased after treatment in the HSSF bed in all weather conditions. The average concentrations for dry weather were close to background concentrations for BOD$_5$ and TN (7.8 mg L$^{-1}$ and 4.5 mg L$^{-1}$, respectively), while the COD concentrations were much higher than that of the background and equaled 48.7 mg L$^{-1}$. During rainy or snowmelt events, the outflow concentrations of these pollutants were much higher in comparison to dry weather. For rainy weather, the outflow concentrations were BOD$_5$, 12.5 mg L$^{-1}$; COD, 81.8 mg L$^{-1}$; and TN, 6.8 mg L$^{-1}$, while for snowmelt the concentrations were much higher: BOD$_5$, 30.0 mg L$^{-1}$; COD, 112.6 mg L$^{-1}$; and TN, 6.8 mg L$^{-1}$. During dry weather, the TW exhibited very effective removal of TN which amounted to 47.7 % of discharged load of nitrogen, while for COD, TSS, and MSS the highest removal efficiency was observed during rainy weather. During snowmelt weather, the system served mainly as a BOD$_5$, VSS, and TN trap (Table 17.1).

The concentrations of TSS during rainy weather decrease significantly after retention in an HSSF bed, and moreover, the share of VSS in TSS increases (Fig. 17.4). This

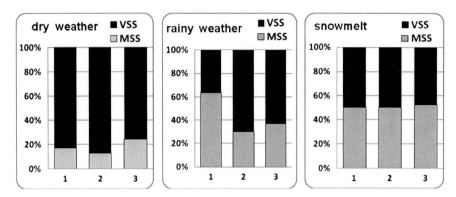

**Fig. 17.4** The composition of suspended solids in dry and rainy weather periods and snowmelt period at three sampling points

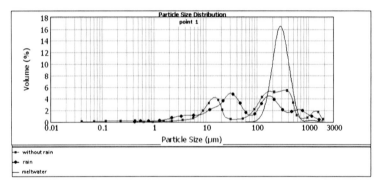

**Fig. 17.5** Particle size distribution for the sampling point 1 – inflow to the retention tank at TW Swelina

could suggest that during storm events, both types of accumulated organic matter, or more likely some microorganisms like diatoms, are washed out from the HSSF bed. Such a phenomenon is quite common in systems with biofilms which prefer stable hydraulic loads. According to Albuquerque et al. (2009), the release of microorganisms from an HSSF bed has no impact on further efficiency of removal of the system. Although the quality of discharged treated wastewater in such cases very often does not meet the requirements, the microorganisms washed out from TWs do not cause any problem to the natural environment (Kadlec and Wallace 2009).

### 17.3.2 Particle Size Distribution During Stormwater Treatment in TW Swelina

Figures 17.5, 17.6, and 17.7 illustrate the size distributions of particles in inflowing water to all three subsequent stages at TW Swelina during dry, wet, and snowmelt periods.

# 17 Treatment Wetland for Overflow Stormwater Treatment: The Impact of... 255

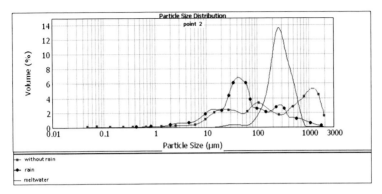

**Fig. 17.6** Particle size distribution for the sampling point 2 – inflow to the HSSF reed bed at TW Swelina

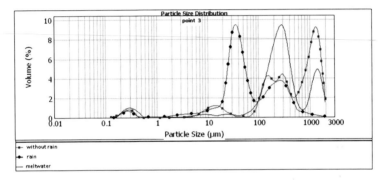

**Fig. 17.7** Particle size distribution for the sampling point 3 – outflow from the HSSF bed at TW Swelina

Granulometric analyses indicated that the size of particles inflowing to a retention pond varied widely. The effective diameters of solids at SP-1 varied from 0.05 to 2,000 μm, both for rainy and dry weather. In case of rainy weather, a higher share of solids in the range from 50 to 2,000 μm was observed, while during dry weather conditions, finer solids (2–1,000 μm) prevailed. In case of snowmelt, the observed range of particle diameters was from 100 to 1,000 μm. These findings are consistent with results reported by Burszta-Adamiak et al. (2011) for stormwater in large cities in Poland.

The granulometric analyses of particle size distribution at SP-2 indicated that the share of particles with diameter <80 μm decreased during rainy weather, while the share of particles with diameters from 400 to 2,000 μm increased. Also, at SP-3 (after HSSF) during dry periods, particle diameters from 10 to 100 μm were prevalent, while during rainy weather, the share of smaller particles decreased below 10 %, and larger particles diameters (80–2,500 μm) dominated. The size of some particles in treated water (after SP-3) was similar to the size of some water algae, for instance, diatoms, suggesting that these microorganisms may be present in treated water as a consequence of being washed out from the HSSF. According to many authors, HSSF beds are sensitive to clogging with suspended solids, especially mineral particles (Vymazal and Kropfelova 2008; Knowles et al. 2011). During the 19-year operation period, the substrate of the HSSF bed in TW Swelina became clogged. This was partly due to unstable hydraulic and organic loadings. The surface flow was observed during visits to study the facility. Analyses of results indicated low efficiency of TSS removal and consequently also other pollutants (Figs. 17.2 and 17.3). According to a recent study by Meyer et al. (2013), it is recommended to apply effective mechanical pretreatment for combined sewer overflow (CSO) system before it is discharged to TWs. In the Italian TW systems for CSO, the final stage of treatment takes place in a pond.

## 17.4 Conclusions

1. Pollutants delivered together with stormwater and waters of the Swelina Stream were characterized by a wide range of particle diameters typical for both colloidal pollutants and suspended solids.
2. The applied treatment system consisting of a pond and a vegetated bed (HSSF) is not effective in removing suspended solids.
3. Both the composition and the size of particles of suspended solids in the treated water can suggest that in the outflow from the wetlands, there are microorganisms (diatoms) which are not dangerous for the environment.
4. According to the latest studies, a surface flow TW should be used instead of an HSSF bed, which is sensitive to clogging.

**Acknowledgments** This scientific research has been carried out as a part of the project "Innovative resources and effective methods of safety improvement and durability of buildings and transport infrastructure in the sustainable development" financed by the European Union from the European Fund of Regional Development based on the Operational Program of the Innovative Economy.

The Provincial Fund for Environmental Protection and Water Management sponsored the Malvern Instruments Ltd. 2000 laser granulometer purchase.

## References

Albuquerque, A., Arendacz, M., Gajewska, M., Obarska-Pempkowiak, H., Randersoen, P., & Kowalik, P. (2009). Removal of organic matter and nitrogen in an horizontal subsurface flow (HSSF) constructed wetland under transient loads. *Water Science and Technology, 60*(7), 1677–1682.

APHA. (2005). *Standard methods for the examination of water and wastewater* (21st ed.). Washington, DC: American Public Health Association.

Burszta-Adamiak, E., Łomotowski, J., Kuśnierz, M., & Smolińska, B. (2011). Oczyszczanie wód z zawiesin w systemach hydrofitowych (Removal of suspended solids in treatment wetlands). *Gaz, Woda i Technika Sanitarna, 12*, 483–485. (In Polish).

Carleton, J. N., Grizzard, T. J., Godrej, A. N., & Post, H. E. (2001). Factors affecting the performance of stormwater treatment wetlands. *Water Research, 35*(6), 1552–1562.

EC (European Commission). (2000). *Water framework directive 2000/60/EC of the European Parliament and of the Council of 23 October 2000 establishing a framework for Community action in the field of water policy.* Brussels: Official Journal of the European Community.

Gasperi, J., Garnaud, S., Rocher, V., & Moiller, R. (2011). Priority substances in combined sewer overflows: Case study of the Paris sewer network. *Water Science and Technology, 63*(5), 853–858.

Kadlec, R. H., & Knight, R. L. (1996). *Treatment wetlands.* Boca Raton: CRC Press.

Kadlec, R. H., & Wallace, S. (2009). *Treatment wetlands* (2nd ed., p. 1116). Boca Raton/New York: CRC Press/Taylor & Francis Group.

Knowles, P. R., Dotro, G., Nivala, J., & Garcia, J. (2011). Clogging in subsurface-flow treatment wetlands: Occurrence and contributing factors. *Ecological Engineering, 37*(2), 99–112.

Królikowski, A., & Królikowska, J. (2009). Ocena wpływu współczynników spływu i opóźnienia na przepływy obliczeniowe w sieci kanalizacji deszczowej (Assessment of the impact of runoff coefficients and delays on the flow calculation of storm water drainage network). *Annual Set of Environment, 11*, 163–171.

Li, Y., Deletic, A., & Fletcher, T. D. (2007). Modeling wet weather sediment removal by stormwater constructed wetlands: Insights from a laboratory study. *Journal of Hydrology, 338*, 285–296.

Meyer, M., Molle, P., Esser, D., Troesch, S., Masi, F., & Dittmer, U. (2013). Constructed wetlands for combined sewer overflow treatment – Comparison of German, French and Italian approaches. *Water, 5*, 1–12.

Obarska-Pempkowiak, H., Gajewska, M., Wojciechowska, E., & Stosik, M. (2011). Constructed wetland systems for aerial runoff treatment in the Gulf of Gdańsk region. *Rocznik Ochrony Środowiska, 13*(1), 173–185.

Polish standards according limits for discharged sewage and environmental protection from July, 24 2006 (No 137 item 984) and January, 28 2009 (No 27 item 169).

Revitt, D. M., Shutes, R. B. E., Jones, R. H., Forshaw, M., & Winter, B. (2004). The performances of vegetative treatment systems for highway runoff during dry and wet conditions. *The Science of the Total Environment, 334–335*, 262–270.

Sansalone, J. J., & Tribouillard, T. (1999). Variation in characteristics of abraded roadway particles as a function of particle size – Implications for water quality and drainage. *Journal of Transportation Research Record, 1690*, 153–163.

Shutes, R. B. E., Revitt, D. M., Lagerberg, I. M., & Barraud, V. C. E. (1999). The design of vegetative constructed wetlands for the treatment of highway runoff. *The Science of the Total Environment, 235*, 189–197.

Vymazal, J., & Kropfelova, L. (2008). *Wastewater treatment in constructed wetlands with horizontal sub-surface flow.* Dordrecht: Springer.

Welker, A. (2007). Occurrence and fate of organic pollutants in combined sewer systems and possible impacts on receiving waters. *Water Science and Technology, 56*, 141–148.

# Chapter 18
# Treatment Wetlands in Rural Areas of Poland for Baltic Sea Protection

Katarzyna Kołecka, Magdalena Gajewska, and Hanna Obarska-Pempkowiak

**Abstract** Water and wastewater management, especially in rural areas, greatly affects loads of nutrients discharged to the Baltic Sea. In Poland, this management is unbalanced because of the dispersed development and in many locations construction of sewerage system is uneconomic or even impossible. For this reason, a significant part of sewage from single-family houses in rural areas must be discharged into domestic sewage systems. The aim of this paper was to analyze the data concerning the Baltic Sea catchment area as well as to present some current results of research on wastewater management mainly in rural areas. Small on-site wastewater treatment plants are usually defined as facilities serving up to 50 people. Depending on the law, their maximum capacity is set at 5 $m^3 d^{-1}$ (according to the Polish Act of Water Law). In rural areas, treatment wetlands seem to be especially useful as on-site wastewater treatment plants. The study indicates the high efficiency of pollutants removal in treatment wetlands, namely, hybrid treatment wetlands with at least two stages of treatment. High treatment efficiency is very important for the protection of the Baltic Sea. The reliability assessment based on Weibull distributions allows for the risk assessment of exceeding the limit values of pollutants in treated wastewater. Very good reliability was observed for TN and COD (303 and 292 day per year, respectively), while reliability for $BOD_5$ and TSS was weaker (219 and 153 day per year, respectively). Proper design and operation of treatment wetland facilities are important in order to achieve a good quality of effluent in the interests of Baltic Sea protection.

**Keywords** Baltic Sea • Wastewater management • On-site wastewater treatment plants • Treatment wetlands

K. Kołecka • M. Gajewska (✉) • H. Obarska-Pempkowiak
Faculty of Civil and Environmental Engineering, Gdansk University of Technology, Gdańsk, Poland
e-mail: mgaj@pg.gda.pl

## 18.1 Introduction

In Poland, about 99.7 % of the territory belongs to the Baltic Sea catchment area. Four catchments of the Baltic Sea belong to the Vistula basin (representing 55.7 % of the country territory) and the Oder basin (33.9 %), coastal river basin (9.3 %), and Neman basin (0.8 %), respectively (Pastuszak 2009).

The Baltic Sea is completely surrounded by land; moreover, it is a relatively shallow regional sea with an average depth of 53 m. Only a few shallow straits connect the Baltic Sea to the North Sea. Its total area is 415,300 km$^2$, and the volume of water is 21,547 km$^3$. The volume of river water flowing into the sea is only about 2 % of its total volume. The Baltic Sea belongs to the brackish seas, due to the low salinity (mean 7 psu – practical salinity units). Annually, only 3 % of the water volume is exchanged. The time period of total water exchange is very long and equals 25–30 years. As a result, dissolved pollutants accumulate and contribute to eutrophication (HELCOM 2009; GIOŚ 2009). The factors responsible for the eutrophication are nutrients represented by nitrogen and phosphorus. This natural process is significantly accelerated by the anthropogenic impact.

Water and wastewater management, especially in rural areas, plays an important role in loads of nutrients discharged to the Baltic Sea. In Poland, this management is unbalanced because of the dispersed development and in many places construction of a sewerage system is uneconomic or even impossible. For this reason, a significant part of sewage from single-family houses in rural areas must be discharged into domestic sewage systems.

The aim of this chapter was to present an analysis of data concerning the Baltic Sea catchment and to give a state-of-the-art report on the results of research on wastewater management, mainly in rural areas.

## 18.2 Influence of Rural Areas for Quality of Surface Water in Poland

In the Baltic Sea, 56 % of nitrogen and 49 % of phosphorus compounds originate from diffuse pollution, whereas only 10 % of nitrogen and 25 % of phosphorus came from point sources (HELCOM 2005). The natural background consists of 31.0 % nitrogen and 25.0 % phosphorus. In recent years, the share of loads from point sources in the overall load of nutrients discharged into the Baltic Sea has strongly decreased. In this area, there has been no significant improvement in several years (Forsberg 1991).

Fertilizers used in agriculture are a significant source of the nutrients load flowing into the rivers and then into the sea. In Poland, the use of nitrogen in mineral fertilizers is still lower than in most EU countries. However, on average, 56 % of crop production in Poland is fertilized with nitrogen.

Rivers are the main source of nutrients for the Baltic Sea. In Poland, the main problem is created by two rivers: Vistula and Oder. It is estimated that approximately 75 % of the total nitrogen (TN) load and 95–99 % of the total phosphorus (TP) load are supplied to the Baltic Sea by rivers. The rest of nutrients load comes from the atmosphere. Approximately 25 % of the nitrogen load comes from atmospheric deposition, which is the second major source of pollution (HELCOM 2004).

The loads of TN and TP discharged by the Polish rivers, as well as the Polish point sources directly polluting the Baltic Sea, represent about 26 % of TN load and 37 % of TP load discharged into the Baltic Sea by all Baltic countries (Pastuszak 2012).

In 10 years, $BOD_5$ load discharged by rivers into the Baltic Sea from Polish territory decreased about 49 % (reaching the level of 137,000 t year$^{-1}$). In the same time, total nitrogen load decreased about 70 % (from 260.5 to 77.9 thousand t year$^{-1}$) and total phosphorus load about 52 % (from 15.5 to 7.4 thousand t year$^{-1}$). In Poland, loads of nutrients discharged into the Baltic Sea per 1 ha amount to about 3.2 kg TN and 0.22 kg TP, and they are one of the lowest amounts among countries surrounding the sea basin. This is mainly the result of significant investment in municipal wastewater treatment plants, elimination of various types of industrial so-called hot-spot sources, and the implementation of the Best Management Practice (BMP). At the same time, relatively low flows of rivers during the last few years have been important for reduction of pollutant loads discharged to the Baltic Sea (GIOŚ 2010).

## 18.3 Community Management of Wastewater in Rural Areas

Water and wastewater management, especially in Polish rural areas, is very much responsible for the loads of nutrients flowing into the Baltic Sea. In rural areas, the water supply systems without sewerage systems dominate. In the urban and rural areas of Poland, the average number of people connected to 1 km of sewerage system are 400 and 74, respectively (Błażejewski 2012). The length of the sewerage system and the number of beneficiaries still have been growing. However, in the last years, only 27.8 % of the population of rural areas was connected to the sewerage system, while the percentage of people using the water supply systems was much higher and amounted to 75.7 % (Fig. 18.1).

Despite an increasing number of wastewater treatment plants (WWTP), many sewerage systems are not connected to WWTP. The reason is that almost two-thirds of the Polish rural population (above ten million) live in areas with dispersed development. During the last 13 years, the number of village WWTPs has increased from 433 to 2,213, but the percentage of rural population served by municipal WWTPs is only 25.7 % (GUS 2011, 2012).

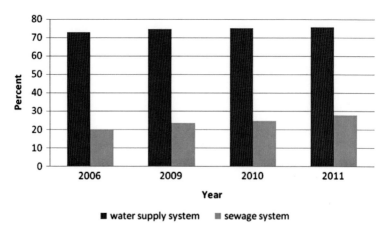

**Fig. 18.1** Number of people as percentage of Polish population in rural areas using water supply system and sewerage system (Data from GUS 2012)

Not all residential areas are connected to a sewage system. Some wastewater is discharged into individual sewage disposal system (septic tanks and small wastewater treatment plants). The number of septic tanks decreased from 2,433,000 in 2009 to 2,359,000 in 2011. During this time, the number of small wastewater treatment plants increased from 62,000 in 2009 by 39.8 % to 103,000 in 2011 (GUS 2012). It is estimated that for 3.8 million people, about 700,000 of small on-site wastewater treatment plants is needed. The estimated volume of treated sewage in these plants will be 384,800 $m^3 d^{-1}$ (Obarska-Pempkowiak et al. 2012).

Due to the dispersed development of Polish rural areas, small on-site wastewater treatment plants seem to be an option to solve the problem of wastewater management in Poland. It is likely that it will improve the quality of the Baltic Sea water.

## 18.4 Types of Small On-Site Wastewater Treatment Plants

Small on-site wastewater treatment plants are usually defined as systems serving up to 50 people. Depending on the law, their maximum capacity is 5 $m^3 d^{-1}$ according to the Polish Act of Water Law (Water Law 2001), and they are called single-family treatment plants. According to the Regulation of the Ministry of the Environment of 24 July 2006 (Environment 2006), wastewater from such facilities should only meet limits for $BOD_5$ (80 % load removal) and TSS (20 % load removal). Local treatment plants are the facilities for < 2,000 PE and must meet discharge limits (Table 18.1).

The most popular natural solutions in Polish rural areas are drainage (not recommended by the European Union) or sand filters. Less popular technologies in the rural areas are biological trickling filters and activated sludge reactors. The drainage is laid down below the ground, across which the effluent from septic tanks

**Table 18.1** The required discharged water quality of local treatment plants for less than 2,000 PE (Dz.U.no 137 item 984 year 2006)

| Parameter | mg L$^{-1}$ | g pe$^{-1}$ d$^{-1}$ |
|---|---|---|
| BOD$_5$ | ≤40 | 4.8 |
| COD | ≤150 | 18 |
| TSS | ≤50 | 6.0 |
| TN | ≤30 | 3.6 |
| TP | ≤5 | 0.6 |

is evenly spread. It can be used in permeable soils, and the distance from the ground should be at least 1.5 m. The disadvantage of this system is a large area requirement from 20 to 60 m$^2$/PE (personal equivalent), and the potential risk of contamination of groundwater (Environment 2006). The sand filter works similarly to the drainage. However, it is used when soil is impermeable or groundwater is located too low. The required filter area is smaller than for the drainage. This system requires the use of an expensive sand fraction, and it does not guarantee elimination of odors. Small activated sludge reactors can provide effective removal of organic pollutants as well as nitrogen and phosphorus compounds. However, the cost of construction is higher, and maintenance and operation requires professional skills. Biofilters (trickling filters) can be located in tanks with a construction similar to septic tanks. However, biofilters require a long start-up time, they are energy consuming, and secondary clarifiers are necessary to eliminate detached microbial biomass.

A relatively new method in Poland is the use of wetland treatment systems with macrophytes of woody species such as willows. The construction and operation costs are competitive with conventional solutions, and the natural look of these systems makes it easy to incorporate them into the rural landscape. In addition, the owners (usually farmers) do not have any problems in operating these systems (GUS 2012; Rosen 2002; Brzostowski et al. 2008).

## 18.5 Experiences with Treatment Wetlands in Poland

### *18.5.1 Single-Stage Facilities with Horizontal Flow*

Since the end of the 1980s, in Poland, dozens of treatment wetlands (TWs) with horizontal subsurface flow (HSSF) have been built as individual or single-family treatment systems. They were built among others in the vicinity of Ciechanów, Ostrołęka, and Lublin. The treatment wetlands have been tested for several years in the last decade (Obarska-Pempkowiak et al. 2012). The obtained results of the monitoring indicate that the efficiency of BOD$_5$ removal ranged between 25.6 % and 99.1 % with an average value of 62.4 %. The inflow loads varied between 11.2 and 115 kg ha$^{-1}$ d$^{-1}$. The efficiency of total nitrogen removal was lower and varied between 22.4 and 84.2 % (mean 44.5 %), with inflow loads varying between 8.5 and

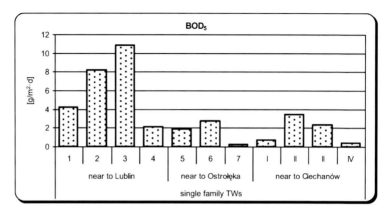

**Fig. 18.2** Removal of BOD$_5$ load in monitored single-family treatment wetlands (TWs) with horizontal subsurface flow

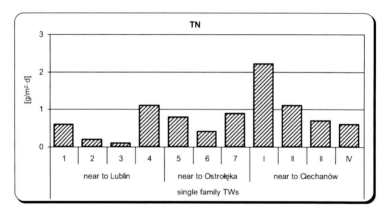

**Fig. 18.3** Removal of total nitrogen (TN) load in monitored single-family treatment wetlands (TWs) with horizontal subsurface flow

34 kg N ha$^{-1}$ d$^{-1}$. The removal of BOD$_5$ and total nitrogen loads is presented in Figs. 18.2 and 18.3, respectively.

Some of the treatment wetlands did not perform properly. The main reason for improper functioning was the incorrect operation of septic tanks which resulted in elevated concentrations of suspended solids and fats discharged into filtration beds that suffered from clogging.

In 2009, a research team from the Gdansk University of Technology built three single-family HSSF TWs in the community of Stężyca. For pretreatment, a primary settling tank and prefilter filled with Pollytag material were used. The pollutants removal efficiency in these facilities was acceptable although the treatment wetlands were fed unevenly. Average removal efficiency was 56.0 % for TSS, 82 % for BOD$_5$, and 59 % for TN. The lower total suspended solids removal was affected by very high TSS concentration in the influent (Obarska-Pempkowiak et al. 2012).

**Table 18.2** The efficiency of pollutants removal (in %) in three single-family TWs with VSSF in Sokoły

| Parameter | VSSF I | VSSF II | VSSF III |
|---|---|---|---|
| $BOD_5$ | 94.9 | 85.5 | 98.1 |
| COD | 79.4 | 31.5 | 93.4 |
| TSS | 41.5 | 33.7 | 94.8 |
| TN | 85.1 | 85.2 | 72.6 |
| TNK | 94.9 | 80 | 92.7 |
| $NH_4^+$-N | 95.7 | 94.8 | 98.5 |

## 18.5.2 Single-Stage Facilities with Vertical Flow

Since 2004, the community of Sokoły has built more than 600 single-family TWs using single-stage beds with vertical flow of sewage (VSSF).

The results of monitoring showed that these facilities were characterized by very high efficiency of pollutants removal. For $BOD_5$, removal efficiency varied between 86 and 98% and for COD from 79 to 94 %. VSSF beds were also found to effectively remove nitrogen (Table 18.2).

In the community of Stężyca, in addition to the above-mentioned HSSF facilities, three facilities with vertical flow and three facilities with two vertical beds working sequentially were built. In all facilities, septic tanks for pretreatment of wastewater were constructed. In these treatment wetlands, an average efficiency of pollutants removal was 75 and 78 % for TSS, 79 and 83 % for $BOD_5$, and 55 and 59.0 % for total nitrogen, respectively (Obarska-Pempkowiak et al. 2012). The results revealed that the most effective configuration was that with two sequentially working vertical flow beds.

## 18.5.3 Hybrid Treatment Wetlands

Hybrid treatment wetlands (HTWs) are applied because they require less area per population equivalent and provide more stable performance. HTWs intended as the local small on-site wastewater treatment plants in Darżlubie, Wiklino, Wieszyno, Sarbsk, and Schodno in Pomerania region in Poland were built in the early 1990s (Table 18.3). The treated wastewater was discharged directly to the Baltic Sea. After mechanical treatment in primary settling tanks, the wastewater was always pumped into HSSF as a first stage of HTWs. The mentioned systems differed from one another in the order and number of treatment stages (Gajewska and Obarska-Pempkowiak 2011). The efficiency of $BOD_5$ and nitrogen removal in HTWs is presented in Table 18.4.

The highest efficiency of $BOD_5$ removal was found in Wiklino, while the lowest in Schodno. The highest efficiency for TN removal was also found in Wiklino and the lowest in Darżlubie. The efficiency of nitrogen removal for four out of five facilities exceeded 60 %. This confirms that the configuration of alternating vertical and horizontal beds effectively removes nitrogen from the wastewater.

**Table 18.3** Characteristics of the hybrid treatment wetland systems

| Plant | Sarbsk | Wiklino | Wieszyno | Schodno | Darżlubie |
|---|---|---|---|---|---|
| Q,m$^3$day$^{-1}$ | 29.7 | 18.6 | 24.5 | 2.2 in winter | 56.7 |
| | | | | 8.9 in summer | |
| Configuration/area, m$^2$ | HF/1610 | HF I/1050 | HF I/600 | HF I/416 | HF I/200 |
| | VF/520 | VF/624 | VF/300 | VF I/307 | C b/400 |
| | ∑ 2,130 | HF II/540 | HF II/600 | HF II/432 | HF II/500 |
| | | ∑ 2,214 | ∑ 1,500 | VF II/180 | VF/250 |
| | | | | ∑ 1,300 | HF III/1000 |
| | | | | | ∑ 3,350 |

*HF* horizontal flow, *VF* vertical flow, *Cb* cascade bed

**Table 18.4** The efficiency of BOD$_5$ and total nitrogen removal in HTWs

| | Wiklino | Sarbsk | Wieszyno | Darżlubie | Schodno |
|---|---|---|---|---|---|
| BOD$_5$ | 95.9 | 95.1 | 86.9 | 82.1 | 78.5 |
| TN | 79.2 | 67.9 | 62.6 | 23.4 | 61.3 |

Values in %

### 18.5.4 Reliability Analysis of Hybrid Treatment Wetlands

HTWs were more stable and effective than single-stage facilities. For this reason, further considerations apply to these types of facilities. Reliability of pollutant removal in treatment wetlands in temperate climate is always a question of post-growing season. Long investigation (from 1998 to 2009) carried out in these four HTWs allows for the assessment of the removal efficiency and its fluctuation during both growing (May to October) and post-growing seasons (November to April) (Fig. 18.4). In all five HTWs, nitrogen compounds expressed as TN were removed with higher efficiency during growing period. The differences in TN removal efficiency did not exceed 12 % which is in accordance with the data given by many authors for maximum TN bioaccumulations in plants (Gajewska and Obarska-Pempkowiak 2011; Kadlec and Wallace 2009; Vymazal 1999, 2007). On the other hand, the TP removal efficiency was up to 11 % higher in post-growing period. Total suspended solids were generally more effectively trapped during the growing season similar to organic matter.

Another question is the reliability of the facility that means to ensure effective treatment in order to meet the effluent requirements imposed by the Regulation of the Minister of Environment from 24 July 2006 (Environment 2006). In the case of wastewater treatment plants, an important factor is working without any disturbances, which can be measured by how often during a year the effluent fulfills the requirements. This parameter could be estimated by statistical assessment with Weibull distributions. The reliability of the facility is expressed by the relationship:

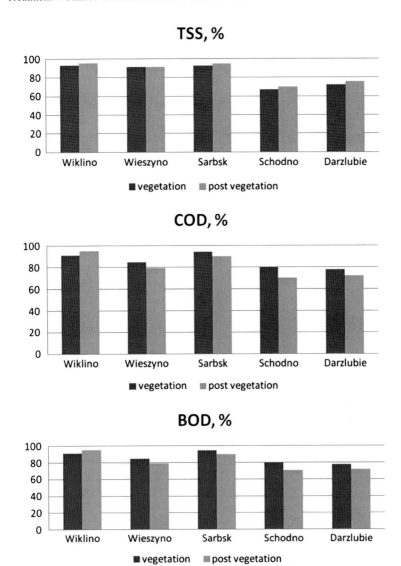

Fig. 18.4 (continued)

(reliability = 1 − Weibull distribution function). Figure 18.5 shows the distribution function of the pollutant concentrations in effluent from HTWs.

Although monitored HTWs were very effective in TSS removal, the effluent met the discharge requirements only in 42 % of cases (Figs. 18.4 and 18.5). Organic matter requirements in effluent were met for $BOD_5$ in 60 % and COD in 80 % of cases. Similar very high reliability of TN (up to 80 % of cases) was achieved for all four HTWs. This high reliability of TN removal is very important because Poland is

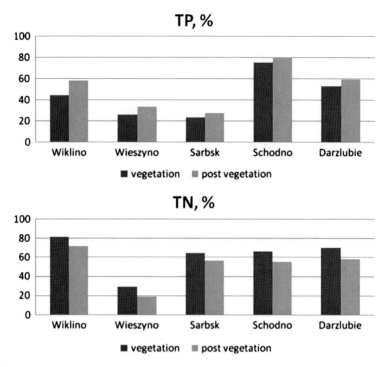

**Fig. 18.4** The efficiency of pollutant removal in HTWs in growing and post-growing periods in Pomerania region, Poland

an area susceptible to eutrophication and wastes are discharged directly into the Baltic Sea.

There are two main reasons to fail the requirements by analyzed HTWs; both are connected with design processes. It is presumed that most of TSS and particulate organics will be trapped in septic tanks, but the facility operators do not follow the rules of emptying them at least twice a year. In addition, in Polish design standards, it is recommended to calculate 150 l d$^{-1}$ population equivalent; however, from field measurements, it has been found that the real water consumption is only about 100 l d$^{-1}$. As a consequence, the HTWs were fed with much higher concentrations of pollutants, and in spite of very effective treatment processes, they could not meet the requirements in some cases.

A comparison of the mean values of organic mass removal rate (as BOD$_5$) and TN MRR confirms that hybrid facilities are more effective in comparison with single-stage ones (Table 18.5).

18 Treatment Wetlands in Rural Areas of Poland for Baltic Sea Protection

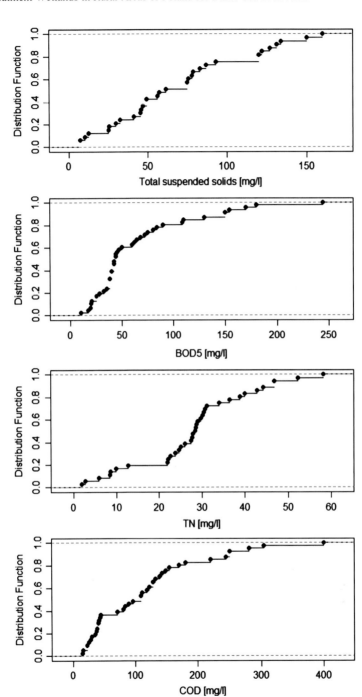

**Fig. 18.5** Cumulative distributions functions of TSS, COD, $BOD_5$, and TN in effluent from HTWs in Pomerania region in Poland

**Table 18.5** The mean values of BOD$_5$ and TN mass removal rates in single-stage and hybrid facilities

|  | BOD$_5$ (g m$^{-2}$ d$^{-1}$) | TN (g m$^{-2}$ d$^{-1}$) |
|---|---|---|
| Hybrid TWs | 5.90 | 4.01 |
| Local single-stage TWs | 0.99 | 0.50 |

## 18.6 Conclusions

Based on our analysis of the data concerning the Baltic Sea catchment area as well as state-of-the-art results of research on wastewater management, mainly in rural areas, the following conclusions can be formulated:

1. The reason of significant pollution of the Baltic Sea and its susceptibility to eutrophication is a long time of water exchange which is equal to 25–30 years.
2. Due to the dispersed development of Polish rural areas, small wastewater treatment plants seem to be an option to solve the problem of wastewater management in Poland.
3. Treatment wetlands technology seems to be especially useful in rural areas as a method for sewage management.
4. The study indicates the high efficiency of pollutants removal in treatment wetlands. Especially effective are hybrid treatment wetlands with at least two treatment stages; this is very important for the protection of the Baltic Sea.
5. The evaluation of reliability based on Weibull distributions allows for an assessment of the risk of exceeding the limit values of pollutants in treated wastewater.
6. Very good reliability of TN and COD (303 and 292 d year$^{-1}$, respectively) and weaker reliability for BOD$_5$ and TSS (219 and 153 d year$^{-1}$, respectively) effluent quality have been observed.
7. Proper design and operations of treatment wetland facilities are very important to achieve a good quality of effluent for Baltic Sea protection.

## References

Błażejewski, R. (2012). Stan i możliwości rozwoju infrastruktury wodociągowo-kanalizacyjnej w Polsce. (The condition and development opportunities of water and sewage infrastructure in Poland). *Gas, Water and Sanitary Engineering*, 2, 49–51.

Brzostowski, N., Hawryłyszyn, M., Karbowski, D., & Paniczko, S. (2008). *Przydomowe oczyszczalnie ścieków. Poradnik.* (On-site wastewater treatment plants. Guide). Narev, Bialystok, Poland: Podlasie Nature Station.

Environment. (2006). *Ministry Regulation according to the limits for discharged sewage and environmental protection from 24 July 2006.* Dz.U.no 137 item 984.

Forsberg, C. (1991). Eutrofizacja Morza Bałtyckiego. Środowisko Morza Bałtyckiego. (Eutrophication of the Baltic Sea. *Environment of the Baltic Sea* (3), electronic version.

Gajewska, M., & Obarska-Pempkowiak, H. (2011). Efficiency of pollutant removal by five multistage 3 constructed wetlands in a temperate climate. *Environment Protection Engineering, 37*, 27–37.

GIOŚ. (2009). *Program Państwowego Monitoringu Środowiska na lata 2010 – 2012*. (Programme of National Environmental Monitoring for the years 2010–2012). Warszawa.

GIOŚ. (2010). *Raport o stanie środowiska w Polsce 2008*. (Report on monitoring status in Poland in 2008). Warszawa: Biblioteka Monitoringu Środowiska.

GUS. (2011). *Ochrona Środowiska 2010* (Environment Protection 2010). Warsaw: GUS.

GUS (2012). *Infrastruktura komunalna w 2011 r.* (Municipal infrastructure in 2011). Wyd. Warsaw: GUS.

HELCOM. (2004). *The fourth Baltic Sea pollution load compilation (PLC-4)* (Baltic Sea Environment Proceedings No. 93). Helsinki: Baltic Marine Environment Protection commission.

HELCOM. (2005). *Nutrient pollution to the Baltic Sea in 2000, 2003* (Baltic Sea environment proceedings no. 100). Helsinki: Helsinki Commission.

HELCOM. (2009). *Baltic facts and figure*. http://www.helcom.fi/environment2/nature/en_GB/facts

Kadlec, R. H., & Wallace, S. (2009). *Treatment wetlands* (2nd ed.). Boca Raton: CRC Press.

Obarska-Pempkowiak, H., Gajewska, M., Wojciechowska, E., & Ostojski, A. (2012). *Oczyszczalnie w ogrodzie*. (Wastewater treatment plants in the garden). Warsaw: Wydawnictwo Seidel-Przywecki.

Pastuszak, M. (2009). *Charakterystyka zlewiska Morza Bałtyckiego. Udział Polskiego rolnictwa w emisji związków azotu i fosforu do Bałtyku* (Characteristics of the Baltic Sea catchment. The share of Polish agriculture in the emissions of nitrogen and phosphorus to the Baltic Sea.) (pp. 17–37). Pulawy: Research Institute – Institute of Soil Science and Plant Cultivation-State Research Institute – Fertilizer Research Institute.

Pastuszak, M. (2012). *Temporal and spatial differences in emission of nitrogen and phosphorus from Polish territory to the Baltic Sea*. Gdynia-Puławy: National Marine Fisheries Research Institute – Institute of Soil Science and Plant Cultivation-State Research Institute – Fertilizer Research Institute.

Rosen, P. (2002). *Przydomowe oczyszczalnie ścieków* (On-site wastewater treatment plants). Warsaw: Central Information Centre of Building.

Vymazal, J. (1999). Nitrogen removal in constructed wetland with horizontal sub-surface flow – Can we determine the key process? In J. Vymazal (Ed.), *Nutrient cycling and retention in natural and constructed wetlands* (pp. 1–17). Leiden: Backhuys Publishers.

Vymazal, J. (2007). Removal of nutrients in various types of constructed wetlands. *Science of the Total Environment, 380*(1–3), 48–65.

Water Law. (2001). Water Law from 18.July 2001 (Dz. U. no. 115, item 1229).

# Chapter 19
# Long-Term Performance of Constructed Wetlands with Chemical Dosing for Phosphorus Removal

Gabriela Dotro, Raul Prieto Fort, Jan Barak, Mark Jones, Peter Vale, and Bruce Jefferson

**Abstract** Fourteen full-scale wastewater treatment plants (WwTPs) owned and operated by Severn Trent Water, a major UK water utility, were assessed to determine the long-term efficiency of using chemical dosing upstream of biofilm processes followed by horizontal flow wetlands to remove phosphorus and any residual metals. Results showed that the flowsheets used are an effective strategy to remove BOD, solids, ammonia, nitrate, and phosphorus, with all sites meeting their discharge consents for the past 10 years. Wetland removal rates of up to 6.7, 3.8, 0.24, 2.4, and 0.5 mg m$^{-2}$ d$^{-1}$ were found for solids, nitrate, ammonia, phosphorus, and iron, respectively. Under flooded conditions, settling mechanisms dominated as evidenced by average surface sludge accumulation rates of 2 cm year$^{-1}$. Subsurface flow resulted in lower or no sludge accumulation on top of the gravel. The average concentration of P and Fe in the sludge per site was in the range of 17–48 and 9–178 g kg$^{-1}$, respectively, much higher than sediment concentrations found in non-dosed wetlands and river sediments. Analysis of suspended, colloidal, and dissolved fractions showed a change in phase for P from particulate to dissolved after the wetland, suggesting release. Phosphorus release was observed under low P conditions in wastewater (<1 mg L$^{-1}$) and high sludge P levels, with release rates of 12–52 44 mg m$^{-2}$ d$^{-1}$ in the field and 44 mg m$^{-2}$ d$^{-1}$ under laboratory conditions. Iron release was observed under controlled experiments but not detected in the field, indicating additional transformation reactions take place in full-scale wetlands. Further research is required to provide adequate data for

---

G. Dotro (✉)
School of Applied Sciences, Cranfield University, Cranfield, Bedfordshire MK43 0AL, UK

Waste Water Research and Development, Severn Trent Water, Coventry, West Midlands CV1 2LZ, UK
e-mail: g.c.dotro@cranfield.ac.uk

R.P. Fort • J. Barak • B. Jefferson
School of Applied Sciences, Cranfield University, Cranfield, Bedfordshire MK43 0AL, UK

M. Jones • P. Vale
Waste Water Research and Development, Severn Trent Water, Coventry, West Midlands CV1 2LZ, UK

modifying the design and improving the operation and maintenance recommendations to ensure wetlands continue to deliver environmental benefit at WwTPs required to remove phosphorus.

**Keywords** Phosphorus • Iron • Aluminium • Metals • Reed beds • Sludge

## 19.1 Introduction

Nutrient enrichment can degrade water bodies by causing oxygen depletion and altering the competitive balance between different aquatic plant and animal species. Phosphorus (P) is considered one of the main contributors, and, as such, P concentrations in freshwater have been proposed to be maintained as near to background levels as possible (Mainstone and Parr 2002). Of particular interest are municipal wastewater treatment plant (WwTPs) as they discharge predominately bioavailable orthophosphate and they are point sources and therefore easier to regulate and manage (Mainstone et al. 2000). Consequently, technology upgrades on existing WwTPs that can help achieve lower effluent P concentrations are of interest. Conventional WwTPs can use aluminium or iron dosing for chemical removal of P, or the conditions within biological treatment processes can be manipulated to enable enhanced biological P removal. The latter requires exposure of biomass to anaerobic, anoxic, and aerobic conditions in a particular sequence and performs better with influents that have significant volatile fatty acid contents. In small WwTPs, the biological treatment stage is typically performed by biofilm technologies, making biological P removal impractical and chemical P removal the only reliable option.

Constructed wetlands have been used for over two decades throughout the world for the removal of organic matter, nitrogen, and suspended solids. There are more than 115,000 wetlands in Europe for wastewater treatment with the vast majority of those used for domestic wastewater treatment (Shutes 2012). The removal of P with wetlands has long been recognised as a limitation of the technology, with removal being limited at start-up and dropping off as the systems mature and saturation of sorption sites occurs (Kadlec and Wallace 2009). One approach is the use of chemical salts, either iron or aluminium based, for reaction with P upstream and use of the wetland for storage and final solids polishing (Brix and Arias 2005; Lauschmann et al. 2013). However, limited long-term, full-scale data is available, thus limiting the uptake of the approach. Severn Trent Water, a major water utility in the UK, has developed a flowsheet where this strategy has been employed since 2005, using a horizontal flow wetland as a tertiary polishing step trapping any residual iron, aluminium, or phosphorus.

Fourteen full-scale WwTPs with chemical dosing for the removal of phosphorus upstream of horizontal flow wetlands were assessed in this study. The role of the wetlands is analysed in terms of treatment performance by system age, sludge storage capacity, and reliability of long-term pollutant storage. The implications of using this approach to achieve tighter effluent discharge levels and future research needs are identified based on the findings of this study.

## 19.2 Materials and Methods

### 19.2.1 Sites and Treatment Flowsheets

Site visits were conducted in order to determine the current condition, any operational issues, the dosing strategy, and the wetland characteristics of each of the 14 sites (Table 19.1). Three types of secondary biological treatment technologies were found: rotating biological contactors (RBCs), trickling filters (TFs), and submerged aerated filters (SAFs). In all cases, the wetlands were used for tertiary treatment (Fig. 19.1). Site 2 is the only aerated wetland of the sample set, with aeration provided by means of perforated tubing at the bottom of the horizontal flow bed, connected to an air compressor. The chemical dose used in all but one site was commercial grade ferric sulphate ($Fe_2(SO_4)_3$ $8H_2O$, 12.5 %), with one site using aluminium sulphate instead due to improved reactivity found during initial jar testing (*data not shown*). This site (site 13) also requires alkalinity dosing to enable nitrification to take place, achieved through addition of sodium carbonate. Mixing of the chemical was either performed in line with a slow drip into the main flow (relying on flow turbulence) or direct into the RBC biozone (relying on the disc rotation to provide mixing).

### 19.2.2 Sampling Records

Historic performance records were obtained from Severn Trent Water's databases. Influent to the wetlands (secondary effluent) and final effluent (exiting the wetlands) were queried for 5-day biochemical oxygen demand (BOD), total suspended solids (TSS), phosphorus (TP), orthophosphates (OP), total iron (Fe), ammonium ($NH_4$–N), nitrites ($NO_2^-$), total oxidised nitrogen (TON), pH, and daily flow. Records were split into before and after start of dosing, and sample sizes were approximated to be equal with a minimum of 5 years of data for the "before" period. For the six sites where data for influent and effluent to the wetlands was available, paired t-tests were conducted at the 0.05 level of significance.

Table 19.1 Summary of site characteristics

| Site ID | Population equivalent | Average flow (m³ day⁻¹) | Secondary biological treatment | Constructed wetlands Area (m²) | Area/PE (m²/PE) | Hydraulic load (mm day⁻¹) | Residence time (hours)[a] | Dosing point | Dosing start date | Years since refurbishment at time of sampling | Consent[b] (mg L⁻¹) BOD | TSS | NH₃ | TP | TFe |
|---|---|---|---|---|---|---|---|---|---|---|---|---|---|---|---|
| Site 1 | 526 | 145 | RBC | 364 | 0.7 | 399 | 14 | B | 03/2012 | 0.25 | 40 | 60 | n/a | 2 | 4 |
| Site 2 | 275 | 67 | RBC | 419 | 1.5 | 160 | 36 | B | 03/2013 | 0.25 | 25 | 45 | 15 | 2 | 4 |
| Site 3 | 795 | 136 | RBC | 480 | 0.6 | 283 | 20 | B | 03/2012 | 0.25 | 30 | 50 | n/a | 2 | 4 |
| Site 4 | 222 | 72 | SAF | 187 | 0.8 | 385 | 15 | A | 03/2007 | 1 | 25 | 45 | n/a | 2 | 4 |
| Site 5 | 415 | 148 | RBC | 412 | 1.0 | 358 | 16 | B | 04/2005 | 1 | 20 | 40 | 10 | 2[c] | 4[c] |
| Site 6 | 1,834 | 830 | RBC | 3,000 | 1.6 | 277 | 21 | B | 03/2010 | 2 | 15 | 30 | 5 | 2 | 4 |
| Site 7 | 1,873 | 520 | TF | 1,250 | 0.7 | 416 | 14 | A | 01/2009 | 3 | 15 | 30 | 5 | 1 | 4 |
| Site 8 | 1,013 | 353 | RBC | 1,175 | 1.2 | 300 | 19 | B | 03/2008 | 4 | 25 | 45 | 5/10[d] | 2 | 4 |
| Site 9 | 607 | 612 | RBC | 912 | 1.5 | 671 | 9 | B | 03/2007 | 5 | 15 | 25 | 5/10[d] | 2 | 4 |
| Site 10 | 1,876 | 655 | TF | 1,400 | 0.7 | 468 | 12 | C | 03/2007 | 5 | 15 | 25 | 5/10[d] | 2 | 4 |
| Site 11 | 4,864 | 2,160 | TF | 2,250 | 0.5 | 960 | 6 | C | 03/2007 | 5 | 15 | 30 | 5 | 2 | 4 |
| Site 12 | 1,297 | 409 | RBC | 1,625 | 1.3 | 252 | 23 | B | 03/2006 | 6 | 25 | 40 | 10 | 2 | 4 |

| Site 13 | 1,839 | 528 | TF | 2,150 | 1.2 | 246 | 23 | A | 03/2005 | 7 | 15 | 25 | 10/20[d] | 2[e] | 1[e] |
|---|---|---|---|---|---|---|---|---|---|---|---|---|---|---|---|
| Site 14 | 1,317 | 353 | RBC | 1,450 | 1.1 | 243 | 24 | B | 04/2005 | 7 | 20 | 45 | 5 | 2[c] | 4[c] |

[a]Theoretical value, with a gravel porosity of 0.4
[b]Total phosphorus consent is a rolling annual average, whereas total iron is an absolute (100 %) consent, and suspended solids, BOD, and ammonia are 95 % percentile
[c]Summer only consent (i.e. April to September)
[d]Summer/winter values
[e]Aluminium sulphate dosing

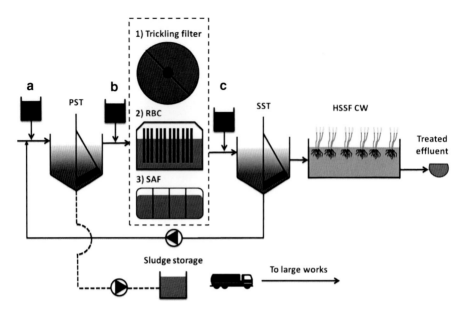

**Fig. 19.1** Flowsheets used for phosphorus removal via chemical dosing in the 14 WwTPs. The *letters* indicate the dosing points for chemical addition, i.e. (**a**) before primary settling tanks, (**b**) after primary settling tanks, or (**c**) before secondary settling tanks

## 19.2.3 Wastewater Samples

Wastewater samples were obtained as 24 h time proportional composites (200 mL per hour; ISCO 3700, Teledyne Isco, UK) from the inlet and outlet of the studied wetlands at sites 1, 3, and 12 during a 2-week sampling period in 2012 and 4-week sampling at sites 1, 2 (aerated wetland), and 13 (Al dosed) in 2013. Site 1 presented the unique opportunity to monitor water quality before (2012) and after (2013) surface sludge was removed as part of partial refurbishment operations. All samples were processed to separate suspended, colloidal, and dissolved fractions as described by Martin (2010). Briefly, samples were progressively filtered through three separate membranes of pore sizes: 1.2 μm, 100 kilodalton (KDa), and 1 KDa. The pore sizes of the last two membranes refer to the atomic mass unit. Suspended particles corresponded to particles bigger than 1.2 μm, with the colloidal phase between 1.2 μm and 1 KDa and the dissolved phase below 1 KDa. Membranes used for collection of suspended solids were made of glass microfiber binder-free (Whatman papers grade GF/C, Whatman[TM], GE Healthcare, UK). Suspended solids fractionation was conducted in standard vacuum filtration equipment following (APHA 2005) procedures. Membranes used to fractionate the samples for colloidal and dissolved phases were made from a low protein-binding modified polyethersulfone (Omega[TM], PALL Life Sciences, UK). The membranes were washed with sequences of ultrapure water, 0.1 M NaOH and 0.1 M HCl, before and after use. The colloidal and dissolved fractionations were conducted in an

Amicon 8400 stirred pressure cell, driven by nitrogen gas at a pressure of 3.7 atm. After fractionation, each filtrate was analysed for total phosphorus, orthophosphates, and total iron using analytical test kits (cuvette tests LCK349, LCK 321, and LCK 320; Hach Lange Ltd., UK).

In addition, on-site testing of oxidation reduction potential (ORP) and dissolved oxygen (DO) concentration in the sediment-water interface was conducted at the time of sample collection. Measurements were conducted in three transects along the length of each bed at inlet, middle, and outlet zones. Redox potential and DO readings were performed in the upper layer of the granular media, below the water table and within the first 10–20 cm. Measurements were taken using a standard epoxy liquid electrode for redox potential with platinum disc sensor (Intellical, Hach Lange Ltd., UK) and a stainless steel lumiphore sensor for dissolved oxygen (Intellical, Hach Lange Ltd., UK) with a probe reader HQ40D digital two-channel multimeter (Hach Lange Ltd., UK).

## 19.2.4 Sludge Samples

Samples of sludge accumulated over the surface of the gravel were collected for analysis of iron and phosphorus content. A sample of 300 mL of sludge was taken per sample point, from the inlet, middle, and outlet. Samples were stored in sealed plastic bags for transport and processed on the same day. Sludge was weighed and oven dried at 105 °C over 48 h. Total solids and water content were measured following (APHA 2005) procedures. Once dried, each sample was ground for digestion. Each sample was extracted with a hydrochloric/nitric acid mixture using a microwave digestion system. The extract was then clarified and made up to volume with deionised water. Iron was determined by atomic absorption, whereas phosphorus content was determined by a spectrometric measurement in solution. Analytical replication was carried out to ensure the reliability of results. The procedure was based on US EPA Method 3051 (US EPA 1994).

## 19.2.5 Release Experiments

Phosphorus and iron release from wetland sludge was tested in batch reactors. Samples of surface sludge and wastewater were collected from the studied wetland at site 3, as well as spot wastewater samples of the wetland influent for use as matrix in the reactor. Influent wastewater was used in order to ensure similar environmental conditions such as pH, dissolved oxygen, and redox potential. A reactor was filled with 2 L of sludge and 2 L of influent water and monitored during a 24 h incubation, which is similar to the retention time in the wetland. Initial concentrations of total phosphorus, orthophosphate, and total iron were measured in the water before filling the reactor. During the incubation time, reactors were kept at room temperature, approximately 18 °C. Water samples and probe measurements of ORP, DO, and pH were taken every hour during working hours and at the end of the experiment (i.e. 24 h later). The total volume of water removed from the reactor by the end of the experiment was calculated to never exceed 5 % of the initial volume. Samples were filtered through 0.45 μm pore size using sterile syringe filters (Millex$^{TM}$ syringe filters, Merck Millipore Corp., UK) and analysed for total phosphorus, orthophosphate, and total iron using analytical test kits as described for water samples. Soluble orthophosphate and total iron release rates were calculated as changes in mass in the liquid phase as a function of time divided by the area of the sediment in the reactor (i.e. 0.035 m$^2$) as per standard procedures.

## 19.3 Results and Discussion

### 19.3.1 Sites Characterisation

#### 19.3.1.1 Flowsheet Performance for Sanitary Determinands

All chemically dosed sites achieved effluent qualities that outperformed the regulatory requirements (Tables 19.1 and 19.2). With the exception of two sites where 95 % percentile values before and after chemical dosing were 21 and 22 and 14 and 16 mg L$^{-1}$, respectively, addition of the chemical had a positive impact on final effluent solids in all flowsheets. Effluent BOD values were consistently low at an average of < 5 mg L$^{-1}$ in all sites before and after addition of the chemicals, thus making statistical analysis irrelevant when considering the error of the method itself. Phosphorus values before chemical addition could reach 13 mg L$^{-1}$ and were significantly lower after chemical addition with the highest average at 1.7 mg L$^{-1}$ in site 14. Effluent iron concentrations generally increased after the addition of chemical salts ($p < 0.05$), as expected. Iron data before start of dosing is, however, limited. All sites provided a degree of nitrification, even when not designed to do so (i.e. sites 1, 3, and 4). The results indicate that, in general, chemical dosing improves the performance of the overall treatment flowsheet for all sanitary determinands with the exception of iron which tends to increase above baseline levels.

#### 19.3.1.2 Wetland Characteristics and Performance

All systems were designed to be operated as subsurface horizontal flow wetlands, in agreement with the design criteria set out in Cooper et al. (1996). However, inspection of the sites revealed the water level was, in all but three, well above the gravel surface (Fig. 19.2). Indeed, the water level in four of the sites was close to the emergency overflow level, leaving only ~ 5 cm of water storage within the bed during an intense rainfall or increased flow period.

19 Long-Term Performance of Constructed Wetlands with Chemical Dosing for...     281

**Table 19.2** Overall treatment works performance before and after dosing started

| Site ID | BOD Before | BOD After | n | TSS Before | TSS After | n | NH₄–N Before | NH₄–N After | n | TP Before | TP After | n | Fe Before | Fe After | n |
|---|---|---|---|---|---|---|---|---|---|---|---|---|---|---|---|
| Site 1 (2012) | 2 (8) | 2 (4) | 57/29 | 3 (13.5) | 3 (10) | 57/29 | 1.3 (9.8) | 0.7 (2.7) | 41/29 | 4.6 (6.13) | 1.6 (2.9) | 21/29 | 0.11 (0.11) | 0.33 (1.0) | 2/26 |
| Site 2 | 9 (19) | 1 (1) | 166/9 | 16 (32) | 3 (6) | 166/9 | 3.6 (10) | 0.1 (0.1) | 163/9 | 3.6 (9.8) | 0.2 (0.3) | 36/9 | 0.11 (0.11) | 0.16 (0.3) | 16/9 |
| Site 3 | 3 (14) | 2 (7) | 53/40 | 6 (21) | 8 (22) | 53/43 | 3.6 (11) | 4.8 (12.8) | 37/40 | 6 (11) | 1.5 (3.8) | 20/42 | – | 0.7 (3.1) | –/40 |
| Site 4 | 4 (10) | 4 (10) | 126/130 | 6 (40) | 10 (22) | 126/129 | 17.9 (39.8) | 17.8 (28.8) | 129/115 | 4.3 (8.0) | 1 (3.7) | 13/91 | 0.30 (0.65) | 1.8 (4.0) | 11/65 |
| Site 5 | 2 (5) | 2 (6) | 68/166 | 5 (19) | 5 (16) | 67/165 | 2.1 (5.8) | 1.6 (8.5) | 68/168 | 4.8 (9.4) | 1.3 (4.5) | 47/72 | – | 0.5 (2.0) | –/40 |
| Site 6 | 1 (2) | 1 (4) | 81/59 | 3 (10) | 4 (9) | 82/60 | 0.34 (1.1) | 1.0 (2.7) | 81/61 | 4.4 (6.0) | 1.4 (2.6) | 56/15 | 0.10 (0.19) | 0.17 (0.7) | 12/51 |
| Site 7 | 4 (6) | 2 (6) | 85/151 | 5 (14) | 7 (16) | 85/155 | 0.8 (2.1) | 1.4 (4.8) | 84/169 | 5.7 (7.9) | 0.6 (1.1) | 58/159 | – | 0.9 (2.8) | –/141 |
| Site 8 | 3 (7) | 2 (4) | 61/77 | 7 (22) | 4 (8) | 61/78 | 2.5 (8.0) | 0.7 (2.3) | 61/78 | 5.8 (8.5) | 0.6 (1.6) | 61/65 | – | 0.2 (0.8) | –/47 |
| Site 9 | 4 (8) | 2 (5) | 47/110 | 8 (18) | 5 (15) | 84/111 | 0.8 (2.1) | 0.6 (2.4) | 82/109 | 2.4 (6.0) | 1.1 (2.3) | 74/76 | 0.7 (1.4) | 0.5 (1.2) | 20/85 |
| Site 10 | 2 (8) | 2 (6) | 71/171 | 3 (9) | 4 (9) | 71/174 | 1.7 (6.3) | 2.5 (7.2) | 71/177 | 5.0 (7.0) | 0.6 (1.6) | 63/40 | – | 0.4 (0.9) | –/76 |
| Site 11 | 7 (12) | 5 (13) | 60/135 | 15 (27) | 13 (26) | 60/136 | 0.5 (1.3) | 0.6 (1.9) | 60/136 | 3.1 (5.2) | 0.9 (2.0) | 59/20 | – | 1.4 (3.3) | –/46 |
| Site 12 | 10 (15) | 2 (6) | 66/122 | 14 (38) | 4 (13) | 70/123 | 3.3 (7.4) | 0.9 (3.3) | 70/123 | 8.2 (13.0) | 1.5 (7.7) | 59/91 | – | 0.11 (0.19) | –/47 |

(continued)

**Table 19.2** (continued)

| Site ID | BOD Before | BOD After | BOD n | TSS Before | TSS After | TSS n | NH₄–N Before | NH₄–N After | NH₄–N n | TP Before | TP After | TP n | Fe Before | Fe After | Fe n |
|---|---|---|---|---|---|---|---|---|---|---|---|---|---|---|---|
| Site 13 | 2 | 2 | 65/139 | 3 | 2 | 65/136 | 0.4 | 1.6 | 65/147 | 5.8 | 0.3 | 57/113 | – | 0.13[a] | –/73 |
|  | (2) | (3) |  | (7) | (6) |  | (0.9) | (6.2) |  | (9.6) | (0.9) |  | (0.3) | (0.23) |  |
| Site 14 | 8 | 2 | 64/144 | 19 | 6 | 64/148 | 2.9 | 1.1 | 65/150 | 7.3 | 1.7 | 47/117 | – | 0.15 | –/70 |
|  | (17) | (5) |  | (41) | (15) |  | (5.3) | (5.6) |  | (10.4) | (4.9) |  |  | (0.3) |  |

Values shown are averages with 95th percentile values in *brackets*
Sample sizes in the n columns are shown as number of samples before/after
[a]Values are for aluminium, as the site is Al dosed

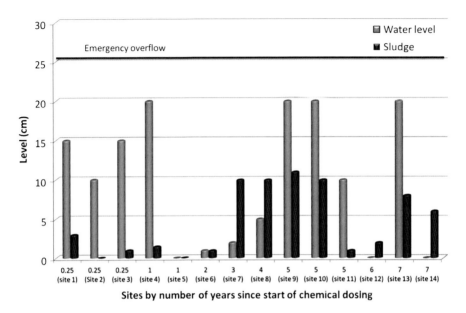

**Fig. 19.2** Water levels across all chemically dosed sites

Concomitant with mechanisms of surface deposition or settling in a wetland operated with surface flow, the quantities of surface sludge over the gravel surface were expected to increase with time since beginning of chemical dosing strategies. The rate of accumulation of surface sludge varied between 0 and 3, averaging 2 cm year$^{-1}$ across all sites. A significantly higher rate was observed in the site with known history of operational issues with solids retention upstream of the treatment wetland (i.e. sites 1 and 7). In addition, the lowest rates of surface sludge accumulation were observed at the two sites operating with the water level below the surface of the gravel (sites 5, 12, and 14). The lack of sludge accumulation in site 2 is potentially related to the aerated nature of the bed. A side-by-side comparative trial conducted by Severn Trent in flooded, tertiary horizontal wetlands with and without aeration showed improved mixing in aerated flooded wetlands which resulted in the absence of surface sludge accumulation during the first year of operation (Butterworth et al. 2013), which agrees with these findings. Site 13 is the site with aluminium dosing and appears to accumulate less sludge than the iron-dosed sites sampled in this study. The lower accumulation rate of sludge in site 11 was unexpected. Whilst this could be due to sludge loss during high flow periods, it is more likely the operator did a partial refurbishment on the surface of the bed that was unrecorded. The average rate of accumulation is similar to those reported for first-stage French vertical flow systems receiving screened sewage (i.e. highly biodegradable solids) with resting periods incorporated in their operation (Molle et al. 2005).

Six sites had records of water quality entering and exiting the wetlands (Fig. 19.3). In terms of nitrogen species, total oxidised nitrate was depleted in all

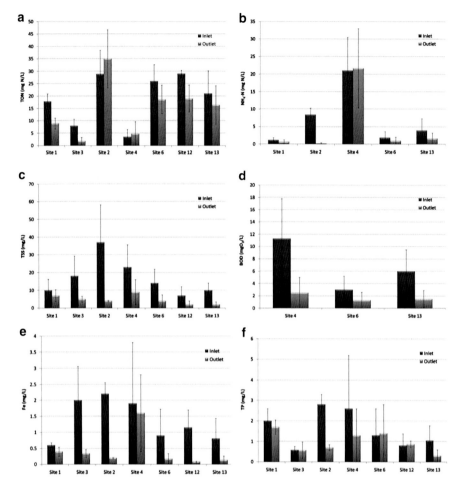

**Fig. 19.3** Wastewater quality feeding and existing six of the wetlands with upstream chemical dosing: (**a**) nitrate, (**b**) ammonia, (**c**) total suspended solids, (**d**) BOD, (**e**) total iron, and (**f**) total phosphorus

systems except site 2, which is an aerated wetland site (i.e. designed to nitrify rather than denitrify), and site 4 (which has nitrate addition at the inlet of the wetland to prevent septic conditions). The rates of nitrate removal were 1.3–3.8 g m$^{-2}$ d$^{-1}$, significantly higher than the 0.08 g m$^{-2}$ d$^{-1}$ reported rates for other HSSF wetlands (Vymazal 2007). This is probably due to the highly nitrified influent received by these systems, enabling nitrate removal from concentrations of 20 to less than 15 mg L$^{-1}$. Ammonia removal was limited with rates of 0, 0.2, 0.24, 0.46, and 1.3 in sites 4, 1, 6, 13, and 2 g m$^{-2}$ d$^{-1}$, respectively, with the highest removal rate in the artificially aerated system, in agreement with current literature (Kadlec and Wallace 2009).

Solids removal rates ranged between 1.2 and 6.7 g m$^{-2}$ d$^{-1}$, with the highest rates corresponding to the highest loading rates. This is in agreement with current understanding of solids removal performance of horizontal flow wetlands (Kadlec and Wallace 2009) and close to the removal rates of 8 g m$^{-2}$ d$^{-1}$ for secondary beds (Vymazal 2009). A positive effect was also found for total iron, with removal rates averaging 0.1–0.5 g m$^{-2}$ d$^{-1}$. This is significantly higher than rates of 0.04–0.014 g m$^{-2}$ d$^{-1}$ reported for (non-dosed) domestic HSSF wetlands (Kröpfelová et al. 2009). Loading rates onto the systems were also significantly higher, which suggests a similar relationship to suspended solids in all sites except site 4. The latter showed a removal rate of 0.1 g Fe m$^{-2}$ d$^{-1}$ in spite of having the highest loading rate of 0.8 g Fe m$^{-2}$ d$^{-1}$ and in spite of having the highest TSS removal rate at 6.7 g m$^{-2}$ d$^{-1}$. This suggests that the iron in this system was in either dissolved or colloidal form rather than particulate. The effect of particle size on removal efficiency was further explored during the detailed fractionation analysis (*see below*).

Phosphorus removal in the six sites ranged between −0.3 (release) and 2.4 g m$^{-2}$ d$^{-1}$ (storage). Sites 3, 6, and 12 exhibited events of P release, with higher concentrations of P in the effluent than the influent based on historic records. The loading for P averaged 0.1–0.7 g m$^{-2}$ d$^{-1}$ and corresponded to influent concentrations ranging between 0.4 and 2.8 mg P L$^{-1}$. Release was observed in the sites where the lowest P influent values were recorded. Notably, influent concentrations to the wetlands were typically lower as system age increased, suggesting better dosing practices upstream of the tertiary system as more operational experience was gained at the individual sites.

### 19.3.2 *Characterisation of Phosphorus and Iron Retention and Release from Wetlands*

The concentration of iron and phosphorus in the sludge of all the wetlands surveyed averaged between 9 and 178 g Fe kg$^{-1}$ dry sludge and 17–48 g P kg$^{-1}$ dry sludge, respectively (Fig. 19.4). Phosphorus sludge concentrations are markedly higher than other constructed wetlands (e.g. (Pant et al. 2001)) and even higher than the concentrations of 2 g kg$^{-1}$ dry matter and 7 g kg$^{-1}$ dry matter reported for river (House and Denison 2002) and lake sediments (Lake et al. 2007), respectively. Iron concentrations were also higher than the 8–21 g kg$^{-1}$ sludge reported from three mature HSSF for domestic wastewater treatment in the Czech Republic (Vymazal and Švehla 2013) and closer to the 101–160 g kg$^{-1}$ dry matter in lakes affected by copper mining activities (Samecka-Cymerman and Kempers 2001).

Whilst the quantities of surface sludge accumulation were expected to increase with system age, the relative concentration of iron and phosphorus was expected to remain the same in a Fe: P ratio between 1.5: 1 and 2.5:1 as per the dosing strategy employed on these sites. Analysis of the surface sludge, however, reveals relatively

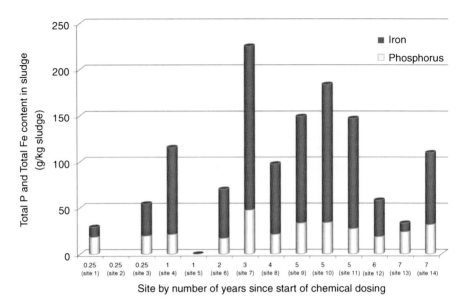

**Fig. 19.4** Iron and phosphorus content of surface sludge accumulation in chemically dosed wetlands. The *numbers of years* represent the time since start-up of chemical dosing at the time of sampling

constant P concentrations in sludge, whilst iron concentrations change with system age (Fig. 19.4). Site 13 achieves P removal via aluminium dosing and thus illustrates P concentrations in sludge at a mature site without Fe accumulation. In this site, the P concentration is consistent with system age, whilst the Fe concentration represents background iron accumulation rates. The ratios of Fe to P range between 1.8 and 4.4, suggesting greater storage capacity for iron when compared to phosphorus and supporting the indications that phosphorus release takes place in some sites, during specific field conditions of an intermittent nature.

To better understand potential release from wetlands, fractionation of water samples entering and leaving four of the wetlands was conducted (Fig. 19.5). Iron was predominantly in the particulate form entering the systems and, as such, was efficiently retained by all the sampled systems. Small quantities of colloidal iron were also captured, whereas the dissolved fraction passed through the system unchanged, as expected in gravel media with limited sorption capacity. The highest changes in concentrations were observed in site 2 and can be directly related to the areal mass loading rate applied.

In three of the six datasets, phosphorus was observed to experience a shift from the particulate to the dissolved fraction. This is indicative of pollutant storage and subsequent release under specific environmental conditions. Phosphorus release from sediments has been observed in different water bodies including rivers, impoundments, and natural wetlands. This phenomenon depends on different environmental conditions and media characteristics such as ORP, DO, pH, temperature, phosphorus concentration in water, and aluminium and iron concentrations in

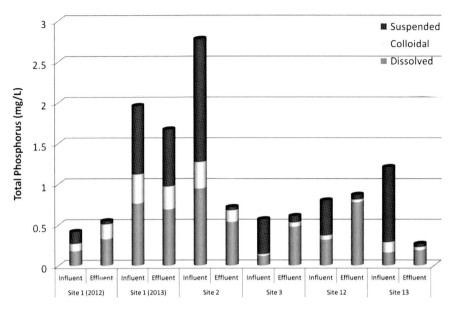

**Fig. 19.5** Total iron (**a**) and total phosphorus (**b**) fractionation in influent and effluent wastewater samples. Site 1 was partially refurbished after sampling in 2012 so that 2013 results represent the first year of operation since refurbishment

sediment. Concentrations of dissolved phosphorus and dissolved iron in water column and pore water at Eh −100 mV have been observed to be up to 10 times higher than those under aerobic conditions (Ann et al. 1999). Most research studies suggest reduction of $Fe^{3+}$ to the more soluble form $Fe^{2+}$ to be the main contributing factor to P release. Microbial reduction of iron takes place when redox potential is between 100 and −100 mV (Reddy and DeLaune 2008). The use of iron as an electron acceptor by microorganisms causes the breakdown of iron-bound phosphates with subsequent release of dissolved phosphates and iron into the water column (Gomez et al. 1999; Lai and Lam 2009; Lake et al. 2007; Olila et al. 1997). The competition between nitrate reducers and ferric iron reducers for electron donors is known for soil systems and anaerobic sediments. The presence of nitrates has been shown to minimise the dissolution of ferric P compounds since nitrates are preferred over ferric iron as electron acceptors by facultative microbes (Søndergaard et al. 2000).

Site 1 exhibited release during 2012 (before refurbishment) and retention during 2013 (after refurbishment). The main changes from 1 year to the next were (a) surface sludge accumulation, with sludge removed at the end of the sampling campaign in 2012, and (b) influent concentrations onto the wetlands increased from 0.5 mg P $L^{-1}$ in 2012 to 1.8 mg P $L^{-1}$ in 2013. Comparison of other key parameters

**Table 19.3** Environmental conditions and nitrate availability in site 1 prior and post refurbishment

| Sample period | DO (mg L$^{-1}$) | ORP (mV) | pH | Temperature (°C) | NO$_3$–N in (mg L$^{-1}$) | NO$_3$–N out (mg L$^{-1}$) |
|---|---|---|---|---|---|---|
| Before refurbishment (2012) | 0.1–1.2 | 160 to −68 | 7.9 | 17.5 | 22.5 | 16.4 |
| After refurbishment (2013) | 0.1–2.1 | 160 to −152 | 7.8 | 18.4 | 17.8 | 8.9 |

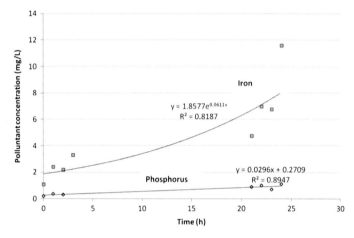

**Fig. 19.6** Iron and phosphorus release during 24 h of incubation in anoxic conditions

believed to influence P release, namely, nitrate, DO, pH, temperature, and ORP showed that conditions within the top layer of the wetland were similar in both years, suggesting it is more likely P release occurred due to change in sludge concentrations and/or influent wastewater P concentrations than a release triggered by anaerobic conditions within the beds (Table 19.3).

Under laboratory conditions, iron release occurred, and its rate was significantly higher than the phosphorus release rates under both laboratory and field conditions (Fig. 19.6). Initial iron concentrations were 1.1 mg Fe L$^{-1}$ and, by the end of the 24 h period, were close to 12 mg Fe L$^{-1}$ in the liquid phase. Phosphorus, on the other hand, started from a concentration of 0.2 mg P L$^{-1}$ and finished at 1.1 mg P L$^{-1}$, indicating a release rate of 44.6 mg m$^{-2}$ d$^{-1}$. Phosphorus flux found in this study was higher than those reported for sediments from other water bodies (Table 19.4), and, when release occurred, it was similar under laboratory and field conditions.

**Table 19.4** Phosphorus release rates from sediment to water in different types of water bodies

| Type of wetland | Values | P release (mg m$^{-2}$ day$^{-1}$) | Conditions | Reference |
|---|---|---|---|---|
| Horizontal flow, iron dosed for P removal from domestic wastewater | Mean ± SD | 52 | Flooded, diurnal composite sampling | This study |
|  | Mean ± SD | 21 | Flooded, diurnal composite sampling |  |
|  | Mean ± SD | 12 | Flooded, diurnal composite sampling |  |
|  | Mean | 44 | Anaerobic (lab reactor) |  |
| River Seine downstream Paris (France) | Range | 14.3–29.3 | Anaerobic conditions | Garban et al. (1995) |
| Constructed wetland treating water from a hyper eutrophic lake (USA) | Mean ± SD | 1.5 ± 0.6 | Continuously flooded | Olila et al. (1997) |
|  |  | 33.4 ± 4.5 | 3 weeks drained and re-flooded |  |
|  |  | 334 ± 14 | 6 weeks drained and re-flooded |  |
| River used as drinking water source (South Korea) | Mean | 5.7 | Aerobic conditions | Kim et al. (2003) |
|  |  | 15 | Anaerobic conditions |  |
| Lower St. Johns River (USA) | Mean | 0.13 | Aerobic conditions | Malecki et al. (2004) |
|  |  | 4.77 | Anaerobic conditions |  |
| Small impoundment in a river (USA) | Range | 0.4–3.6 | Aerobic conditions | Haggard and Soerens (2006) |
|  |  | 14.5–15.5 | Anaerobic conditions |  |
| Eutrophic marshes (Hong Kong) | Mean ± SD | 9.4 ± 3.3 | Aerobic conditions | Lai and Lam (2009) |
|  |  | 31.7 ± 5.8 | Anaerobic conditions |  |

## 19.4 Implications for the Use of Wetlands for P and Fe Polishing

The findings from this study agree with current understanding of wetlands as effective solids filters, retaining solids of any nature provided they are in true particulate form rather than in colloidal fractions. In conventional applications, it is desirable to keep the water flowing through the media to allow sufficient contact time with the biofilm (for biological and sorption processes) and passing through the gravel pores for filtration of solids. Whilst the studied wetlands were designed as conventional applications, the surveys showed most of them were operated with surface flow. This can be advantageous in that stored sludge is easily accessible for periodic removal, but operating the wetlands with surface flow shifts the main removal mechanism from filtration or physical entrapment to settling. For effective settling to take place, different design criteria should be used to take into account the type of particle settling, its rate, and resuspension requirements to avoid periodic release of stored solids during higher flow conditions such as storm events. Furthermore, a system designed to store P and Fe solids on the surface of the gravel at the rates found in this study would only last, at most, 12 years before refurbishment was required to avoid the top of the surface reaching the overflow pipe level (i.e. 25 cm). Consideration into the appropriate freeboard necessary for a system designed for this purpose becomes one of the key criteria, as is currently the case for surface flow wetlands for applications such as storage of ochre from mine drainage operations (Sheoran and Sheoran 2006). Alternatively, the horizontal wetlands could be operated under subsurface flow and thus minimise the risk of sludge resuspension and loss through the overflow pipes. Unfortunately, phosphorus release was observed regardless of where the water level was in the wetlands, so this strategy can only limit the risk of failure due to sludge loss but not necessarily biogeochemical cycling.

Phosphorus and iron cycling in constructed wetland systems is complex, and whilst there is information from the natural wetlands literature, an improved understanding of their chemistry under wastewater treatment conditions is needed to incorporate the appropriate features in the design, operation, and maintenance guidance for treatment wetland operators. From the 14 sites studied, three presented events of phosphorus release which appeared to be related more to sludge and wastewater P concentrations than DO and nitrate availability within the beds. Phosphorus release rates appear to be linear, and, as such, hydraulic residence time could potentially be one of the design parameters to adjust for limiting final effluent P concentrations under at-risk conditions such as seasonally low influent P concentrations or high accumulation of P-containing surface sludge. Further research is required to better profile Fe, P, and Al cycling in constructed wetlands for wastewater treatment within the context of chemical addition for phosphorus removal.

## 19.5 Conclusions

Constructed wetlands have been successfully used as part of the treatment strategies in 14 sites in the UK that have chemical addition for phosphorus removal. Four main conclusions have been reached:

- The flowsheets used – consisting of primary settling, iron or aluminium addition, and a secondary biological stage in the form or RBCs, trickling filters, or SAFs – followed by a tertiary horizontal flow wetland are an effective method to remove BOD, solids, ammonia, nitrate, and phosphorus, with all sites meeting their discharge consents for the past 10 years.
- Constructed wetlands are an effective method to capture residual solids. Under flooded operation, settling mechanisms dominate as evidenced by surface sludge accumulation. Subsurface flow results in lower or no sludge accumulation on top of the gravel.
- The concentration of P and Fe in sludge and the fractionation of water quality show phosphorus release under low P in wastewater and high sludge levels. The relative impact of sludge Fe and P concentrations versus the wastewater concentration is, as yet, unknown.
- Further research is required to provide adequate data for modifying the design and to improve the operation and maintenance guidance to ensure wetlands continue to deliver environmental benefit at small WwTPs required to remove phosphorus.

# References

Ann, Y., Reddy, K. R., & Delfino, J. J. (1999). Influence of redox potential on phosphorus solubility in chemically amended wetland organic soils. *Ecological Engineering, 14*(1–2), 169–180.

APHA, A., & WEF. (2005). In A. Eaton, L. Clesceri, E. Rice, & A. Greenberg (Eds.), *Standard methods for the examination of water and wastewater* (21st ed.). Washington, DC: American Public Health Association.

Brix, H., & Arias, C. A. (2005). The use of vertical flow constructed wetlands for on-site treatment of domestic wastewater: New Danish guidelines. *Ecological Engineering, 25*(5), 491–500.

Butterworth, E., Dotro, G., Jones, M., Richards, A., Onunkwo, P., Narroway, Y., & Jefferson, B. (2013). Effect of artificial aeration on tertiary nitrification in a full-scale subsurface horizontal flow constructed wetland. *Ecological Engineering, 54*, 236–244.

Cooper, P. F., Job, J. D., Green, B., & Shutes, R. B. E. (1996). *Reed beds and constructed wetlands for wastewater treatment*. Swindon: WRc.

Garban, B., Ollivon, D., Poulin, M., Gaultier, V., & Chesterikoff, A. (1995). Exchanges at the sediment-water interface in the river seine, downstream from Paris. *Water Research, 29*(2), 473–481.

Gomez, E., Durillon, C., Rofes, G., & Picot, B. (1999). Phosphate adsorption and release from sediments of brackish lagoons: PH, O2 and loading influence. *Water Research, 33*(10), 2437–2447.

Haggard, B. E., & Soerens, T. S. (2006). Sediment phosphorus release at a small impoundment on the Illinois river, Arkansas and Oklahoma, USA. *Ecological Engineering, 28*(3), 280–287.

House, W. A., & Denison, F. H. (2002). Total phosphorus content of river sediments in relationship to calcium, iron and organic matter concentrations. *Science of the Total Environment, 282-283*, 341–351.

Kadlec, R., & Wallace, S. (2009). *Treatment wetlands* (2nd ed.). Boca Raton: CRC Press.

Kim, L., Choi, E., & Stenstrom, M. K. (2003). Sediment characteristics, phosphorus types and phosphorus release rates between river and lake sediments. *Chemosphere, 50*(1), 53–61.

Kröpfelová, L., Vymazal, J., Švehla, J., & Štíchová, J. (2009). Removal of trace elements in three horizontal sub-surface flow constructed wetlands in the Czech republic. *Environmental Pollution, 157*(4), 1186–1194.

Lai, D. Y. F., & Lam, K. C. (2009). Phosphorus sorption by sediments in a subtropical constructed wetland receiving stormwater runoff. *Ecological Engineering, 35*(5), 735–743.

Lake, B. A., Coolidge, K. M., Norton, S. A., & Amirbahman, A. (2007). Factors contributing to the internal loading of phosphorus from anoxic sediments in six Maine, USA, lakes. *Science of the Total Environment, 373*(2–3), 534–541.

Lauschmann, R. E., Lechner, M., Ertl, T., & Langergraber, G. (2013). Experiences with pre-precipitation of phosphorus in a vertical flow constructed wetland in Austria. *Water Science and Technology, 67*(10), 2337–2341.

Mainstone, C. P., & Parr, W. (2002). Phosphorus in rivers — Ecology and management. *Science of the Total Environment, 282–283*, 25–47.

Mainstone, C. P., Parr, W., & Day, M. (2000). *Phosphorus and river ecology, tackling sewage inputs*. Peterborough: English Nature and Environment Agency.

Malecki, L. M., White, J. R., & Reddy, K. R. (2004). Nitrogen and phosphorus flux rates from sediment in the lower St. Johns River estuary. *Journal of Environmental Quality, 33*(4), 1545–1555.

Molle, P., Liénard, A., Boutin, C., Merlin, G., & Iwema, A. (2005). How to treat raw sewage with constructed wetlands: An overview of the French systems. *Water Science and Technology, 51*(9), 11–21.

Olila, O. G., Reddy, K. R., & Stites, D. L. (1997). Influence of draining on soil phosphorus forms and distribution in a constructed wetland. *Ecological Engineering, 9*(3–4), 157–169.

Pant, H. K., Reddy, K. R., & Lemon, E. (2001). Phosphorus retention capacity of root bed media of sub-surface flow constructed wetlands. *Ecological Engineering, 17*(4), 345–355.

Reddy, K. R., & DeLaune, R. D. (2008). *Biogeochemistry of wetlands*. Boca Raton: CRC Press.

Samecka-Cymerman, A., & Kempers, A. J. (2001). Concentrations of heavy metals and plant nutrients in water, sediments and aquatic macrophytes of anthropogenic lakes (former open cut brown coal mines) differing in stage of acidification. *Science of the Total Environment, 281*(1–3), 87–98.

Sheoran, A. S., & Sheoran, V. (2006). Heavy metal removal mechanism of acid mine drainage in wetlands: A critical review. *Minerals Engineering, 19*(2), 105–116.

Shutes, B. (2012). *Myths, reality, and future potential of constructed wetlands*. In Proceedings of 13th IWA international conference on wetland systems for water pollution control. November 2012. Perth, Australia.

Søndergaard, M., Jeppesen, E., & Jensen, J. P. (2000). Hypolimnetic nitrate treatment to reduce internal phosphorus loading in a stratified lake. *Lake and Reservoir Management, 16*(3), 195–204.

US EPA. (1994). *Method 3051: Microwave assisted acid digestion of sediments, sludges, soils and oils*. Washington, DC: US Environmental Protection Agency.

Vymazal, J. (2007). Removal of nutrients in various types of constructed wetlands. *Science of the Total Environment, 380*, 48–65.

Vymazal, J. (2009). The use constructed wetlands with horizontal sub-surface flow for various types of wastewater. *Ecological Engineering, 35*(1), 1–17.

Vymazal, J., & Švehla, J. (2013). Iron and manganese in sediments of constructed wetlands with horizontal subsurface flow treating municipal sewage. *Ecological Engineering, 50*, 69–75.

# Chapter 20
# Use of the Macrophyte *Cyperus papyrus* in Wastewater Treatment

Njenga Mburu, Diederik P.L. Rousseau, Johan J.A. van Bruggen, and Piet N.L. Lens

**Abstract** *Cyperus papyrus*, commonly referred to as papyrus, belongs to the *Cyperaceae* family and is one of the most prolific emergent macrophytes in African subtropical and tropical wetlands. Botanical studies have shown that stands of papyrus are capable of accumulating large amounts of nutrients and have a high standing biomass. Its $C_4$ photosynthetic pathway makes *C. papyrus* highly productive with dry weight biomass generation rates of up to 6.00 kg m$^{-2}$ years$^{-1}$ and nutrient uptake rates of up to 7.10 kg ha$^{-1}$days$^{-1}$ and 0.24 kg ha$^{-1}$days$^{-1}$ of, respectively, nitrogen and phosphorus. *C. papyrus* plants take about 6–9 months to mature with a highly reliable natural regrowth and replenishment on a site after harvesting. Studies featuring side-by-side investigations with unplanted controls show that *C. papyrus* has mostly a positive effect on the treatment of wastewater. The ability of *C. papyrus* to use nutrients from the wastewater and the incorporation of heavy metals and organics into its phytomass, added to its easy management by regular harvesting, makes it one of the most suitable plants to be used in wastewater phytoremediation in tropical areas. Therefore, it continues to be an excellent candidate for application as a macrophyte in the constructed wetland wastewater treatment technology. As such, determining the potential scope of the performance of *C. papyrus* is vital for the optimal application and design of *C. papyrus*-based constructed wetland systems. This work collates growth, productivity and performance information from various independent studies incorporating the *C. papyrus* macrophyte.

N. Mburu
UNESCO-IHE Institute for Water Education, P. O. Box 3015, 2601 DA Delft, The Netherlands

Department of Civil Engineering, Dedan Kimathi University of Technology, P.O Box 657, 10100 Nyeri, Kenya

D.P.L. Rousseau (✉)
Department of Industrial Biological Sciences, Ghent University Campus Kortrijk, Graaf Karel de Goedelaan 5, 8500 Kortrijk, Belgium
e-mail: diederik.rousseau@ugent.be

J.J.A. van Bruggen • P.N.L. Lens
UNESCO-IHE Institute for Water Education, P. O. Box 3015, 2601 DA Delft, The Netherlands

**Keywords** *Cyperus papyrus* • Tropics • Harvesting • Biomass • Tropical wetlands

## 20.1 Introduction

In the subtropical and tropical climate, *C. papyrus* is one of the most interesting macrophytes because it is amongst the most productive plants in wetlands (Kansiime et al. 2005; Heers 2006; Perbangkhem and Polprasert 2010). This plant has a high potential of producing biomass from solar energy, which is one of the recommended criteria for the selection of macrophytes in tropical areas with abundant sunshine for use in constructed wetlands (Perbangkhem and Polprasert 2010). The papyrus vegetation has been shown to actively improve wastewater quality through contribution to the removal of organic compounds (Vymazal and Kröpfelová 2009), heavy metals (Sekomo 2012), pathogens (Okurut 2000; Kansiime and Nalubega 1999) and excess nutrients such as nitrogen and phosphorus (Kansiime et al. 2007a, b, Perbangkhem and Polprasert 2010). This could be attributed to their ecological characteristics of high phytomass, well-developed root system and high photosynthetic rate (Jones 1988; Muthuri et al. 1989; Kansiime et al. 2007a). Various authors have presented experimental research findings on aspects of the characteristics of *C. papyrus* that have an influence on water quality improvement, both in the natural and constructed wetlands. This chapter collates these findings by answering the following questions related to the application and management of the macrophyte *C. papyrus* in wastewater treatment:

(a) What is the growth habitat of *C. papyrus* macrophytes, its morphology and physical effects on water quality improvement including the surface area for attachment of microbial growth?
(b) What is the influence of the metabolism of *C. papyrus on* water quality improvement (i.e. the plant nutrients (nitrogen and phosphorus) uptake potential, the root oxygen leakage and the biomass productivity)?
(c) What is the harvesting practice and the regeneration capacity for *C. papyrus* after harvesting?

## 20.2 Influence of Macrophyte on Pollutant Bioconversion and Removal in Treatment Wetlands

The biogeochemical cycling and storage of nutrients, organic compounds and metals in natural wetlands is mimicked in constructed wetlands, through the use of plants, porous media and associated microorganisms (Sonavane et al. 2008; Hunter et al. 2001). The presence of emergent macrophytes is one of the most conspicuous features of constructed wetlands, and their presence distinguishes constructed wetlands from unplanted soil filters or lagoons (Vymazal 2011; Greenway 2007). Their positive role on the performance of constructed wetlands has been well established in numerous studies measuring treatment with and without plants

(Kadlec and Wallace 2009; Brisson and Chazarenc 2009; Akratos and Tsihrintzis 2007; Yang et al. 2007).

Generally, the performance of wetlands for wastewater treatment depends on the growth potential and ability of macrophytes to develop sufficient root systems for microbial attachment and material transformations and to incorporate nutrients into plant biomass that can be subsequently harvested for nutrient removal (Vymazal and Kröpfelová 2009; Kyambadde et al. 2004). However, empirical exploitation of plants is a common practice. Availability, expected water quality, normal and extreme water depths, climate and latitude, maintenance requirements and project goals are amongst the variables that determine the selection of plant species for constructed wetlands (Stottmeister et al. 2003).

While there is a recognition that the improvement of water quality in treatment wetland applications is primarily due to microbial activity (Faulwetter et al. 2009; Kadlec and Wallace 2009), experience has shown that wetland systems with vegetation or macrophytes have a higher efficiency of water quality improvement than those without plants (Coleman et al. 2001; Tanner 2001; Brisson and Chazarenc 2009). The emphasis of constructed wetland technology to date has been on soft tissue emergent plants including *Cyperus papyrus*, *Phragmites*, *Typha* and *Schoenoplectus* (Kadlec and Wallace 2009; Okurut 2000). These are fast-growing species that have lower lignin contents and are adaptable to variable water depths. The productivity of emergent macrophytes is the highest amongst the aquatic plant communities in the tropics as well as in temperate regions. Emergent macrophytes are characterised by a photosynthetic aerial part above the water surface and a basal part rooted in the water substrate.

Emergent macrophytes find application in both surface and subsurface flow configurations of constructed wetlands. The significance of the plants used for wastewater purification has been emphasised by previous researchers (Brix 1997; Peterson and Teal 1996; Gersberg et al. 1983). Vymazal (2011) summarised the various roles played by emergent macrophytes in different configurations of constructed wetlands (Table 20.1).

Macrophyte plants, in addition to their site-specific roles (i.e. attenuation of light, water current and wind velocity, aesthetic appearance, etc.), are essential in the wetland treatment systems because they have properties that foster many wastewater treatment processes (Kyambadde 2005; Kadlec and Wallace 2009). Aquatic plants can absorb inorganic (nutrients, metals, etc.) and organic pollutants (aromatics, hydrocarbons, etc.) from wastewater and incorporate them into their own structure (Haberl et al. 2003), thus providing a storage and a release of nutrients through the plant growth cycle (NAS-NRC 1976; Shetty et al. 2005). They create favourable conditions for microbes that contribute to the processing of pollutants by influencing the oxygen supply to the water, providing attachment surfaces, providing carbon and electron donor via carbon content of litter and root exudates (Kadlec and Wallace 2009; Brix and Schierup 1989). Further, aquatic plants promote stable residual accretions in the wetland (Vymazal 2007; Greenway 2007). These residuals contain pollutants as part of their structure or in absorbed form and hence represent a burial process of contaminants (Kadlec and Wallace

**Table 20.1** Major role of macrophytes in constructed treatment wetlands (Vymazal 2011)

| Macrophyte property | Role in treatment process |
|---|---|
| Aerial plant tissue | Light attenuation – reduced growth of photosynthesis |
| | Influence of microclimate – insulation during winter |
| | Reduced wind velocity – reduced risk of resuspension |
| | Aesthetic-pleasing appearance of the system |
| | Storage of nutrients |
| Plant tissue in water | Filtering effect – filter out large debris |
| | Reduced current velocity – increased rate of sedimentation, reduced risk of resuspension |
| | Excretion of photosynthetic oxygen – increased aerobic degradation |
| | Uptake of nutrients |
| | Provision of surface for periphyton attachment |
| Roots and rhizomes in the sediment | Stabilising the sediment surface – less erosion |
| | Prevention of the medium clogging in vertical flow systems |
| | Provision of surface for bacterial growth |
| | Release of oxygen increases degradation (and nitrification) |
| | Uptake of nutrients |
| | Release of antibiotics, phytometallophores and phytochelatins |

With kind permission from Springer Science+Business Media: Hydrobiologia, Plants used in constructed wetlands with horizontal subsurface flow: a review, Volume 674, 2011, Jan Vymazal, Table 1

2009). These facts have been exploited in constructed wetland systems which have been widely used during the past decades for the treatment of wastewater because of their good efficacy to improve water quality at low operational costs (Neralla et al. 2000; Vymazal 1999; Rousseau et al. 2004; Molle et al. 2005; Zurita et al. 2008; Perbangkhem and Polprasert 2010). The natural wetlands too have been shown to have potential as a sink and buffering site for organic and inorganic pollutants (Sekomo et al. 2010; Mannino et al. 2008; Buchberger and Shaw 1995; Muthuri and Jones 1997).

Wetland vegetation enhance evapotranspiration (and a corresponding increase in the hydraulic retention time) which can be explained by a net biomass productivity accompanied by transpiration (Kyambadde et al. 2005; Kansiime and Nalubega 1999). Emergent macrophyte vegetation tends to increase rates of water loss through evapotranspiration when compared to rates of evaporation from bodies of open water (Jones and Humphries 2002).

At present there is no clear evidence that treatment performance is superior or different between the common emergent wetland species used in treatment wetlands (IWA 2000; Zhu et al. 2010). Even so some soft tissue emergent macrophytes including *Phragmites* sp., *Schoenoplectus* sp., *Typha* sp. and *Carex* (true sedge) are well known for their potentials in constructed wetlands treating wastewater and faecal sludge, and their performances are well documented, especially for the high latitudes and temperate climate regions (Ciria et al. 2005; Stein et al. 2006;

Fennessy et al. 1994; Coleman et al. 2001). These macrophytes are, however, not found in all regions of the world and efforts are being made worldwide to select candidate macrophytes to be exploited locally in constructed wetlands (Brisson and Chazarenc 2009; Perbangkhem and Polprasert 2010; Azza et al. 2000; Huang et al. 2010; Yang et al. 2007).

## 20.3 *Cyperus papyrus* Macrophyte

### 20.3.1 *History and Growth Habitat*

*Cyperus papyrus*, commonly called papyrus or paper plant, is a member of the *Cyperaceae* sedge family, a group of plants closely related to grasses (Michael 1983). The *Cyperaceae* family has about 75 genera and more than 4,000 species, which are for a large part perennial rhizomateous herbs growing in moist places. *C. papyrus* has a long history of being harvested and has been used over millennia, such as for the manufacture of the first paper by the ancient Egyptians (Terer et al. 2012). It once grew wild throughout the Nile Valley (Egypt, Ethiopia) and can still be found in the swamps and marshes of Central, East and Southern Africa (El-Ghani et al. 2011; Chale 1987; van Dam et al. 2011; Boar et al. 1999). It was widely cultivated in Egypt for its many uses: boats, rope, food (boiled pith and rhizomes), sandals, boxes, mats, sails, blankets, cloth, mummy wrappings, firewood (dried rhizomes), medicine and building materials as well as writing materials (scrolls *papyri*) (Leach and Tait 2000). It is the largest of the sedges and a monocot that is native to riverbanks and mouths, lakeshores, floodplains and wet soil areas of North and tropical Africa. Outside Africa, it is thought to be native to the Hula Valley in Israel where it reaches its northernmost limits. It has been introduced and naturalised in the Mediterranean (Sicily, Malta), the USA (Florida) and India (Terer et al. 2012).

*C. papyrus* is the dominant species of many swamps in East and Central Africa and can be found growing in both lentic and lotic freshwater environments with stable hydrological regimes (permanently flooded). It cannot cope with rapid water level changes and water flow (Jones 1988; Jones and Muthuri 1985; Serag 2003; Kresovich et al. 1982). Due to its rhizomatic root structure, it can also be found floating with a mat-like root structure (up to 1.5 m thick) in open waters as deep as 3–4 m (Azza et al. 2000; Kansiime and Nalubega 1999; Thompson et al. 1979). As a member of the sedge family, it does not hold economic importance as a crop plant; nevertheless in some regions, it still finds application in weaving mats, baskets, screens and even sandals (van Dam et al. 2011; Osumba et al. 2010; Morrison et al. 2012).

A substantial number of sedges are weeds, invading crop fields in all climates of the world. Sedges do, however, have a considerable ecological importance. They are of extreme importance to primary production as well as an integral part of the hydrologic cycle (Saunders et al. 2007). Today, the most important uses of papyrus wetlands are those of ecological resources and services (Maclean et al. 2011; van Dam et al. 2007).

The *C. papyrus* wetland soils and plants may absorb or adsorb heavy metals, pathogens, inorganic forms of nitrogen, phosphorus, other nutrients and trace elements, and the rhizomes of the plant prevent soil erosion and trap polluted sediments from inflowing water. Consequently, *C. papyrus* has found application in both constructed and natural wetlands for water quality improvements. Thus, even in modern times papyrus may have an important role in cleaning up wastewater pollution from industrial, municipal and domestic sources as captured in the listed selection of studies in Table 20.2. Side-by-side investigations with unplanted controls show that the macrophyte *C. papyrus* has mostly a positive effect, i.e. supports higher treatment efficiency for the removal of organics (COD, $BOD_5$), faecal coliforms, heavy metals and nutrients such as nitrogen and phosphorus (Abira 2007; Okurut 2000; Kyambadde et al. 2005; Nyakang'o and van Bruggen 1999; Sekomo 2012).

In constructed wetland applications, *C. papyrus* has been found to establish well from rhizome fragment propagules and also to adapt well to wastewater conditions (Mburu et al. 2013; Abira 2007; Okurut 2000). This characteristic of vegetative reproduction via rhizomes and rapid recovery after damage to aboveground growth is shared by other effective invasive macrophytes such as *Phragmites* (Meyerson et al. 2000).

## *20.3.2* Cyperus papyrus *Morphology*

*C. papyrus* has its culm or stem (often triangular) growing to an average height of 3–5 m above ground and taking 6–9 months to mature (Gaudet 1977; Kresovich et al. 1982; Terer et al. 2012; Muthuri and Jones 1997). The culm has a large proportion of a spongy aerenchyma on its inside, and to a small extent, it is capable of photosynthesis (Okurut 2000). It is topped by characteristically large, spherical-shaped (finely dissected bracteoles), reproductive umbels (it bears flowers) that serve also as the main photosynthetic surface of the plant (Mnaya et al. 2007; Jones and Humphries 2002). The rhizomes and the roots together form a mat-like structure that is the base for swamp development. *C. papyrus* can grow well in the subtropical and tropical climate and is amongst the most productive plants of wetlands (Heers 2006; Kansiime et al. 2005; Perbangkhem and Polprasert 2010).

*C. papyrus* is considered to be unique due to its $C_4$ photosynthetic pathway in spite of the fact that it grows in a wetland ecosystem, which appears an unlikely habitat for $C_4$ species (Jones 1988, 1987; Jones and Humphries 2002; Saunders et al. 2007). Plants utilising the $C_4$ photosynthetic pathway show higher potential efficiencies in the use of intercepted radiation, water and nitrogen for the production of dry matter than do other photosynthetic types (Piedade et al. 1991). $C_4$ species are most numerous in tropical and warm, temperate semi-arid zones, where their

Table 20.2 Application and investigations of *Cyperus papyrus* macrophyte for water quality removal

| Location | Set-up | Area | Type of (waste)water | Removal efficiency | Reference |
|---|---|---|---|---|---|
| Kenya | Experimental CW | $3.2 \times 1.2 \times 0.8$ m | Pulp and paper mill | [a]TN:75 %, [a]BOD: 90 %, [a]TSS: 81 %, [a]COD: 52 %, Phenols: 73–96 % | Abira (2007) |
| Uganda | Experimental CW | 60 m$^2$ | Urban sewage | BOD, NH$_4^+$-N, P: 68.6–86.5 % | Kyambadde et al. (2005) |
| Kenya (Kahawa swamp) | Natural wetland | 3.7 ha | Domestic sewage | DO, NH$_4^+$-N, ortho-P: 77–85 % | Chale (1985) |
| Uganda (Nakivubo swamp) | Natural wetland | 2.5 km$^2$ | Urban sewage | [a] NH$_4^+$-N:89.4 %, [a] ortho-P: 80 %, [a]COD: 70 % | Kansiime et al. (2003) |
| Uganda | Experimental CW | 320 m$^2$ | Domestic sewage | TSS: 80 %, [a]coliforms: 4log units | Okurut (2000) |
| Congo (Upemba wetlands) | Rectangular transects, $2 \times 10$–20 m | 250 m$^2$ | Natural | | Thompson et al. (1979) |
| Kenya | Operational CW | 0.5 ha | Domestic sewage | | Nyakang'o and van Bruggen (1999) |
| Thailand | Experimental CW | 3 m$^2$ | Domestic sewage | | Perbangkhem and Polprasert (2010) |
| Kenya (lake Naivasha) | Quadrants $3 \times 3$ m (lake-) | | Natural | | Muthuri et al. (1989) |
| Tanzania (Rubondo Island) | Quadrants $0.5 \times 0.25$ m | 1.41 km$^2$ | Natural | | Mnaya et al. (2007) |
| Kenya (Lake Naivasha) | Quadrants $0.5 \times 0.5$ m | 150 km$^2$ | Natural | | Boar (2006) |
| Kenya (Lake Victoria shore) | Quadrants $2 \times 2$ m | 60 km stretch | Natural | | Osumba et al. (2010) |
| Rwanda (Nyabugogo wetland) | 13 sampling sites | 60 ha | Municipal/industrial (heavy metals) | Cd: 4.2, Cr: 45.8, Cu: 29.7, Pb: 56.1 All values in mg kg$^{-1}$ | Sekomo et al. (2010) |

(continued)

Table 20.2 (continued)

| Location | Set-up | Area | Type of (waste)water | Removal efficiency | Reference |
|---|---|---|---|---|---|
| Uganda (Natete wetland) | Transects (325 m, 350 m long) | 1 km$^2$ | Municipal sewage | $NH_4^+$-N: 21 %, $NO_3^-$-N: 98 %, TN: 45 % | Kanyiginya et al. (2010) |
| Egypt (Nile Delta) | 50 sites (located by GPS) | 2,250 km$^2$ | Natural | | El-Ghani et al. (2011) |
| Cameroon | Yard scale | 1 m$^2$ | Faecal sludge | | Kengne et al. (2008) |
| USA | Microcosm | 0.49 × 0.35 m | Synthetic sewage | | Morgan et al. (2008) |
| China | Microcosm | 0.26 m$^2$ | Synthetic wastewater | | Wang et al. (2008) |

*CW* constructed wetland
[a]Maximum values

greater water-use efficiency appears to be a key selective advantage (Piedade et al. 1991). The $C_4$ photosynthetic pathway makes *C. papyrus* highly productive with dry weight biomass generation of up to 6.28 kg m$^{-2}$ years$^{-1}$ (Terer et al. 2012).

Aerobic conditions in the roots and rhizomes of *C. papyrus* are maintained by oxygen transport from the atmosphere through the aerenchyma of the culms (Li and Jones 1995). Aerenchymous plant tissue is an important adaptation to flooding in wetland plants through which transport of gases to and from the roots through the vascular tissues of the plant above water and in contact with the atmosphere takes place (Singer et al. 1994). Li and Jones (1995) reported a diffusive oxygen transport between the rhizomes and the culms of 2.45 and 3.29 mol m$^{-3}$ of oxygen at day- and night-time, respectively. This provides an aerated root zone and thus lowering the plant's reliance on external oxygen diffusion through water and soil (Kadlec and Wallace 2009).

In the tropical swamps *C. papyrus* establishment, growth and mortality occur concurrently through the year so that there is little temporal change in the standing crop (Muthuri et al. 1989). The culms can be divided into different age classes; some authors have used classification based on three age classes, namely, juvenile, with unopened umbels; mature, with opened green umbels; and senescent with more than half of the umbels brown (achlorophyllous) (Muthuri et al. 1989), while others have identified six age classes, namely, young elongated culm with closed umbel, elongated culm with umbel just opening, fully elongated culm and fully expanded umbel, fully elongated culm and fully expanded umbel but older, senescing culm (>10 % achlorophyllous) and dead culm (>80 % achlorophyllous) (Muthuri and Jones 1997). Culm density is controlled by density-dependent mortality (Thompson et al. 1979).

## 20.3.3 Cyperus papyrus *Biomass Productivity*

In both natural and constructed wetlands with *C. papyrus*, vegetative and reproductive parts above the ground level and their root systems comprise a substantial part of the wetland biomass (Fig. 20.1). Emergent macrophytes in swamps and marshes are amongst the most productive plant communities (Muthuri et al. 1989). *C. papyrus* vegetation is highly productive and under favourable temperatures, hydrological regime and nutrient availability estimates of aerial standing live biomass (including scale leaves, culm and umbel) often exceeding 5,000 g (dry weight) m$^{-2}$ (Saunders et al. 2007).

The productivity of natural papyrus wetlands is found to be variable (Table 20.3) and controlled by different factors such as climate, nutrient availability and the prevailing general hydrological conditions (Okurut 2000). Differences in aerial biomass of papyrus in various sites have been attributed to prevailing climatic conditions. Some studies have noted a trend of an increase in standing biomass of papyrus swamps with increase in altitude. Nevertheless, this trend has not been

**Fig. 20.1** Photos of the aboveground vegetative and reproductive parts of *Cyperus papyrus* in a constructed wetland at Juja, Kenya

found to hold true by all authors (Muthuri et al. 1989; Thompson et al. 1979; Mnaya et al. 2007). Unlike other emergent aquatic plants (Table 20.4), its high productivity rates and standing/harvestable biomass make *C. papyrus* have a high nutrient removal potential more so in wetlands receiving a high nutrient load. The harvesting of biomass presents a potential for biological nutrient removal (Kyambadde et al. 2005; Kansiime and Nalubega 1999).

Estimating biomass or primary productivity in tropical swamps, which have relatively stable biomass, requires measurements of population dynamics and the life cycle of individual shoots (Muthuri et al. 1989), unlike in the temperate ecosystems, where common methods of estimating primary productivity include measurements of peak biomass, maximum minus minimum biomass or methods which account for death and decomposition between harvests (Sala and Austin 2000; Muthuri et al. 1989).

### 20.3.3.1 Aboveground Biomass

The aerial organs of the *papyrus* (umbel, culm and scale leaves) contribute about 50 % of the total plant biomass (Thompson et al. 1979), in which the largest proportion is in culms (Muthuri et al. 1989). The high aerial biomass of the papyrus (Table 20.3) is unlike many other perennial emergent macrophytes such as *Typha latifolia* (890–2,500 g m$^{-2}$), *Scirpus validus* (2,355–2,650 g m$^{-2}$) and *Phragmites australis* (1,110–5,500 g m$^{-2}$) (Kadlec and Wallace 2009) that have a large proportion of their biomass in the form of roots and rhizomes (Muthuri et al. 1988). High aerial primary production indicates that less carbohydrate is assimilated in the rhizomes; hence, the living culms act as storage organs. This function is normally for the rhizome (Tanner 1996). Boar et al. (1999) established that biomass allocation to the various *C. papyrus* tissues is directly related to the fertility of the growing media such that least investment in roots and rhizome indicates plenty of nutrient supply. However, some studies have found that an

**Table 20.3** Biomass production of *Cyperus papyrus* growing in different types of wetlands

| Study/site | Biomass production (g dry mass m$^{-2}$) Below ground | Above ground | References |
|---|---|---|---|
| Pilot-scale FWS CW/Uganda | 1,250 | 2,250 | Kyambadde et al. (2005) |
| Natete wetland, Kampala, Uganda | 1,288 ± 8.3 | 1,019 ± 13.8 | Kanyiginya et al. (2010) |
| Man-made swamp/Kahawa swamp, Kenya | 4,955[a] | | Chale (1987) |
| Lake Naivasha, Kenya | | 3,245 | Jones and Muthuri (1985) |
| Flooded river valley, Busoro, Rwanda | | 1,384 | Jones and Muthuri (1985) |
| Nakivubo wetland (two sites), Uganda | | 883–1,156 | Kansiime et al. (2003) |
| Pilot-scale-constructed wetland, Uganda | | 3,529–5,844 | Kansiime et al. (2003) |
| Constructed wetland, Thailand | 2,200–3,100[a] | | Perbangkhem and Polprasert (2010) |
| Constructed wetland, Uganda | 16,900–18,700[a] | | Kansiime et al. (2005) |
| Natural wetland, Lake Naivasha, Kenya | | 2,731 | Muthuri et al. (1989) |
| Natural wetland, Lake Naivasha, Kenya | | 4,652 | Terer et al. (2012) |
| Rubondo Island, Lake Victoria, Tanzania | 4,144 ± 452 | 5,789 ± 435 | Mnaya et al. (2007) |
| Lake Naivasha, Kenya | 6,950 ± 860[a] | | Boar (2006) |
| Nakivubo wetland, Uganda | 6,700[a] | | Kansiime et al. (2007a) |
| Kirinya wetland, Uganda | 7,200[a] | | Kansiime et al. (2007a) |
| Nakivubo wetland, Uganda | 1,158 | 2,480 | Mugisha et al. (2007) |
| Kirinya wetland, Uganda | 4,343 | 3,290 | Mugisha et al. (2007) |

*FWS CW* free water surface-constructed wetland
[a]Total biomass

important fraction of the plants' biomass is stored in the belowground stands of the papyrus (Kengne et al. 2008; Kanyiginya et al. 2010; Saunders et al. 2007).

### 20.3.3.2 Belowground Biomass

The belowground biomass of *C. papyrus* consists of an interlaced but permeable root mat with a rhizomatic structure (Kansiime and Nalubega 1999). Measurements

**Table 20.4** Biomass production of other macrophytes growing in different types of wetlands

| Macrophyte | Study/site | Biomass production (g dry mass m$^{-2}$) Below ground | Above ground | Reference |
|---|---|---|---|---|
| Colocasia esculenta | Nakivubo wetland, Uganda | 1,236 | 2,024 | Mugisha et al. (2007) |
| Colocasia esculenta | Kirinya wetland, Uganda | 1,697 | 2,463 | Mugisha et al. (2007) |
| Miscanthus violaceus | Nakivubo wetland, Uganda | 870 | 1,190 | Mugisha et al. (2007) |
| Miscanthus violaceus | Kirinya wetland, Uganda | 1,470 | 1,680 | Mugisha et al. (2007) |
| Phragmites mauritianus | Nakivubo wetland, Uganda | 745 | 1,790 | Mugisha et al. (2007) |
| Phragmites mauritianus | Kirinya wetland, Uganda | 1,452 | 3,030 | Mugisha et al. (2007) |
| Phragmites australis | Tidal salt marsh, N. America | | 727–3,663 | Meyerson et al. (2000) |
| Phragmites australis | Freshwater marsh, N. America | | 980–2,642 | Meyerson et al. (2000) |

of rhizomes and root mass in the papyrus vegetation involve excavation to the maximum depth to which the roots are found (Muthuri et al. 1989). In natural swamps, the rooting mat has been estimated to contribute up to 30–52 % of the total biomass (Okurut 2000; Boar et al. 1999). The belowground biomass (i.e. the root and rhizomes) surface area provides attachment sites which are conducive for the proliferation of bacterial biomass. The roots and rhizomes influence the wastewater residence time, trapping and settling of suspended particles, surface area for pollutant adsorption, uptake, assimilation in plant tissues and oxygen for organic and inorganic matter oxidation in the rhizosphere (Kansiime and Nalubega 1999; Okurut 2000; Kyambadde et al. 2004). For example, the nature and density of the rooting biomass can greatly influence the extent of faecal bacteria removal via sedimentation and attachment processes. This influence was demonstrated in the studies of Kansiime and Nalubega (1999) in a natural wetland where faecal coliform counts were consistently higher in zones dominated by the *Miscanthidium violaceum* macrophyte, than in zones dominated by *C. papyrus*. The rooting mat of the former was tight and compact and thus had a reduced total surface area. In contrast, the papyrus mat is hollow and interwoven giving it a larger surface area for entrapment and attachment of faecal coliforms (Okurut 2000). Sekomo (2012) established that *C. papyrus* plants play an important role in metal retention. The *C. papyrus* root system was the most important part of the plant in heavy metal retention, followed by the umbel and finally the stem.

## 20.3.4 Nutrient Uptake and Storage

The removal of soluble inorganic nitrogen and phosphorus via absorption from either the water column or the sediment and storage in plant tissue is a direct mechanism of nutrient sequestration (Greenway 2007). Table 20.5 shows values for nutrient uptake of *C. papyrus* under different set-ups. The difference in the uptake rates may be attributed to nutrient availability under the experimental conditions and/or the growth phase of the macrophyte. A comparison of nutrient concentrations in plants, soil and water column in the Natete wetland (Kampala, Uganda) found that *C. papyrus* stored the highest amounts of nutrients as compared to soil and water (Kanyiginya et al. 2010). Plants take up nutrients as a requirement for their growth. These nutrients accumulate in plant parts which present an opportunity to remove excess nutrients from wetland systems through harvesting the aerial plant phytomass (Kansiime et al. 2007a). In this regard, plants with high rates of net primary productivity and higher nutrient uptake are preferred in wetlands subject to wastewater inputs (Kansiime et al. 2007a, b).

The nutrient elements essential for plant growth would be removed in proportion to their compositional ratios in the particular species (Boyd 1970). For *C. papyrus*, Chale (1987) found the nitrogen concentrations of the various plant organs were 4.8 % roots, 8.4 % rhizomes, 4.5 % scales, 4.8 % culms and 6.2 % umbels on dry weight basis. As compared to phosphorus, the concentrations were 0.09 % roots, 0.11 % rhizomes, 0.09 % scales, 0.10 % culms and 0.13 % umbels. A high content of nutrients is observed for the aerial biomass of papyrus, an indication of active translocation and storage of nutrients to parts of the plant where they are needed for primary growth, e.g. synthesis of amine acids and enzymes (Kyambadde et al. 2005; Muthuri and Jones 1997; Kansiime et al. 2007b). Significantly higher

**Table 20.5** *Cyperus papyrus* nutrient uptake rates under varying experimental set-ups

| Type of wastewater | P uptake (kg ha$^{-1}$ days$^{-1}$) | N uptake (kg ha$^{-1}$ days$^{-1}$) | Reference |
|---|---|---|---|
| Septic tank effluent (constructed wetland, Uganda) | 0.24 | 7.1 | Okurut (2000) |
| Natural wetland (Lake Naivasha) | 0.06 | 1.18 | Muthuri et al. (1989) |
| Municipal sewage (Nakivubo wetland) | 0.21 | 1.3 | Kansiime et al. (2003) |
| Natural wetland (Upemba swamps) | 0.06 | 1.18 | Thompson et al. (1979) |
| Domestic wastewater (constructed wetland) | 0.14 | 3.01 | Brix (1994) |
| **Other macrophytes** | | | |
| *Phragmites australis* (infiltration wetland) | 0.22 | 2.14 | Okurut (2000) |
| *Eichhornia crassipes* (diverse wastewaters) | 0.2–2.0 | 1.6–6.6 | Okurut (2000) |

amounts of nutrients are sequestered in papyrus umbels and culms compared to roots/rhizomes portions (Kyambadde et al. 2005). Similar observations have been made by Mugisha et al. (2007) who established that photosynthetic organs of *C. papyrus* (culm and umbel) generally had a higher nutrient content than other organs (roots and rhizome) at the Nakivubo and Kirinya wetlands at the shores of Lake Victoria in Uganda. Nevertheless, nutrients in papyrus plants decrease with the age of the plant as the nutrients are translocated to the metabolically active juvenile plants for growth (Mugisha et al. 2007).

Okurut (2000) found nutrient removal from wastewater via plant uptake to show variability at different growth phases. The growth rate of *Cyperus papyrus* is the highest in juvenile plants and the lowest in mature plants, and the nitrogen uptake rate by *Cyperus papyrus* is the highest in juvenile plants and the lowest in mature plants. Uptake was correlated with the biomass yields exhibited in the different phases. The total nitrogen content was the highest in the juvenile plants and decreased with increasing age. This enables the plant to recycle nutrients from the old portions to new growth (Boyd 1970). Generally, (a) the rate of nutrient uptake by macrophytes is limited by its growth rate and the concentration of nutrients within the plant tissue, and (b) nutrient storage is dependent on the plant tissue nutrient concentration and potential for biomass accumulation (Greenway 2007).

## 20.4 Wastewater Treatment with *Cyperus papyrus*

*Cyperus papyrus* plays an important role in the water quality enhancement, the effects of which can be readily observed in terms of dissolved oxygen (DO), pH and redox potential of their surroundings and the attenuation of pollutant parameter profiles from influent to effluent (Huang et al. 2010; Okurut 2000). For constructed wetlands to be effective in water pollution control, they must function as a "pollutant" sink for organics, sediments, nutrients and metals, i.e. these pollutants must be transformed, degraded or removed from the wastewater and stored within the wetland either in the sediment or the plants. Although there is still debate about the relative importance of macrophytes versus microbes in nutrient removal, plant biomass still accounts for substantial removal and storage of nitrogen and phosphorus (Brix 1997; IWA 2000).

Macrophytes can contribute directly through uptake (nutrients and heavy metals), sedimentation, adsorption or phytovolatilisation or indirectly to pollutant removal in constructed wetlands. Indirect processes are related to biofilm growth around roots, evapotranspiration and the pumping of oxygen towards the rhizosphere that changes the redox conditions (Imfeld et al. 2009; Carvalho et al. 2012; Kadlec and Wallace 2009). Some of these mechanisms are addressed in the sections below.

## 20.4.1 Root Oxygen Release into the Rhizosphere

Papyrus-dominated wetlands like all other natural wetlands are characterised by low dissolved oxygen concentrations (Okurut 2000). The main reason for this state is that surface aeration and photosynthetic oxygen transfer mechanisms are poor or non-existent due to the dense plant canopy. On the other hand, oxygen leakage to the rhizosphere is important in constructed wetlands with subsurface flow for aerobic degradation of oxygen-consuming substances and nitrification (Brix 1994). The photosynthetic characteristics of wetland species can affect their ability to provide oxygen, and this ultimately influences their disposal efficiencies.

The peak photosynthetic quantum efficiency, i.e. the amount of $CO_2$ that is fixed or the amount of $O_2$ that is released via assimilation when the photosynthetic apparatus absorbs one photon (Huang et al. 2010) for *C. papyrus*, has been reported to range between 26 and 40 μmol $CO_2$ m$^{-2}$ s$^{-1}$ (Jones 1987, 1988; Saunders and Kalff 2001). In their work on plant photosynthesis and its influence on removal efficiencies in constructed wetlands, Huang et al. (2010) published the photosynthetic rates of five wetland plants, *Phragmites, Ipomoea, Canna, Camellia* and *Dracaena*, at light saturation. These ranged between 11.6 and 31.32 μmol $CO_2$ m$^{-2}$ s$^{-1}$. *C. papyrus* presents a comparable potential for oxygen production via photosynthesis. Kansiime and Nalubega (1999) estimated oxygen release rates of 0.017 g m$^{-2}$ days$^{-1}$ by C. *papyrus* plants. The oxygen released is available for microbiota within the rhizosphere.

## 20.4.2 Surface for Microorganism's Attachment

In natural and constructed wetlands, macrophyte root structures provide microbial attachment sites. In an experimental microcosm set-up, Gagnon et al. (2007) found that microbes were present on substrates and roots as an attached biofilm and abundance was correlated to root surface throughout depth. Indeed planted wastewater treatment systems outperform unplanted ones, mainly because plants stimulate belowground microbial populations (Gagnon et al. 2007). Plant species root morphology and development seem to be key factors influencing microbial–plant interactions. Kyambadde et al. (2004) measured a higher root surface and microbial density in a constructed wetland planted with *C. papyrus* (average root surface area 208.6 cm$^2$) compared to *Miscanthidium violaceum* (average root surface area 72.2 cm$^2$). *C. papyrus* and *Miscanthidium violaceum* differed in their root recruitment rate and root number in a microcosm-constructed wetland. The root recruitment rate per constructed wetland unit was 77 and 32 roots per week for *C. papyrus* and *Miscanthidium violaceum*, respectively, and *C. papyrus* had more adventitious roots and larger root surface area than *Miscanthidium violaceum* (Kyambadde et al. 2004). Further, *C. papyrus* seems to promote greater nitrogen removal efficiencies, through nitrification and denitrification rates of bacteria associated with it roots (Morgan et al. 2008).

### 20.4.3 Evapotranspiration

The average daily water vapour flux from the papyrus vegetation through canopy evapotranspiration in a wetland located near Jinja (Uganda) on the Northern shore of Lake Victoria was approximated by Saunders et al. (2007) as 4.75 kg $H_2O$ $m^{-2}$ $days^{-1}$ (=4.75 mm $days^{-1}$), which was approximately 25 % higher than water loss through evaporation from open water (approximated as 3.6 kg $H_2O$ $m^2$ $days^{-1}$). Jones and Muthuri (1985) reported an evapotranspiration rate of 12.5 mm $days^{-1}$ at the fringing papyrus swamp on Lake Naivasha, while Kyambadde et al. (2005) reported $24.5 \pm 0.6$ mm $days^{-1}$ for a subsurface horizontal flow wetland in Kampala (Uganda). Evapotranspiration rates vary sharply since they depend on numerous factors influencing the ecosystem's prevailing microclimate, as listed by Kadlec and Wallace (2009). For example, common reed transpiration rates oscillate between 4.7 and 12.4 mm $days^{-1}$ depending on meteorological conditions (Holcová et al. 2009). Evapotranspiration (ET) by plants can significantly affect the hydrological balance of treatment wetlands. The water lost through ET concentrates pollutants within the wetland, while the volume reduction results in longer hydraulic retention times (Kadlec and Wallace 2009). For low-loaded systems or systems with longer retention times, such evapotranspiration rates can exceed the influent wastewater flow, leading to a zero discharge.

### 20.4.4 Cyperus papyrus *Harvesting and Regeneration Potential*

In order to achieve a permanent nutrient removal from wetland systems, *C. papyrus* harvesting is encouraged, but this requires careful timing (Kiwango and Wolanski 2008). Total nitrogen in aerial biomass of *C. papyrus* decreases from the juvenile plants to older plants (Mugisha et al. 2007). Hence, to minimise internal nutrient cycling and eventual export of the nutrients from wetland systems, sustainable harvesting of aerial biomass at different growth stages can be used as a strategy to reduce nutrients, especially in wastewater treatment wetlands. The regeneration potential of *C. papyrus,* i.e. the inherent capacity for natural regrowth and replenishment, on a site after a disturbance has been found to be highly reliable (Osumba et al. 2010). However, overharvesting (within less than one 6-month growing season) of *C. papyrus* can reduce this regeneration potential leading to weak spatial connectivity, papyrus stand fragmentation and increased landscape patchiness in natural wetlands (Osumba et al. 2010). Modelling studies of papyrus wetlands by van Dam et al. (2007) have proposed a harvesting rate between 10 and 30 % of the total biomass per year. At higher harvesting rates, nutrient uptake and retention by papyrus does not increase proportionally because of reduction in plant biomass, leading to lower uptake (van Dam et al. 2007). Muthuri et al. (1988) established a ceiling aerial biomass of 2,731 g m$^{-2}$ after 6 months, at a previously harvested section of a swamp at Lake Naivasha (Kenya), while for the undisturbed sections of the swamp, an aerial biomass of 3,602 g m$^{-2}$ was recorded. Water levels after harvesting are thought to affect biomass yield. Osumba et al. (2010) found that flooded sites give the least regenerated biomass yields.

## 20.5 Conclusion

This literature survey reveals that the macrophyte *C. papyrus* has found application in constructed wetlands for remediating a variety of pollutants in wastewater from different sources. The majority of the application of the *C. papyrus* macrophyte in constructed wetlands is found in the developing tropical countries, where papyrus is occurring locally. The macrophyte possesses a robust morphology and metabolism, and it is easy to establish and manage, thus making constructed wetlands incorporating *C. papyrus* wetland vegetation a promising wastewater treatment option for wider application. The production and harvesting of vegetation biomass from these treatment wetlands can provide a permanent route for the removal of nutrients, with economic benefits for communities that engage in the trade of papyrus products.

## References

Abira, M. A. (2007). *A pilot constructed treatment wetland for pulp and paper mill wastewater: Performance, processes and implications for the Nzoia river, Kenya*. PhD thesis, UNESCO-IHE, DELFT, Netherlands.

Akratos, C. S., & Tsihrintzis, V. A. (2007). Effect of temperature, HRT, vegetation and porous media on removal efficiency of pilot-scale horizontal subsurface flow constructed wetlands. *Ecological Engineering, 29*(2), 173–191.

Azza, N. G. T., Kansiime, F., Nalubega, M., & Denny, P. (2000). Differential permeability of papyrus and Miscanthidium root mats in Nakivubo swamp, Uganda. *Aquatic Botany, 67*(3), 169–178.

Boar, R. R. (2006). Responses of a fringing Cyperus papyrus L. swamp to changes in water level. *Aquatic Botany, 84*(2), 85–92.

Boar, R. R., Harper, D. M., & Adams, C. S. (1999). Biomass allocation in *Cyperus papyrus* in a tropical wetland, Lake Naivasha, Kenya. *Biotropica, 31*(3), 411–421.

Boyd, C. E. (1970). Vascular aquatic plants for mineral nutrient removal from polluted waters. *Economic Botany, 24*(1), 95–103.

Brisson, J., & Chazarenc, F. (2009). Maximizing pollutant removal in constructed wetlands: Should we pay more attention to macrophyte species selection? *Science of the Total Environment, 407*(13), 3923–3930.

Brix, H. (1994). Functions of macrophytes in constructed wetlands. *Water Science and Technology, 29*(4), 71–78.

Brix, H. (1997). Do macrophytes play a role in constructed treatment wetlands? *Water Science and Technology, 35*(5), 11–17.

Brix, H., & Schierup, H.-H. (1989). The use of aquatic macrophytes in water pollution control. *Ambio, 18*(2), 100–107.

Buchberger, S. G., & Shaw, G. B. (1995). An approach toward rational design of constructed wetlands for wastewater treatment. *Ecological Engineering, 4*(4), 249–275.

Carvalho, P. N., Basto, M. C. P., & Almeida, C. M. R. (2012). Potential of *Phragmites australis* for the removal of veterinary pharmaceuticals from aquatic media. *Bioresource Technology, 116*, 497–501.

Chale, F. M. M. (1985). Effects of a cyperus papyrus L. swamp on domestic waste water. *Aquatic Botany, 23*(2), 185–189.

Chale, F. (1987). Plant biomass and nutrient levels of a tropical macrophyte (Cyperus papyrus L.) receiving domestic wastewater. *Aquatic Ecology, 21*(2), 167–170.

Ciria, M. P., Solano, M. L., & Soriano, P. (2005). Role of macrophyte *Typha latifolia* in a constructed wetland for wastewater treatment and assessment of its potential as a biomass fuel. *Biosystems Engineering, 92*(4), 535–544.

Coleman, J., Hench, K., Garbutt, K., Sexstone, A., Bissonnette, G., & Skousen, J. (2001). Treatment of domestic wastewater by three plant species in constructed wetlands. *Water, Air, & Soil Pollution, 128*(3), 283–295.

El-Ghani, M. A., El-Fiky, A. M., Soliman, A., & Khattab, A. (2011). Environmental relationships of aquatic vegetation in the fresh water ecosystem of the Nile Delta, Egypt. *African Journal of Ecology, 49*(1), 103–118.

Faulwetter, J. L., Gagnon, V., Sundberg, C., Chazarenc, F., Burr, M. D., Brisson, J., Camper, A. K., & Stein, O. R. (2009). Microbial processes influencing performance of treatment wetlands: A review. *Ecological Engineering, 35*(6), 987–1004.

Fennessy, S. M., Cronk, J. K., & Mitsch, W. J. (1994). Macrophyte productivity and community development in created freshwater wetlands under experimental hydrological conditions. *Ecological Engineering, 3*(4), 469–484.

Gagnon, V., Chazarenc, F., Comeau, Y., & Brisson, J. (2007). Influence of macrophyte species on microbial density and activity in constructed wetlands. *Water Science and Technology, 56*(3), 249–254.

Gaudet, J. J. (1977). Uptake, accumulation, and loss of nutrients by papyrus in tropical swamps. *Ecology, 58*(2), 415–422.

Gersberg, R. M., Elkins, B. V., & Goldman, C. R. (1983). Nitrogen removal in artificial wetlands. *Water Research, 17*(9), 1009–1014.

Greenway, M. (2007). The role of macrophytes in nutrient removal using constructed wetlands environmental bioremediation technologies. In S. Singh & R. Tripathi (Eds.), *Environmental bioremediation technologies* (pp. 331–351). Berlin Heidelberg: Springer.

Haberl, R., Grego, S., Langergraber, G., Kadlec, R. H., Cicalini, A. R., Dias, S. M., Novais, J. M., Aubert, S., Gerth, A., Thomas, H., & Hebner, A. (2003). Constructed wetlands for the treatment of organic pollutants. *Journal of Soils and Sediments, 3*(2), 109–124.

Heers, M. (2006). *Constructed wetlands under different geographic conditions: Evaluation of the suitability and criteria for the choice of plants including productive species.* Master thesis, Hamburg University of Applied Sciences, Germany Faculty of Life Sciences, Department of Environmental Engineering.

Holcová, V., Šíma, J., Edwards, K., Semančíková, E., Dušek, J., & Šantrůčková, H. (2009). The effect of macrophytes on retention times in a constructed wetland for wastewater treatment. *International Journal of Sustainable Development and World Ecology, 16*(5), 362–367.

Huang, J., Wang, S.-h., Yan, L., & Zhong, Q.-s. (2010). Plant photosynthesis and its influence on removal efficiencies in constructed wetlands. *Ecological Engineering, 36*(8), 1037–1043.

Hunter, R. G., Combs, D. L., & George, D. B. (2001). Nitrogen, phosphorous, and organic carbon removal in simulated wetland treatment systems. *Archives of Environmental Contamination and Toxicology, 41*(3), 274–281.

Imfeld, G., Braeckevelt, M., Kuschk, P., & Richnow, H. H. (2009). Monitoring and assessing processes of organic chemicals removal in constructed wetlands. *Chemosphere, 74*(3), 349–362.

IWA. (2000). *Constructed wetlands for pollution control: Processes, performance, design and operation* (Scientific and technical report no 8). London: International Water Association.

Jones, M. B. (1987). The photosynthetic characteristics of papyrus in a tropical swamp. *Oecologia, 71*(3), 355–359.

Jones, M. B. (1988). Photosynthetic responses of C3 and C4 wetland species in a tropical swamp. *Ecology, 76*, 253–262.

Jones, M. B., & Humphries, S. W. (2002). Impacts of the C4 sedge Cyperus papyrus L. on carbon and water fluxes in an African wetland. *Hydrobiologia, 488*(1), 107–113.

Jones, M. B., & Muthuri, F. M. (1985). The canopy structure and microclimate of papyrus (*Cyperus papyrus*) swamps. *Journal of Ecology, 73*(2), 481–491.

Kadlec, R. H., & Wallace, S. (2009). *Treatment wetlands* (2nd ed.). Boca Raton: CRC Press.

Kansiime, F., & Nalubega, M. (1999). *Wastewater treatment by a natural wetland: The Nakivubo swamp Uganda -processes and implications*. PhD thesis, UNESCO-IHE, DELFT, Netherlands.

Kansiime, F., et al. (2003). The effect of wastewater discharge on biomass production and nutrient content of Cyperus papyrus and Miscanthidium violaceum in the Nakivubo wetland, Kampala, Uganda. *Water Science and Technology, 48*(5), 233–240.

Kansiime, F., Oryem-Origa, H., & Rukwago, S. (2005). Comparative assessment of the value of papyrus and cocoyams for the restoration of the Nakivubo wetland in Kampala, Uganda. *Physics and Chemistry of the Earth Parts A B C, 30*(11–16), 698–705.

Kansiime, F., Kateyo, E., Oryem-Origa, H., & Mucunguzi, P. (2007a). Nutrient status and retention in pristine and disturbed wetlands in Uganda: Management implications. *Wetlands Ecology and Management, 15*(6), 453–467.

Kansiime, F., Saunders, M., & Loiselle, S. (2007b). Functioning and dynamics of wetland vegetation of Lake Victoria: An overview. *Wetlands Ecology and Management, 15*(6), 443–451.

Kanyiginya, V., et al. (2010). Assessment of nutrient retention by Natete wetland Kampala, Uganda. *Physics and Chemistry of the Earth Parts A/B/C, 35*(13–14), 657–664.

Kengne, I. M., Akoa, A., Soh, E. K., Tsama, V., Ngoutane, M. M., Dodane, P. H., & Koné, D. (2008). Effects of faecal sludge application on growth characteristics and chemical composition of *Echinochloa pyramidalis* (Lam.) Hitch. and Chase and *Cyperus papyrus* L. *Ecological Engineering, 34*(3), 233–242.

Kiwango, Y. A., & Wolanski, E. (2008). Papyrus wetlands, nutrients balance, fisheries collapse, food security, and Lake Victoria level decline in 2000–2006. *Wetlands Ecology and Management, 16*(2), 89–96.

Kresovich, S., Wagner, C. K., Scantland, D. A., Groet, S. S., & Lawhon, W. T. (1982). *The utilization of emergent aquatic plants for biomass energy systems development* (Solar Energy Research Institute Task No. 3337.01 WPA No. 274–81). Golden: Solar Energy Research Institute.

Kyambadde, J. (2005). *Optimizing processes for biological nitrogen removal in Nakivubo wetland, Uganda*. PhD thesis, Royal Institute of Technology, Stockholm.

Kyambadde, J., Kansiime, F., Gumaelius, L., & Dalhammar, G. (2004). A comparative study of *Cyperus papyrus* and *Miscanthidium violaceum*-based constructed wetlands for wastewater treatment in a tropical climate. *Water Research, 38*(2), 475–485.

Kyambadde, J., Kansiime, F., & Dalhammar, G. (2005). Nitrogen and phosphorus removal in substrate-free pilot constructed wetlands with horizontal surface flow in Uganda. *Water, Air, and Soil Pollution, 165*(1–4), 37–59.

Leach, B., & Tait, J. (2000). Papyrus. In *Ancient Egyptian materials and technology* (pp. 227–253).

Li, M., & Jones, M. B. (1995). $CO_2$ and $O_2$ transport in the aerenchyma of *Cyperus papyrus* L. *Aquatic Botany, 52*(1–2), 93–106.

Maclean, I., Boar, R., & Lugo, C. (2011). A review of the relative merits of conserving, using, or draining papyrus swamps. *Environmental Management, 47*(2), 218–229.

Mannino, I., Franco, D., Piccioni, E., Favero, L., Mattiuzzo, E., & Zanetto, G. (2008). A cost-effectiveness analysis of seminatural wetlands and activated sludge wastewater-treatment systems. *Environmental Management, 41*(1), 118–129.

Mburu, N., Tebitendwa, S., Rousseau, D., van Bruggen, J., & Lens, P. (2013). Performance evaluation of horizontal subsurface flow–constructed wetlands for the treatment of domestic wastewater in the tropics. *Journal of Environmental Engineering, 139*(3), 358–367.

Meyerson, L. A., Saltonstall, K., Windham, L., Kiviat, E., & Findlay, S. (2000). A comparison of Phragmites australis in freshwater and brackish marsh environments in North America. *Wetlands Ecology and Management, 8*(2/3), 89–103.

Michael, J. (1983). Papyrus: A new fuel for the third world. *Magazine, 99*(1370), 418–421.

Mnaya, B., Asaeda, T., Kiwango, Y., & Ayubu, E. (2007). Primary production in papyrus (*Cyperus papyrus* L.) of Rubondo Island, Lake Victoria, Tanzania. *Wetlands Ecology and Management, 15*(4), 269–275.

Molle, A., Liénard, C., Boutin, G. M., & Iwema, A. (2005). How to treat raw sewage with constructed wetlands: An overview of the French systems. *Water Science and Technology, 51*(9), 11–21.

Morgan, J. A., Martin, J. F., & Bouchard, V. (2008). Identifying plant species with root associated bacteria that promote nitrification and denitrification in ecological treatment systems. *Wetlands, 28*(1), 220–231.

Morrison, E., Upton, C., Odhiambo-K'oyooh, K., & Harper, D. (2012). Managing the natural capital of papyrus within riparian zones of Lake Victoria, Kenya. *Hydrobiologia, 692*(1), 5–17.

Mugisha, P., Kansiime, F., Mucunguzi, P., & Kateyo, E. (2007). Wetland vegetation and nutrient retention in Nakivubo and Kirinya wetlands in the Lake Victoria basin of Uganda. *Physics and Chemistry of the Earth Parts A B C, 32*(15–18), 1359–1365.

Muthuri, F. M., & Jones, M. B. (1997). Nutrient distribution in a papyrus swamp: Lake Naivasha, Kenya. *Aquatic Botany, 56*(1), 35–50.

Muthuri, F. M., Jones, M. B., & Imbamba, S. K. (1989). Primary productivity of papyrus (Cyperus papyrus) in a tropical swamp; Lake Naivasha, Kenya. *Biomass, 18*(1), 1–14.

NAS-NRC. (1976). *Making aquatic weeds useful: Some perspectives for developing countries*. Washington, DC: National Academy of Sciences.

Neralla, S., Weaver, R. W., Lesikar, B. J., & Persyn, R. A. (2000). Improvement of domestic wastewater quality by subsurface flow constructed wetlands. *Bioresource Technology, 75*(1), 19–25.

Nyakang'o, J. B., & van Bruggen, J. J. A. (1999). Combination of a well functioning constructed wetland with a pleasing landscape design in Nairobi, Kenya. *Water Science and Technology, 40*(3), 249–256.

Okurut, T. O. (2000). *A pilot study on municipal wastewater treatment using a constructed wetland in Uganda*. PhD dissertation, UNESCO-IHE, Institute for Water Education, Delft, The Netherlands.

Osumba, J. J. L., Okeyo-Owuor, J. B., & Raburu, P. O. (2010). Effect of harvesting on temporal papyrus (Cyperus papyrus) biomass regeneration potential among swamps in Winam Gulf wetlands of Lake Victoria Basin, Kenya. *Wetlands Ecology and Management, 18*(3), 333–341.

Perbangkhem, T., & Polprasert, C. (2010). Biomass production of papyrus (Cyperus papyrus) in constructed wetland treating low-strength domestic wastewater. *Bioresource Technology, 101*(2), 833–835.

Peterson, S. B., & Teal, J. M. (1996). The role of plants in ecologically engineered wastewater treatment systems. *Ecological Engineering, 6*(1–3), 137–148.

Piedade, M. T. F., Junk, W. J., & Long, S. P. (1991). The productivity of the C_4 Grass *Echinochloa polystachya* on the Amazon floodplain. *Ecology, 72*(4), 1456–1463.

Rousseau, D. P. L., Vanrolleghem, P. A., & Pauw, N. D. (2004). Constructed wetlands in Flanders: A performance analysis. *Ecological Engineering, 23*(3), 151–163.

Sala, O. E., & Austin, A. T. (2000). Methods of estimating aboveground net primary productivity. In O. E. Sala, R. B. Jackson, H. A. Mooney, & R. W. Howarth (Eds.), *Methods in ecosystem science* (pp. 31–43). New York: Springer.

Saunders, M. J., & Kalff, J. (2001). Denitrification rates in the sediments of Lake Memphremagog, Canada-USA. *Water Research, 35*(8), 1897–1904.

Saunders, M. J., Jones, M., & Kansiime, F. (2007). Carbon and water cycles in tropical papyrus wetlands. *Wetlands Ecology and Management, 15*(6), 489–498.

Sekomo, C. B. (2012). *Development of a low-cost alternative for metal removal from textile wastewater*. PhD thesis, UNESCO-IHE Institute for Water Education Delft, The Netherlands.

Sekomo, C. B., Nkuranga, E., Rousseau, D. P. L., & Lens, P. N. L. (2010). Fate of heavy metals in an urban natural wetland: The Nyabugogo swamp (Rwanda). *Water, Air, and Soil Pollution, 214*(1), 321–333.

Serag, M. S. (2003). Ecology and biomass production of *Cyperus papyrus* L. on the Nile bank at Damietta, Egypt. *Journal of Mediterranean Ecology, 4*(3–4), 15–24.

Shetty, U. S., Sonwane, K. D., & Kuchekar, S. R. (2005). Water Hyacinth (Eichornia crassipes) as a natural tool for pollution control. *Annali di Chimica, 95*(9–10), 721–725.

Singer, A., Eshel, A., Agami, M., & Beer, S. (1994). The contribution of aerenchymal $CO_2$ to the photosynthesis of emergent and submerged culms of *Scirpus lacustris* and *Cyperus papyrus*. *Aquatic Botany, 49*(2–3), 107–116.

Sonavane, P. G., Munavalli, G. R., & Ranade, S. V. (2008). Nutrient removal by root zone treatment systems: A review. *Journal of Environmental Science & Engineering, 50*(3), 241–248.

Stein, O. R., Biederman, J. A., Hook, P. B., & Allen, C. (2006). Plant species and temperature effects on the k-C* first-order model for COD removal in batch-loaded SSF wetlands. *Ecological Engineering, 26*(2), 100–112.

Stottmeister, U., Wießner, A., Kuschk, P., Kappelmeyer, U., Kästner, M., Bederski, O., Müller, R. A., & Moormann, H. (2003). Effects of plants and microorganisms in constructed wetlands for wastewater treatment. *Biotechnology Advances, 22*(1–2), 93–117.

Tanner, C. C. (1996). Plants for constructed wetland treatment systems – A comparison of the growth and nutrient uptake of eight emergent species. *Ecological Engineering, 7*(1), 59–83.

Tanner, C. C. (2001). Plants as ecosystem engineers in subsurface-flow treatment wetlands. *Water Science and Technology, 44*(11–12), 9–17.

Terer, T., Triest, L., & Muthama Muasya, A. (2012). Effects of harvesting *Cyperus papyrus* in undisturbed wetland, Lake Naivasha, Kenya. *Hydrobiologia, 680*(1), 135–148.

Thompson, K., Shewry, P. R., & Woolhouse, H. W. (1979). Papyrus swamp development in the Upemba Basin, Zaïre: Studies of population structure in *Cyperus papyrus* stands. *Botanical Journal of the Linnean Society, 78*(4), 299–316.

van Dam, A., Dardona, A., Kelderman, P., & Kansiime, F. (2007). A simulation model for nitrogen retention in a papyrus wetland near Lake Victoria, Uganda (East Africa). *Wetlands Ecology and Management, 15*(6), 469–480.

van Dam, A., Kipkemboi, J., Zaal, F., & Okeyo-Owuor, J. (2011). The ecology of livelihoods in East African papyrus wetlands (ECOLIVE). *Reviews in Environmental Science and Biotechnology, 10*(4), 291–300.

Vymazal, J. (1999). Removal of $BOD_5$ in constructed wetlands with horizontal sub-surface flow: Czech experience. *Water Science and Technology, 40*(3), 133–138.

Vymazal, J. (2007). Removal of nutrients in various types of constructed wetlands. *Science of the Total Environment, 380*(1–3), 48–65.

Vymazal, J. (2011). Plants used in constructed wetlands with horizontal subsurface flow: A review. *Hydrobiologia, 674*, 133–156.

Vymazal, J., & Kröpfelová, L. (2009). Removal of organics in constructed wetlands with horizontal sub-surface flow: A review of the field experience. *Science of the Total Environment, 407*(13), 3911–3922.

Wang, C., et al. (2008). Interactive influence of N and P on their uptake by four different hydrophytes. *African Journal of Biotechnology, 7*(19), 3480–3486.

Yang, Q., Chen, Z.-H., Zhao, J.-G., & Gu, B.-H. (2007). Contaminant removal of domestic wastewater by constructed wetlands: Effects of plant species. *Journal of Integrative Plant Biology, 49*(4), 437–446.

Zhu, S.-X., Ge, H.-L., Ge, Y., Cao, H.-Q., Liu, D., Chang, J., Zhang, C.-B., Gu, B.-J., & Chang, S.-X. (2010). Effects of plant diversity on biomass production and substrate nitrogen in a subsurface vertical flow constructed wetland. *Ecological Engineering, 36*(10), 1307–1313.

Zurita, F., Belmont, M. A., De Anda, J., & Cervantes-Martinez, J. (2008). Stress detection by laser-induced fluorescence in Zantedeschia aethiopica planted in subsurface-flow treatment wetlands. *Ecological Engineering, 33*(2), 110–118.

# Chapter 21
# Does the Presence of Weedy Species Affect the Treatment Efficiency in Constructed Wetlands with Horizontal Subsurface Flow?

Jan Vymazal

**Abstract** In the Czech Republic, *Phalaris arundinacea* and *Phragmites australis* are two most frequent plants used in constructed wetlands for wastewater treatment. While *Phragmites* is quite resistant to the invasion of other species, *Phalaris* is commonly invaded by other wetland species. Moreover, in several systems, *Phalaris* was completely overgrown by *Urtica dioica* in the inflow zone and *Epilobium hirsutum* in the outflow zone. In the present study, the effect of *Phalaris* replacement with other species on treatment performance of two constructed wetlands has been evaluated. In constructed wetlands Zbenice and Čistá, *Phalaris* was overgrown after 8 years of operation in 2005 and 2003, respectively. The results in terms of $BOD_5$ and TSS revealed that there was no difference in discharged water quality during the periods before and after *Phalaris* was overgrown.

**Keywords** Macrophytes • Competition • Treatment efficiency • Nitrogen • Organics

## 21.1 Introduction

The presence of macrophytes is one of the most conspicuous features of wetlands, and their presence distinguishes constructed wetlands from unplanted soil filters or lagoons. The most important roles of macrophytes in horizontal subsurface flow constructed wetlands (HF CWs) in cold and temperate climate are as follows: (1) the insulation of the bed during winter, (2) provision of substrate (roots and rhizomes) for the growth of attached bacteria, (3) oxygen release to otherwise anoxic/anaerobic rhizosphere, (4) nutrient uptake and storage, and (5) release of root exudates with antimicrobial properties (e.g., Seidel 1976; Brix 1997; Gagnon et al. 2006).

---

J. Vymazal (✉)
Department of Applied Ecology, Faculty of Environmental Sciences, Czech University of Life Sciences Prague, Kamýcká 129, 165 21 Praha 6, Czech Republic

ENKI, o.p.s, Dukelská 145, 379 01 Třeboň, Czech Republic
e-mail: vymazal@yahoo.com

The early systems in the Czech Republic were planted only with common reed (*Phragmites australis*) based on then available literature (Kickuth 1982; Brix 1987; Cooper 1990). In a few systems, also *Typha latifolia* was added. Since the mid-1990s, Reed canary grass (*Phalaris arundinacea*) has often been used in combination with *Phragmites* (Vymazal and Kröpfelová 2005). The major reasons for the use of *Phalaris* were: (1) excellent germination from the seeds, (2) easy planting, (3) very fast growth creating full cover of the surface during the first growing season if planted in spring, and (4) provision of good insulation during the winter. Since the mid-1990s, many constructed wetlands designed to treat municipal wastewater in the Czech Republic have been planted with a combination of *Phragmites* and *Phalaris* in bands perpendicular to water flow. This combination turned out to be very effective because *Phalaris* creates very soon, contrary to *Phragmites*, a reasonable aboveground biomass which is necessary for winter insulation. In a long-term run, *Phragmites* usually outcompetes *Phalaris*, but this encroachment does not influence the treatment performance of the system (Vymazal 2013; Vymazal and Kröpfelová 2005).

In the literature, there is very little information about the presence of weed species in constructed wetlands (Cooper et al. 2004; Vymazal 2013). In constructed wetlands, however, a weed is considered a species which was not intentionally planted and voluntarily occurs in constructed wetland. The weedy species are usually found in the Czech constructed wetlands as individual plants, and only several species form denser stands within the originally planted species. In the inflow zone, stinging nettle (*Urtica dioica*) is by far the most frequent "weedy" species. In addition, *Urtica* is the only species which became a dominant species in several systems (Fig. 21.1). It has been observed that *Urtica* was present in zones originally planted with *Phragmites* as well as with *Phalaris*, but in *Phalaris* stands, *Urtica* exhibited substantially denser growth. The more vigorous growth of *Urtica* was particularly observed in *Phalaris* zones which were not regularly mowed, and the *Urtica* belowground organs were mostly distributed in the decaying litter of *Phalaris* above the filtration bed surface and, hence, above the water level.

This observation is in accordance with findings of Klimešová and Šrůtek (1995) that *Phalaris* and *Urtica* reacted differently to the lack of oxygen in the soil. While *Urtica* stops growing and starts to decay, *Phalaris* under the anoxic/anaerobic conditions continues to grow. This is also in accordance with findings of Klimešová (1994) that *Urtica* does not survive long-term flooding, especially during the spring and summer. In the Czech constructed wetlands, the wastewater is distributed in a layer of large stones, and therefore, the systems do not suffer too much from ponding which is usually caused by clogging. Another reason that affects the occurrence of *Urtica* in *Phalaris* stands is the fact that *Urtica* exhibits about twice higher maximum relative growth rate than *Phalaris* (Grimme et al. 1988). Klimešová (1995) observed that *U. dioica* was able to establish itself more often in sites shaded by herbaceous plants in the natural floodplains. This ability may favor the establishment of *U. dioica* in constructed wetlands with tall emergent species.

Similarly to the inflow zone, the most frequent "weedy" species in the outflow zone is *Urtica dioica*. However, in the outflow zones, it only seldom created dense stands

**Fig. 21.1** *Urtica dioica* replacing *Phalaris arundinacea* at the inflow zone of CW Čistá in 2009

and was rather found in the form of individual plants. On the other hand, the second most frequent species, *E. hirsutum* creates dense stands with very large individual plants and in several systems outcompeted originally planted *Phalaris* (Fig. 21.2).

The results reported in the literature indicate that mixed vegetation is more effective in pollutant removal as compared to stands of single species (Karathanasis et al. 2003; Fraser et al. 2004). However, most of the studies were relatively short, and it is a question whether all the species in a mixture would survive in a long-term run. It is well known from constructed wetlands (e.g., Vymazal 2013) as well as from natural habitats (e.g., Vymazal et al. 2008) that competition among plant species may be a slow process which can take 5–10 years before any noticeable change in plant species composition is observed.

It has also been shown that plant species selection may influence the treatment efficiency of the system. In Table 21.1, examples of the influence of plant species on treatment performance of HF CWs reported in the literature are shown.

The objective of this paper was to evaluate the potential influence of weedy species on the treatment performance of two constructed wetlands with subsurface flow in the Czech Republic, which have been "invaded" by weeds.

## 21.2 Study Sites and Methods

The effect of replacement of originally planted *Phalaris arundinacea* by weed species was evaluated at constructed wetlands Zbenice and Čistá (Table 21.2). Both species were planted with *Phalaris* and *Phragmites* planted in bands

**Fig. 21.2** *Epilobium hirsutum* replacing *Phalaris arundinacea* at the outflow zone of CW Zbenice in 2008

**Table 21.1** Examples of the effect of various plant species on treatment performance of HF CWs

| Reference | Location | Plant effect |
|---|---|---|
| Finlayson and Chick (1983) | Australia | *Scirpus validus* superior to *P. australis* in removing $NH_4^+$-N and TKN Removal of P: *S. validus* > *Typha* spp. > *P. australis* |
| Gersberg et al. (1986) | USA | *Scirpus lacustris* and *Phragmites australis* better than *Typha latifolia* in $NH_4$-N removal |
| Burgoon et al. (1989) | USA | *Sagittaria latifolia* (91.8 %) more efficient than *T. latifolia* (85.6 %), *Scirpus pungens* (75.5 %), and *P. australis* (67.5 %) in removing TKN |
| Coleman et al. (2001) | USA | *T. latifolia* outperformed *Juncus effusus* and *Scirpus cyperinus* |
| Fraser et al. (2004) | USA | *S. validus* superior to *P. arundinacea* in P and N removal |

perpendicular to wastewater flow. At Čistá, *Phalaris* was overgrown by *Urtica* after 8 years of operation (Fig. 21.1), and at Zbenice, *Epilobium* became dominant species in the outflow zone replacing *Phalaris* also after 8 years of operation (Fig. 21.2).

Both systems have been regularly visited several times every year since they were put in operation. During all visits, the evaluation of vegetation and photodocumentation were carried out. Results of all inflow and outflow concentrations were obtained from the local municipalities. At CWs Zbenice and Čistá, samples in the inflow and outflow were taken four times and six times a year, respectively.

**Table 21.2** Main design parameters of HF CWs Zbenice and Čistá

|  | Start of operation | Designed PE | Area ($m^2$) | HLR (cm $d^{-1}$) | Filtr. medium | Size (mm) |
| --- | --- | --- | --- | --- | --- | --- |
| Čistá | 1995 | 800 | 3,020 | 5.9 | Gravel | 4–8 |
| Zbenice | 1997 | 200 | 1,000 | 2.0 | Gravel | 8–16 |

*PE* population equivalent

## 21.3 Results and Discussion

In Fig. 21.3 the removal of $BOD_5$ at both monitored and constructed wetlands is shown. At Čistá, the average outflow concentration of 5.9 mg $L^{-1}$ during the period 1995–2003 did not significantly differ ($p > 0.05$) from the average outflow concentration of 7.2 mg $L^{-1}$ during the period 2004–2012 after *Urtica* replaced *Phalaris*. At Zbenice, the concentrations before and after compete replacement of *Phalaris* by *Epilobium* were 11.9 and 11.2 mg $L^{-1}$, respectively, with no statistical difference ($p > 0.05$).

Similar results were observed for TSS (Fig. 21.4), and no statistical difference was observed as well. At Čistá, the average outflow TSS concentrations before and after *Phalaris* replacement were 4.1 mg $L^{-1}$ and 4.6 mg $L^{-1}$, respectively, while at Zbenice the respective concentrations were 12.2 mg $L^{-1}$ and 10.4 mg $L^{-1}$.

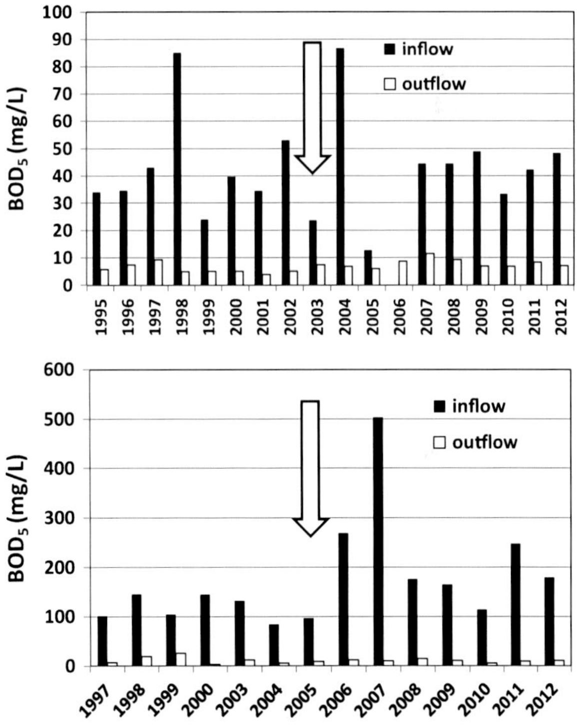

**Fig. 21.3** Removal of $BOD_5$ at constructed wetlands Čistá (*top*) during the period 1995–2012 and Zbenice (*bottom*) during the years 1997–2012. The *arrows* indicate the replacement of *Phalaris* with weed species

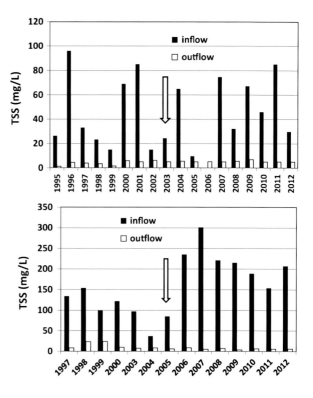

**Fig. 21.4** Removal of TSS at constructed wetlands Čistá (*top*) during the period 1995–2012 and Zbenice (*bottom*) during the years 1997–2012. The *arrows* indicate the replacement of *Phalaris* with weed species

The results may be surprising, especially for the case where *Urtica* replaced *Phalaris* as *Urtica* roots usually do not grow into the filtration zone.

## 21.4 Conclusions

Weedy species are usually restricted to the margins of the filtration beds in HF CWs. In most cases, the occurrence of weedy species is restricted to individual plants with exception of *Urtica dioica* and *Epilobium hirsutum* which may outcompete originally planted species, namely, *Phalaris arundinacea*. However, the results from such systems indicated that the change in vegetation composition does not affect the treatment performance in terms of $BOD_5$ and suspended solids.

## References

Brix, H. (1987). Treatment of wastewater in the rhizosphere of wetland plants – The root zone method. *Water Science and Technology, 19*, 107–118.
Brix, H. (1997). Do macrophytes play a role in constructed treatment wetlands? *Water Science and Technology, 35*(5), 11–17.

Burgoon, P. S., Reddy, K. R., & DeBusk, T. A. (1989). Domestic wastewater treatment using emergent plants cultured in gravel and plastic substrate. In D. A. Hammer (Ed.), *Constructed wetlands for wastewater treatment* (pp. 536–541). Chelsea: Lewis Publishers.

Coleman, J., Hench, L., Garbutt, K., Sextone, A., Bissonnette, G., & Skousen, J. (2001). Treatment of domestic wastewater by three plant species in constructed wetlands. *Water, Air, and Soil Pollution, 128*, 283–295.

Cooper, P. F. (Ed.). (1990). *European design and operation guidelines for reed bed treatment systems* (WRc Rep. UI 17). Prepared for the European Community/European Water Pollution Control Association Emergent Hydrophyte Treatment System Expert Contact Group.

Cooper, D., Griffin, P., & Cooper, P. F. (2004). Factor affecting the longevity of sub-surface horizontal flow systems operating as tertiary treatment for sewage effluent. In A. Liénard (Ed.), *Proceedings of 9th international conference on wetland systems for water pollution control* (pp. 259–266). Lyon: ASTEE and Cemagref.

Finlayson, C. M., & Chick, A. J. (1983). Testing the potential of aquatic plants to treat abattoir effluent. *Water Research, 17*, 415–422.

Fraser, L. H., Carty, S. M., & Steer, D. (2004). A test of four plant species to reduce total nitrogen and total phosphorus from soil leachate in subsurface wetland microcosms. *Bioresource Technology, 94*, 185–192.

Gagnon, V., Chazarenc, F., Comeau, Y., & Brisson, J. (2006). Influence of macrophytes species on microbial density and activity in constructed wetlands. In *Proceedings of 10th international conference of wetland systems for water pollution control* (pp. 1025–1033). Lisbon: MAOTDR.

Gersberg, R. M., Elkins, B. V., Lyon, S. R., & Goldman, C. R. (1986). Role of aquatic plants in wastewater treatment by artificial wetlands. *Water Research, 20*, 363–368.

Grimme, J. P., Hodgson, J. G., & Hunt, R. (1988). *Comparative plant ecology. A functional approach to common British species*. London: Unwin Hyman.

Karathanasis, A. D., Potter, C. L., & Coyne, M. S. (2003). Vegetation effect on fecal bacteria, BOD, and suspended solid removal in constructed wetlands treating domestic wastewater. *Ecological Engineering, 20*, 157–169.

Kickuth, R. (1982). A low-cost process for purification of municipal and industrial waste water. *Der Tropenlandwirt, 83*, 141–154.

Klimešová, J. (1994). The effects of timing and duration of floods on growth of young plants of *Phalaris arundinacea* L. and *Urtica dioica* L.: An experimental study. *Aquatic Botany, 48*, 21–29.

Klimešová, J. (1995). Population dynamics of *Phalaris arundinacea* L. and *Urtica dioica* L. in a floodplain during a dry period. *Wetlands Ecology and Management, 3*, 79–85.

Klimešová, J., & Šrůtek, M. (1995). Vertical distribution of underground organs of *Phalaris arundinacea* and *Urtica dioica* in a floodplain: A comparison of two methods. *Preslia, 67*, 47–53 (in Czech).

Seidel, K. (1976). Macrophytes and water purification. In J. Tourbier & R. W. Pierson (Eds.), *Biological control of water pollution* (pp. 109–122). Philadelphia: Pennsylvania University Press.

Vymazal, J. (2013). Vegetation development in subsurface flow constructed wetlands in the Czech Republic. *Ecological Engineering, 61P*, 575–581.

Vymazal, J., & Kröpfelová, L. (2005). Growth of *Phragmites australis* and *Phalaris arundinacea* in constructed wetlands for wastewater treatment in the Czech Republic. *Ecological Engineering, 25*, 606–621.

Vymazal, J., Craft, C. B., & Richardson, C. J. (2008). Plant community response to long-term N and P fertilization. In C. J. Richardson (Ed.), *The Everglades experiments: Lessons for ecosystem restoration* (Ecological studies 201, pp. 505–527). Dordrecht: Springer Science +Business Media, LLC.

# Index

**A**
Aboveground biomass, 4–9, 11–13, 302–303, 316
Activated sludge, 156, 169, 197, 204, 262, 263
Aerated wetlands, 144, 275, 278, 283, 284
Aerobic conditions, 18, 27, 28, 225, 274, 287, 301
Agricultural runoff, 137–146
Agricultural watershed, 138
Agriculture, 6, 14, 17–30, 51, 52, 58–59, 63, 66, 68, 89, 94–98, 103–120, 123–134, 138–143, 145, 196, 206, 230, 260
Alluvial floodplain forest, 41–53
*Alopecurus pratensis*, 6, 12
Aluminum, 20, 206–222, 225, 274, 275, 278, 282, 283, 286, 291
Ammonium, 5, 59, 61, 62, 65, 156, 159, 162, 207, 275, 284, 291
Anaerobic conditions, 20, 27, 28, 204–207, 212, 222, 225, 288, 316
Arable land, 105, 116–118, 120, 125, 126, 131, 133

**B**
Bacterial community, 159, 161, 162, 167
Baltic Sea, 244, 249, 259–270
Bioaccumulation, 33, 39, 246, 266
Bioavailability, 34–36, 39, 205, 206, 274
Biomass, 1–14, 63, 64, 104, 105, 125, 127–133, 160, 161, 163, 205, 216, 263, 274, 294–296, 301–306, 308, 309, 316

Bog, 90, 95
Boron, 206, 207, 209, 210
Buffers, 58–63, 65–70, 138, 140–142, 145, 175, 195, 197, 296
Bulk density, 21, 26–28, 44–47, 50, 52, 65

**C**
Carbon, 17, 21, 22, 24–26, 28, 29, 36, 42, 44, 46, 47, 52, 67, 83, 89, 104, 119, 132, 164, 200, 204, 207, 209, 212, 213, 217, 223, 225, 275, 295
Carbon sequestration, 41–53
Chemical oxygen demand (COD), 156, 159, 160, 162, 166, 176–179, 183–186, 191–193, 195–198, 204, 207, 209, 212, 213, 230, 232, 233, 238, 241–244, 250, 252, 253, 265, 267, 269, 270, 298
Chlorophyll *a*, 77, 80, 81, 83, 84
Clogging, 144, 151, 157, 158, 160–162, 164, 167–169, 190, 193–195, 200, 229, 235, 256, 264, 316
Coastal plains, 43
Competition, 63, 204, 263, 274, 287, 317
Constructed wetland models, 149–169
Constructed wetlands, 39, 58, 66, 125, 132
Cooling capacity, 120, 124
Copper, 34–39, 213, 285
Culms, 298, 301, 305–306
Cutting régime, 3–6, 8–13
Cyanide, 204, 206, 207, 209, 212, 225
*Cyperus papyrus*, 293–309

## D

Deforestation, 104, 119, 134
Denitrification, 89, 100, 133, 156, 284, 307–308
Diffuse pollution, 60, 66, 138, 260
Domestic sewage, 238, 239, 242, 245, 260
Drained peatland, 58, 66, 68
Dry periods, 75, 138, 140, 166, 167, 251, 252, 254, 255
Dry weather, 94, 252–255

## E

Ecosystem, 13, 25, 28, 42, 53, 95, 105, 113, 119, 298, 302, 308
Ecosystem structure, 73
Electroplating wastewater, 203–225
*Elytrigia repens*, 6, 12
Energy crop, 14
Energy fluxes, 105, 108, 110, 111, 124
Environmental sustainability, 195
*Epilobium hirsutum*, 318, 320
Eutrophication, 42, 138, 143, 244, 260, 268, 270
Evaporative fiction, 103
Evapotranspiration, 104–107, 110, 111, 113, 117–120, 124, 132, 164, 166, 296, 306, 308

## F

Fen, 64, 90, 94, 95, 118
Fertilizer, 3, 5, 6, 9–12, 142, 260
Fill and drain, 203–225
Fishpond, 105, 106, 133
Flooding, 3, 4, 13, 14, 42, 75, 133, 138, 230, 273, 283, 291, 297, 301, 309, 316
Floodplain, 2–4, 7, 12, 13, 41–53, 95, 115, 117, 124, 126, 131, 133, 134, 297, 316
Forest, 41–53, 57–70, 105, 115, 116, 118–120, 124–126, 131, 132, 134, 251
Forest fertilization, 58, 62, 63
Forestry management, 62, 63, 66, 69, 188
Freshwater wetland, 17, 29, 42
Fungicides, 139

## G

Gas emission, 27, 29, 69, 88, 89, 91, 92, 100, 119, 156
Gas flux, 17–30, 91, 94–96
Graminoids, 77, 79, 80, 83
Gravel, 132, 160, 166, 176, 225, 230, 233, 235, 279, 280, 283, 286, 290, 291
Greenhouse gases, 17–30, 67, 69, 89, 104, 119

## H

Harvesting, 4–6, 8, 9, 13, 14, 52, 58, 62, 63, 65, 67, 115, 134, 141, 142, 197, 294, 295, 297, 302, 305, 308–309
Heavy metals, 33, 34, 39, 151, 250, 294, 298, 304, 306
Helophytes, 230, 234, 235
Horizontal flow, 145, 166, 190, 238, 239, 251, 263–265, 274, 275, 280, 283, 285, 290, 291, 308, 315–320
Hydraulic load, 143, 144, 166, 175–187, 192, 234, 254, 256
Hydrological cycle, 104, 105, 119, 120, 298
Hydrological gradient, 75

## I

Immobilisation, 33–39, 63, 156
Iron, 33–39, 206–224, 274, 275, 279, 280, 283–291

## L

Land cover, 105, 106, 115, 119, 125–128, 131, 132, 134
Landscape, 14, 17, 18, 29, 45, 53, 103–120, 123–134, 141, 142, 251, 263, 308
Litter, 64, 295, 316

## M

Macrophytes, 145, 160, 193, 230, 263, 293–309, 315
Manganese, 207–217, 219, 223
Metals, 33–39, 145, 151, 157, 204–209, 212–225, 250, 294, 295, 298, 304, 306
Methane, 2, 17, 18, 21–25, 27–29, 68, 69, 87–100, 224
Methanogenesis, 88, 89, 100, 224
Microcosm, 203–225, 307
Mineralisation, 3, 29, 42, 43, 47, 49–53, 58, 65, 66, 70, 96, 124, 144, 156, 192, 196, 205, 206, 213, 217, 223, 225, 229, 230, 234, 235, 240, 250, 253, 256, 260
Mire, 58, 67, 69, 90, 94, 95
Moisture, 4, 6–7, 12, 21, 27, 44, 88, 89, 91, 94–99, 128, 130, 134
Multi-stage constructed wetland, 189–200

## N

Nanoparticles, 33–39
Natural habitat, 3, 317
Natural marsh, 17

Index 325

Natural wetland, 18, 20, 22–27, 29, 68–69, 132, 286, 290, 294, 296, 298, 304, 307, 308
Nickel, 203, 206–215, 218, 221, 223, 225
Nitrification, 156, 167, 176, 229–235, 246, 247, 275, 280, 307, 308
Nitrogen, 3, 5, 6, 21, 26, 28, 42–44, 46, 47, 49, 52, 58, 89, 91, 96, 128, 151, 156–159, 162, 176, 179, 187, 234, 238, 241, 244–246, 252, 253, 260, 261, 263–266, 274, 275, 279, 283, 294, 298, 305–308
  leaching, 138
  removal, 176, 179, 246, 263, 265, 266, 307
Nitrous oxide, 17, 21, 22, 25, 27, 29, 89
Nutrients, 3–6, 12, 13, 20, 41–53, 57–70, 74, 124, 125, 132–134, 138, 142, 143, 145, 158–159, 164, 166, 191, 193, 239, 244–246, 250, 260, 261, 274, 294, 295, 298, 301, 302, 305–306, 308, 309, 315
  accumulation, 41–53, 63, 66
  additions, 61, 62
  loss, 58, 124, 132, 134

**O**

On-site wastewater treatment plants, 262–263, 265
Organic carbon, 21, 26, 28, 42, 44, 46, 47, 52, 67, 207, 225
Organic load, 175, 176, 178, 185, 192, 195–197, 256
Organic matter, 18, 36, 45–47, 51, 144, 156–159, 166, 195, 196, 204, 206, 212, 223, 229–230, 235, 238, 239, 242–244, 246, 247, 250, 252–254, 266, 274
Organic soil, 3, 4, 7, 8, 13, 52
Oxidation-reduction potential (ORP), 207, 216, 224, 279, 280, 286, 288
Oxygen diffusion, 301
Oxygen release, 164, 166, 307, 315

**P**

Particle size distribution, 47, 250, 254, 256
Pathogens, 139, 145, 294, 298
Peatland, 58, 65, 66, 68, 87–100, 124
Pesticides, 138, 139, 141–145, 151
*Phalaris arundinacea*, 2, 3, 6–12, 106, 316–318, 320
Phosphorus, 5, 6, 42–44, 47, 49, 52, 58, 133, 138, 158, 238, 244, 260, 261, 263, 273–291, 294, 298, 305
Photosynthetically active radiation, 75
*Phragmites australis*, 39, 79, 176, 206, 316

Phytoremediation, 2, 34, 39
Plant community, 42, 50, 73, 75, 295, 301
Plant growth, 235, 295, 305
Plant productivity, 50
Plants, 2, 18, 34, 42, 59, 73, 98, 104, 133, 139, 158, 176, 190, 206, 229, 235, 261, 274, 294, 314
Pollutants, 59, 138, 145–146, 150, 151, 161, 167, 212, 230, 238, 240, 242, 243, 246, 247, 249–256, 260, 261, 263–268, 270, 274, 286, 294–297, 304, 306, 308, 309, 317
Ponds, 43, 132, 133, 142–144, 157, 158, 161, 164, 166, 167, 241, 243, 251, 253, 255, 256, 316
Porosity, 21, 27, 157, 158, 161, 235
Productivity, 1–14, 20, 22, 24, 27–29, 42, 51, 58, 66, 67, 69, 139, 143, 190, 191, 193, 200, 205, 260, 294–298, 301–305, 307, 309

**R**

Rain, 119, 124, 132, 133, 139, 140, 179, 250–255, 280
Redox potential, 164, 168, 279, 287, 306
Reed beds, 189, 192, 229, 255
Reflectance spectra, 73–84, 127
Remote sensing, 74
Removal rate, 143, 144, 195–197, 199, 268, 270, 284, 285
Restoration, 2, 18, 19, 25, 27, 29, 58, 63, 65, 66, 68–70, 132, 133, 190, 251
Restored wetlands, 17–30, 57–70, 119
Retention capacity, 59–62, 66, 69
Rhizomes, 145, 216, 234, 243, 297, 298, 301–306, 315
Rhizosphere, 145, 164, 213, 229, 304, 307, 315
Riverine wetlands, 91
Roots, 38, 89, 104, 143, 145, 157, 158, 164, 165, 168, 216, 224, 234, 243, 294, 295, 297, 298, 301–308, 315, 320
Runoff, 51, 58, 59, 61, 66, 69, 124, 132–134, 137–146, 193, 249–251
Rural areas, 141, 237–239, 242, 246, 259–270

**S**

Sand, 43–46, 50, 144, 157, 166, 176, 189–192, 195, 197, 206, 212, 213, 215, 217, 222, 230, 232, 233, 251, 262, 263
Satellite data, 125, 126, 128
Scanning electron microscopy (SEM), 205, 209, 218, 220, 222

Sediments, 18, 33–39, 41–53, 57–70, 91, 138, 142–144, 158, 164, 195–197, 205, 208, 223, 250, 252–253, 279, 280, 285–289, 298, 304–306
Septic tanks, 190, 239–240, 242, 243, 245, 246, 262–265, 268
Sequential extraction procedure (SEP), 34–37, 39, 203, 205, 207–208, 213–216, 223–224
Single family treatment plant (SFTP), 238
Slough, 18, 19, 43–47, 49, 50, 92
Sludge, 156, 169, 191, 192, 196, 200, 204, 239, 240, 262, 263, 274, 278, 279, 283, 285–288, 290, 291, 296, 300
Snowmelt, 62, 64, 69, 138, 252–255
Soil accretion, 43, 45–47, 49, 50, 53
Solar energy, 74, 103–120, 124, 134, 294
Solar energy dissipation, 124
Standing crop, 301
Stormwater, 249–256
Subsurface flow constructed wetlands, 149–169
Sulphate, 34, 156, 275, 277
Sulphide, 156, 193, 195, 196
Sulphur, 151, 156, 159
Suspended solids (SS), 59, 238, 242–244, 250, 252–254, 264, 266, 274, 275, 277, 278, 284, 285, 320
Sustainable landscape management, 124, 132

T
Temperature, 20, 75, 88, 89, 91, 95, 97, 104–106, 108, 109, 113, 115–120, 123–134, 138, 156, 159, 161, 164–167, 177, 180–182, 204, 208, 250, 280, 288, 301
Thermo-vision pictures, 117
Treatment efficiency, 144, 166, 197, 209, 212, 225, 233, 234, 298, 315–320
Treatment wetlands, 63, 237–247, 249–256, 259–270, 290, 294–297, 308, 309
Tropical wetlands, 293
Tropics, 89, 94–96, 294, 295, 297, 298, 301, 302, 309
Two-stage vertical flow, 175–187

U
*Urtica dioica*, 106, 316, 317, 320

V
Vertical flow (VF), 156, 157, 161, 169, 175–187, 191, 230, 239, 243, 265, 266, 283
Vivianite, 34–39

W
Wastewater, 145, 150, 151, 158, 176, 183, 190–193, 197, 200, 203–225, 229–235, 240–244, 246, 248, 254, 261–263, 265, 270, 278–279, 284, 287, 288, 290, 291, 294–296, 298, 304–306, 308, 309, 316, 318
management, 260, 261, 270
treatment, 2, 138, 189–200, 238, 246, 250, 261–263, 265, 266, 274, 285, 290, 291, 293–309
Weed, 297, 315–320
Wet grassland, 3, 4, 6, 9–14
Wetlands, 2, 17, 34, 42, 58, 89, 105, 124, 138, 150, 175, 191, 230, 238, 250, 274, 294, 315
buffer, 57–70
management, 3, 14, 18, 34, 58, 59
soils, 14, 18, 22–24, 26, 27, 42, 298
Wet meadow, 4–6, 13, 27, 29, 105, 106, 110–113, 117–120, 125, 133, 134
Winery wastewater, 189–200
Wooded buffer zones, 229

Z
Zinc, 34–39, 207–215, 218, 223, 224